Introduction to Statistical Methods for Financial Models

CHAPMAN & HALL/CRC
Texts in Statistical Science Series

Statistical Theory: A Concise Introduction
F. Abramovich and Y. Ritov

Practical Multivariate Analysis, Fifth Edition
A. Afifi, S. May, and V.A. Clark

Practical Statistics for Medical Research
D.G. Altman

Interpreting Data: A First Course in Statistics
A.J.B. Anderson

Introduction to Probability with R
K. Baclawski

Linear Algebra and Matrix Analysis for Statistics
S. Banerjee and A. Roy

Modern Data Science with R
B. S. Baumer, D. T. Kaplan, and N. J. Horton

Mathematical Statistics: Basic Ideas and Selected Topics, Volume I, Second Edition
P. J. Bickel and K. A. Doksum

Mathematical Statistics: Basic Ideas and Selected Topics, Volume II
P. J. Bickel and K. A. Doksum

Analysis of Categorical Data with R
C. R. Bilder and T. M. Loughin

Statistical Methods for SPC and TQM
D. Bissell

Introduction to Probability
J. K. Blitzstein and J. Hwang

Bayesian Methods for Data Analysis, Third Edition
B.P. Carlin and T.A. Louis

Second Edition
R. Caulcutt

The Analysis of Time Series: An Introduction, Sixth Edition
C. Chatfield

Introduction to Multivariate Analysis
C. Chatfield and A.J. Collins

Problem Solving: A Statistician's Guide, Second Edition
C. Chatfield

Statistics for Technology: A Course in Applied Statistics, Third Edition
C. Chatfield

Analysis of Variance, Design, and Regression : Linear Modeling for Unbalanced Data, Second Edition
R. Christensen

Bayesian Ideas and Data Analysis: An Introduction for Scientists and Statisticians
R. Christensen, W. Johnson, A. Branscum, and T.E. Hanson

Modelling Binary Data, Second Edition
D. Collett

Modelling Survival Data in Medical Research, Third Edition
D. Collett

Introduction to Statistical Methods for Clinical Trials
T.D. Cook and D.L. DeMets

Applied Statistics: Principles and Examples
D.R. Cox and E.J. Snell

Multivariate Survival Analysis and Competing Risks
M. Crowder

Statistical Analysis of Reliability Data
M.J. Crowder, A.C. Kimber, T.J. Sweeting, and R.L. Smith

An Introduction to Generalized Linear Models, Third Edition
A.J. Dobson and A.G. Barnett

Nonlinear Time Series: Theory, Methods, and Applications with R Examples
R. Douc, E. Moulines, and D.S. Stoffer

Introduction to Optimization Methods and Their Applications in Statistics
B.S. Everitt

Extending the Linear Model with R: Generalized Linear, Mixed Effects and Nonparametric Regression Models, Second Edition
J. Faraway

Linear Models with R, Second Edition
J. Faraway

A Course in Large Sample Theory
T.S. Ferguson

Multivariate Statistics: A Practical Approach
B. Flury and H. Riedwyl

Readings in Decision Analysis
S. French

Discrete Data Analysis with R: Visualization and Modeling Techniques for Categorical and Count Data
M. Friendly and D. Meyer

Markov Chain Monte Carlo: Stochastic Simulation for Bayesian Inference, Second Edition
D. Gamerman and H.F. Lopes

Bayesian Data Analysis, Third Edition
A. Gelman, J.B. Carlin, H.S. Stern, D.B. Dunson, A. Vehtari, and D.B. Rubin

Multivariate Analysis of Variance and Repeated Measures: A Practical Approach for Behavioural Scientists
D.J. Hand and C.C. Taylor

Practical Longitudinal Data Analysis
D.J. Hand and M. Crowder

Logistic Regression Models
J.M. Hilbe

Richly Parameterized Linear Models: Additive, Time Series, and Spatial Models Using Random Effects
J.S. Hodges

Statistics for Epidemiology
N.P. Jewell

Stochastic Processes: An Introduction, Second Edition
P.W. Jones and P. Smith

The Theory of Linear Models
B. Jørgensen

Pragmatics of Uncertainty
J.B. Kadane

Principles of Uncertainty
J.B. Kadane

Graphics for Statistics and Data Analysis with R
K.J. Keen

Mathematical Statistics
K. Knight

Introduction to Functional Data Analysis
P. Kokoszka and M. Reimherr

Introduction to Multivariate Analysis: Linear and Nonlinear Modeling
S. Konishi

Nonparametric Methods in Statistics with SAS Applications
O. Korosteleva

Modeling and Analysis of Stochastic Systems, Third Edition
V.G. Kulkarni

Exercises and Solutions in Biostatistical Theory
L.L. Kupper, B.H. Neelon, and S.M. O'Brien

Exercises and Solutions in Statistical Theory
L.L. Kupper, B.H. Neelon, and S.M. O'Brien

Design and Analysis of Experiments with R
J. Lawson

Design and Analysis of Experiments with SAS
J. Lawson

A Course in Categorical Data Analysis
T. Leonard

Statistics for Accountants
S. Letchford

Introduction to the Theory of Statistical Inference
H. Liero and S. Zwanzig

Statistical Theory, Fourth Edition
B.W. Lindgren

Stationary Stochastic Processes: Theory and Applications
G. Lindgren

Statistics for Finance
E. Lindström, H. Madsen, and J. N. Nielsen

The BUGS Book: A Practical Introduction to Bayesian Analysis
D. Lunn, C. Jackson, N. Best, A. Thomas, and D. Spiegelhalter

Introduction to General and Generalized Linear Models
H. Madsen and P. Thyregod

Time Series Analysis
H. Madsen

Pólya Urn Models
H. Mahmoud

Randomization, Bootstrap and Monte Carlo Methods in Biology, Third Edition
B.F.J. Manly

Statistical Regression and Classification: From Linear Models to Machine Learning
N. Matloff

Introduction to Randomized Controlled Clinical Trials, Second Edition
J.N.S. Matthews

Statistical Rethinking: A Bayesian Course with Examples in R and Stan
R. McElreath

Statistical Methods in Agriculture and Experimental Biology, Second Edition
R. Mead, R.N. Curnow, and A.M. Hasted

Statistics in Engineering: A Practical Approach
A.V. Metcalfe

Statistical Inference: An Integrated Approach, Second Edition
H. S. Migon, D. Gamerman, and F. Louzada

Beyond ANOVA: Basics of Applied Statistics
R.G. Miller, Jr.

A Primer on Linear Models
J.F. Monahan

Stochastic Processes: From Applications to Theory
P.D Moral and S. Penev

Applied Stochastic Modelling, Second Edition
B.J.T. Morgan

Elements of Simulation
B.J.T. Morgan

Probability: Methods and Measurement
A. O'Hagan

Introduction to Statistical Limit Theory
A.M. Polansky

Applied Bayesian Forecasting and Time Series Analysis
A. Pole, M. West, and J. Harrison

Statistics in Research and Development, Time Series: Modeling, Computation, and Inference
R. Prado and M. West

Essentials of Probability Theory for Statisticians
M.A. Proschan and P.A. Shaw

Introduction to Statistical Process Control
P. Qiu

Sampling Methodologies with Applications
P.S.R.S. Rao

A First Course in Linear Model Theory
N. Ravishanker and D.K. Dey

Essential Statistics, Fourth Edition
D.A.G. Rees

Stochastic Modeling and Mathematical Statistics: A Text for Statisticians and Quantitative Scientists
F.J. Samaniego

Statistical Methods for Spatial Data Analysis
O. Schabenberger and C.A. Gotway

Bayesian Networks: With Examples in R
M. Scutari and J.-B. Denis

Large Sample Methods in Statistics
P.K. Sen and J. da Motta Singer

Introduction to Statistical Methods for Financial Models
T. A. Severini

Spatio-Temporal Methods in Environmental Epidemiology
G. Shaddick and J.V. Zidek

Decision Analysis: A Bayesian Approach
J.Q. Smith

Analysis of Failure and Survival Data
P. J. Smith

Applied Statistics: Handbook of GENSTAT Analyses
E.J. Snell and H. Simpson

Applied Nonparametric Statistical Methods, Fourth Edition
P. Sprent and N.C. Smeeton

Data Driven Statistical Methods
P. Sprent

Generalized Linear Mixed Models: Modern Concepts, Methods and Applications
W. W. Stroup

Survival Analysis Using S: Analysis of Time-to-Event Data
M. Tableman and J.S. Kim

Applied Categorical and Count Data Analysis
W. Tang, H. He, and X.M. Tu

Elementary Applications of Probability Theory, Second Edition
H.C. Tuckwell

Introduction to Statistical Inference and Its Applications with R
M.W. Trosset

Understanding Advanced Statistical Methods
P.H. Westfall and K.S.S. Henning

Statistical Process Control: Theory and Practice, Third Edition
G.B. Wetherill and D.W. Brown

Generalized Additive Models: An Introduction with R, Second Edition
S. Wood

Epidemiology: Study Design and Data Analysis, Third Edition
M. Woodward

Practical Data Analysis for Designed Experiments
B.S. Yandell

Introduction to Statistical Methods for Financial Models

Thomas A. Severini

Northwestern University

Evanston, Illinois, USA

CRC Press
Taylor & Francis Group
Boca Raton London New York

CRC Press is an imprint of the
Taylor & Francis Group, an **informa** business
A CHAPMAN & HALL BOOK

CRC Press
Taylor & Francis Group
6000 Broken Sound Parkway NW, Suite 300
Boca Raton, FL 33487-2742

Printed on acid-free paper

International Standard Book Number-13: 978-1-138-19837-1 (Hardback)

Library of Congress Cataloging-in-Publication Data

Names: Severini, Thomas A. (Thomas Alan), 1959- author.
Title: Introduction to statistical methods for financial models / Thomas A. Severini.
Description: Boca Raton, FL : CRC Press, [2018] | Includes bibliographical references and index.
Identifiers: LCCN 2017003073| ISBN 9781138198371 (hardback) | ISBN 9781315270388 (e-book master) | ISBN 9781351981910 (adobe reader) | ISBN 9781351981903 (e-pub) | ISBN 9781351981897 (mobipocket)
Subjects: LCSH: Finance--Statistical methods. | Finance--Mathematical models.
Classification: LCC HG176.5 .S49 2017 | DDC 332.072/7--dc23
LC record available at https://lccn.loc.gov/2017003073

Visit the Taylor & Francis Web site at
http://www.taylorandfrancis.com

and the CRC Press Web site at
http://www.crcpress.com

To Karla

Contents

Preface

This book provides an introduction to the use of statistical concepts and methods to model and analyze financial data; it is an expanded version of notes used for an advanced undergraduate course at Northwestern University, "Introduction to Financial Statistics." A central theme of the book is that by modeling the returns on assets as random variables, and using some basic concepts of probability and statistics, we may build a methodology for analyzing and interpreting financial data.

The audience for the book is students majoring in statistics and economics as well as in quantitative fields such as mathematics and engineering; the book can also be used for a master's level course on statistical methods for finance. Readers are assumed to have taken at least two courses in statistical methods covering basic concepts such as elementary probability theory, expected values, correlation, and conditional expectation as well as introductory statistical methodology such as estimation of means and standard deviations and basic linear regression. They are also assumed to have taken courses in multivariate calculus and linear algebra; however, no prior experience with finance or financial concepts is required or expected.

The 10 chapters of the book fall naturally into three sections. After a brief introduction to the book in Chapter 1, Chapters 2 and 3 cover some basic concepts of finance, focusing on the properties of returns on an asset. Chapters 4 through 6 cover aspects of portfolio theory, with Chapter 4 containing the basic ideas and Chapter 5 presenting a more mathematical treatment of efficient portfolios; the estimation of the parameters needed to implement portfolio theory is the subject of Chapter 6. The remainder of the book, Chapters 7 through 10, discusses several models for financial data, along with the implications of those models for portfolio theory and for understanding the properties of return data. These models begin with the capital asset pricing model in Chapter 7; its more empirical version, the market model, is covered in Chapter 8. Chapter 9 covers the single-index model, which extends the market model to the returns on several assets; more general factor models are the topic of Chapter 10.

In addition to building on the basic concepts covered in math and statistics courses, the book introduces some more advanced topics in an applied setting. Such topics include covariance matrices and their properties, shrinkage estimation, the use of simulation to study the properties of estimators, multiple testing, estimation of standard errors using resampling, and optimization methods. The discussion of such methods focuses on their use and the interpretation of the results, rather than on the underlying theory.

Data analysis and computation play a central role in the book. There are detailed examples illustrating how the methods presented may be implemented in the statistical software R; the methods described are applied to genuine financial data, which may be conveniently downloaded directly into R. These examples include both the use of R packages when available and the writing of small R programs when necessary. I have tried to provide sufficient details so that readers with even minimal experience in R can successfully implement the methodology; however, those with no R experience will likely benefit from one of the many introductory books or online tutorials available.

Each chapter ends with exercises and suggestions for further reading. The exercises include both questions requiring analytic solutions and those requiring data analysis or other numerical work; in nearly all cases, any R functions needed have been discussed in the examples in the text. Finance and financial statistics are well-studied fields about which much has been written. The books and papers given as suggestions for further reading were chosen based on the expected background of the reader, rather than to reference the most definitive treatments of a topic.

I would like to thank Karla Engel who was instrumental in preparing the manuscript and who provided many useful comments and corrections; it is safe to say that this book would not have been completed without her help. I would like to thank Matt Davison (University of Western Ontario) for a number of valuable comments and suggestions. Several anonymous reviewers made helpful comments at various stages of the project and their contributions are gratefully acknowledged. I would also like to thank Rob Calver and the staff at CRC Press/Taylor & Francis for suggestions and other help throughout the publishing process.

1

Introduction

The goal of this book is to present an introduction to the statistical methodology used in investment analysis and financial econometrics, which are concerned with analyzing the properties of financial markets and with evaluating potential investments. Here, an "investment" refers to the purchase of an asset, such as a stock, that is expected to generate income, appreciate in value, or ideally both. The evaluation of such an investment takes into account its potential financial benefits, along with the "risk" of the investment based on the fact that the asset may decrease in value or even become worthless.

A major advance in the science of investment analysis took place beginning in the 1950s when probability theory began to be used to model the uncertainty inherent in any investment. The "return" on an investment, that is, the proportional change in its value over a given period of time, is modeled as a random variable and the investment is evaluated by the properties of the probability distribution of its return. The methods used in this statistical approach to investment analysis form an important component of the field known as *quantitative finance* or, more recently, *financial engineering*. The methodology used in quantitative finance may be contrasted with that based on *fundamental analysis*, which attempts to measure the "true worth" of an asset; for example, in the case of a stock, fundamental analysis uses financial information regarding the company issuing the stock, along with more qualitative measures of the firm's profitability.

For instance, in the statistical models used in analyzing investments, the expected value of the return on an asset gives a measure of the expected financial benefit from owning the asset and the standard deviation of the return is a measure of its variability, representing the risk of the investment. It follows that, based on this approach, an ideal investment has a return with a large expected value and a small standard deviation or, equivalently, a large expected value and a small variance. Thus, the analysis of investments using these ideas is often referred to as *mean-variance analysis*.

Concepts from probability and statistics have been used to develop a formal mathematical framework for investment analysis. In particular, the properties of the returns on a *portfolio*, a set of assets owned by a particular investor, may be derived using properties of sums of the random variables representing the returns on the individual assets. This approach leads to a methodology for selecting assets and constructing portfolios known as *modern portfolio theory* or *Markowitz portfolio theory*, after Harry Markowitz, one of the pioneers in this field.

A central concept in this theory is the *risk aversion* of investors, which assumes that, when choosing between two investments with the same expected return, investors will prefer the one with the smaller risk, that is, the one with the smaller standard deviation; thus, the optimal portfolios are the ones that maximize the expected return for a given level of risk or, conversely, minimize the risk for a given expected return. It follows that numerical optimization methods, which may be used to minimize measures of risk or to maximize an expected return, play a central role in this theory.

An important feature of these methods is that they do not rely on accurate predictions of the future asset returns, which are generally difficult to obtain. The idea that asset returns are difficult to predict accurately is a consequence of the statistical model for asset prices known as a *random walk* and the assumption that asset prices follow a random walk is known as the *random walk hypothesis*. The random walk model for prices asserts that changes in the price of an asset over time are unpredictable, in a certain sense. The random walk hypothesis is closely related to the *efficient market hypothesis*, which states that asset prices reflect all currently available information. Although there is some evidence that the random walk hypothesis is not literally true, empirical results support the general conclusion that accurate predictions of future returns are not easily obtained.

Instead, the methods of modern portfolio theory are based on the properties of the probability distribution of the returns on the set of assets under consideration. In particular, the mean return on a portfolio depends on the mean returns on the individual assets and the standard deviation of a portfolio return depends on the variability of the individual asset returns, as measured by their standard deviations, along with the relationship between the returns, as measured by their correlations. Thus, the extent to which the returns on different assets are related plays a crucial role in the properties of portfolio returns and in concepts such as diversification.

Of course, in practice, parameters such as means, standard deviations, and correlations are unknown and must be estimated from historical data. Thus, statistical methodology plays a central role in the mean-variance approach to investment analysis. Although, in principle, the estimation of these parameters is straightforward, the scale of the problem leads to important challenges. For instance, if a portfolio is based on 100 assets, we must estimate 100 return means, 100 return standard deviations, and 4950 return correlations.

The properties of the returns on different assets are often affected by various economic conditions relevant to the assets under consideration. Hence, statistical models relating asset returns to available economic variables are important for understanding the properties of potential investments. For instance, the theoretical *capital asset pricing model (CAPM)* and its empirical version, known as the *market model*, describe the returns on an asset in terms of their relationship with the returns on the equity market as a whole, known as the *market portfolio*, and measured by a suitable market index, such as the Standard & Poors (S&P) 500 index. Such models are useful for understanding the nature of the risk associated with an asset, as well as the relationship between the expected return on an asset and its risk. The *single-index model* extends this idea to a model for the correlation structure of the returns on a set of assets; in this model, the correlation between the returns on two assets is described in terms of each asset's correlation with the return on the market portfolio.

The CAPM, the market model, and the single-index model are all based on the relationship between asset returns and the return on some form of a market portfolio. Although the behavior of the market as a whole may be the most important factor affecting asset returns, in general, asset returns are related to other economic variables as well. A *factor model* is a type of generalization of these models; it describes the returns on a set of assets in terms of a few underlying "factors" affecting these assets. Such a model is useful for describing the correlation structure of a set of asset returns as well as for describing the behavior of the mean returns of the assets. The factors used are chosen by the analyst; hence, there is considerable flexibility in the exact form of the model. The parameters of a factor model are estimated using statistical techniques such as regression analysis and the results provide useful information for understanding the factors affecting the asset returns; the results from an analysis based on a factor model are important in analyzing potential investments and constructing portfolios.

Data Analysis and Computing

Data analysis is an important component of the methodology covered in this book and all of the methods presented are illustrated on genuine financial data. Fortunately, financial data are readily available from a number of Internet sources such as finance.yahoo.com and the Federal Reserve Economic Data (FRED) website, fred.stlouisfed.org. Experience with such data is invaluable for gaining a better understanding of the features and challenges of financial modeling.

The analyses in the book use the statistical software R which can be downloaded, free of charge, at www.r-project.org. Analysts often find it convenient to use a more user-friendly interface to R such as RStudio, which is available at www.rstudio.com; however, the examples presented here use only the standard R software. R includes many functions that are useful for statistical data analysis; in addition, it is a programming language and users may define their own functions when convenient. Such user-defined functions will be described in detail and implemented as needed; no previous programming experience is necessary.

There are two features of R that make it particularly useful for analyzing financial data. One is that stock price data may be downloaded directly into R.

The other is that there are many R packages available that extend its functionality; several of these provide functions that are useful for analyzing financial data.

Suggestions for Further Reading

A detailed nontechnical introduction to financial analysis based on statistical concepts is given in Bernstein (2001). Chapter 1 of Fabozzi et al. (2006) gives a concise account of the history of financial modeling. Malkiel (1973) contains a nontechnical discussion of the random walk hypothesis and its implications, as well as many of the criticisms of the random walk hypothesis that have been raised.

For readers with limited experience using R, the document "Introduction to R," available on the R Project website at https://cran.r-project.org/doc/manuals/r-release/R-intro.pdf, is a good starting point. Dalgaard (2008) provides a book-length treatment of basic statistical methods using R with many examples. The "Quick-R" website, at http://www.statmethods.net/index.html, contains much useful information for both the beginner and experienced user.

2

Returns

2.1 Introduction

As discussed in Chapter 1, the goal of this book is to provide an introduction to the statistical methodology used in modeling and analyzing financial data. This chapter introduces some basic concepts of finance and the types of financial data used in this context. The analyses focus on the returns on an asset, which are the proportional changes in the price of the asset over a given time interval, typically a day or month. The statistical foundations for the analysis of such data are presented, along with statistical methods that are useful for investigating the properties of return data.

2.2 Basic Concepts

Consider an asset, such as one share of a particular stock, and let P_t denote the price of the asset at time t, $t = 0, 1, 2, \dots$ so that P_0 is the initial price, P_1 is the price at time 1, P_2 is the price at time 2, and so on. Some assets pay *dividends*, a specified amount at a given time. For example, one share of IBM stock may pay a dividend of \$1.20 each quarter. These dividends make the asset worth more than simply the price. For now, assume that there are no dividends.

The *net return* or, simply, the *return*, on the asset over the period from time $t - 1$ to time t is defined as

$$R_t = \frac{P_t - P_{t-1}}{P_{t-1}} = \frac{P_t}{P_{t-1}} - 1, \quad t = 1, 2, \dots.$$

That is, the return on the asset is simply the proportional change in its price over a given time period; the return is positive if the price increased and is negative if the price decreased.

Example 2.1 Suppose that, for a given asset, $P_0 = 60$, $P_1 = 62.40$, $P_2 = 63.96$, $P_3 = 61.40$, and $P_4 = 66$; assume that all prices are in dollars but, for simplicity, the dollar sign is omitted. Then the returns are

$$R_1 = \frac{62.40 - 60}{60} = 0.040, \quad R_2 = \frac{63.96 - 62.40}{62.40} = 0.025,$$

$$R_3 = \frac{61.40 - 63.96}{63.96} = -0.040, \quad R_4 = \frac{66 - 61.40}{61.40} = 0.075. \qquad \square$$

The *revenue* from holding the asset is given by

$$\text{revenue} = (\text{investment}) \times (\text{return}).$$

Therefore, in Example 2.1, if the initial investment is \$100, the revenue over the period from $t = 0$ to $t = 1$ is

$$100(0.04) = \$4.$$

Normally, we focus on the return rather than on the revenue, which depends on the amount invested.

The *gross return* on the asset over the period from time $t-1$ to time t is

$$\frac{P_t}{P_{t-1}} = 1 + R_t$$

so that, for example, the gross return corresponding to $R_1 = 0.04$ is simply 1.04.

We may be interested in returns over a length of time longer than one period. The return over the time period from time $t-k$ to time t, known as the k-period return at time t, is defined as the proportional change in price over that time period. Let $R_t(k)$ denote the k-period return at time t. Then

$$R_t(k) = \frac{P_t - P_{t-k}}{P_{t-k}} = \frac{P_t}{P_{t-k}} - 1, \quad t = k, k+1, \ldots .$$

Multiperiod returns are related to one-period returns by

$$\begin{aligned} 1 + R_t(k) &= \frac{P_t}{P_{t-k}} = \frac{P_t}{P_{t-1}} \frac{P_{t-1}}{P_{t-2}} \cdots \frac{P_{t-k+1}}{P_{t-k}} \\ &= (1 + R_t)(1 + R_{t-1}) \cdots (1 + R_{t-k+1}). \end{aligned}$$

Note that $1 + R_t(k)$ is the gross return from $t-k$ to t and $1 + R_t, 1 + R_{t-1}, \ldots,$ are the single-period gross returns.

Example 2.2 Using the sequence of prices given in Example 2.1, the two-period return at time 4 is

$$R_3(2) = \frac{P_4 - P_2}{P_2} = \frac{66 - 63.96}{63.96} = 0.032.$$

Recall that $R_3 = -0.040$, and $R_4 = 0.075$. Then

$$R_4(2) = (1 + 0.075)(1 - 0.040) - 1 = 0.032. \qquad \square$$

Log-Returns

It is sometimes convenient to work with *log-returns*, defined by $r_t = \log(1 + R_t), t = 1, 2, \ldots$; note that throughout the book, "log" will denote natural logarithms.

Let $p_t = \log P_t$, $t = 0, 1, \ldots$ denote the log prices. Then the log-returns are defined as

$$r_t = \log(1 + R_t) = \log \frac{P_t}{P_{t-1}} = p_t - p_{t-1}.$$

That is, log-returns are simply the change in the log-prices.

One advantage of working with log-returns is that it simplifies the analysis of multi-period returns. Let $r_t(k)$ denote the k-period log-return at time t. Then, by analogy with the single-period case, $r_t(k) = \log(1 + R_t(k))$ and

$$\begin{aligned} r_t(k) &= \log(1 + R_t(k)) \\ &= \log((1 + R_t)(1 + R_{t-1}) \cdots (1 + R_{t-k+1})) \\ &= \log(1 + R_t) + \log(1 + R_{t-1}) + \cdots + \log(1 + R_{t-k+1}) \\ &= r_t + r_{t-1} + \cdots + r_{t-k+1}; \end{aligned}$$

that is, the k-period log-return at time t is simply the sum of the k single-period log-returns, $r_{t-k+1}, r_{t-k+2}, \ldots, r_t$. Alternatively, because $r_t = p_t - p_{t-1}$, the k-period log-return is the change in the log-price from period $t - k$ to period t,

$$r_t(k) = p_t - p_{t-k}.$$

Example 2.3 Using the sequence of prices given in Example 2.1, $P_0 = 60$, $P_1 = 62.40$, $P_2 = 63.96$, $P_3 = 61.40$, and $P_4 = 66$, the log-prices are given by

$$\begin{aligned} p_0 &= \log(60) = 4.0943, \\ p_1 &= \log(62.40) = 4.1336, \\ p_2 &= \log(63.96) = 4.1583, \\ p_3 &= 4.1174, \end{aligned}$$

and

$$p_4 = 4.1897.$$

It follows that the log-returns are

$$\begin{aligned} r_1 &= p_1 - p_0 = 4.1336 - 4.0943 = 0.0393, \\ r_2 &= p_2 - p_1 = 4.1583 - 4.1366 = 0.0217, \\ r_3 &= p_3 - p_2 = 4.1174 - 4.1583 = -0.0409, \end{aligned}$$

and

$$r_4 = p_4 - p_3 = 4.1897 - 4.1174 = 0.0723.$$

Alternatively, the log-returns may be calculated from the returns; for example $R_1 = 0.04$ so that

$$r_1 = \log(1 + R_1) = \log(1 + 0.04) = \log(1.04) = 0.0392,$$

with the difference between this and our previous result due to round-off error. The three-period log-return at time 4 is

$$r_4(2) = r_3 + r_4 = -0.0409 + 0.0723 = 0.0314;$$

alternatively, using the result from Example 2.2,

$$r_4(2) = \log(1 + R_4(3)) = \log(1 + 0.032) = 0.0315. \qquad \square$$

Dividends

Now suppose that there are dividends. Let D_t represent the dividend paid immediately prior to time t, that is, after time $t-1$ but before time t; for convenience, we will refer to such a dividend as being paid "at time t." Then the gross return from time $t-1$ to time t takes into account the payment of the dividend, along with the change in price; it is defined as

$$1 + R_t = \frac{P_t + D_t}{P_{t-1}}.$$

The net return is given by

$$\begin{aligned} R_t = \frac{P_t + D_t}{P_{t-1}} - 1 &= \left(\frac{P_t}{P_{t-1}} - 1\right) + \frac{D_t}{P_{t-1}} \\ &= \text{(proportional change in price)} \\ &\quad + \text{(dividend as a proportion of price at time } t-1). \end{aligned}$$

Thus, it is possible to make money from an investment in an asset even if the asset's price declines over time.

The multiperiod return from period $t-k$ to period t is defined by an analogy with the no-dividend case:

$$\begin{aligned} 1 + R_t(k) &= (1 + R_t)(1 + R_{t-1}) \cdots (1 + R_{t-k+1}) \\ &= \left(\frac{P_t + D_t}{P_{t-1}}\right)\left(\frac{P_{t-1} + D_{t-1}}{P_{t-2}}\right) \cdots \left(\frac{P_{t-k+1} + D_{t-k+1}}{P_{t-k}}\right). \end{aligned}$$

Example 2.4 Suppose that, as in Example 2.1, $P_0 = 60$, $P_1 = 62.40$, $P_2 = 63.96$, and suppose that there are dividends $D_1 = 2$ and $D_2 = 1$. Then

$$R_1 = \frac{P_1 + D_1}{P_0} - 1 = \frac{62.40 + 2}{60} - 1 = 0.073$$

and

$$R_2 = \frac{P_2 + D_2}{P_1} - 1 = \frac{63.96 + 1}{62.40} - 1 = 0.041.$$

The two-period return at time 2 is

$$R_2(2) = \left(\frac{P_2 + D_2}{P_1}\right)\left(\frac{P_1 + D_1}{P_0}\right) - 1 = (1.073)(1.041) - 1 = 0.117. \qquad \square$$

Note that, when there are dividends, the definition of multiperiod returns assumes that the dividends are reinvested. To see this, consider the following example.

Example 2.5 Consider an asset with prices $P_0 = 8$, $P_1 = 10$, and $P_1 = 12$ and with dividends $D_1 = 2$, $D_2 = 1$. Suppose that our initial investment is \$200. The initial price of the asset is $P_0 = 8$; hence, we buy $200/8 = 25$ shares. The price at time $t = 1$ is $P_1 = 10$; therefore, in time period 1, those shares are worth $(25)(10) = \$250$, plus we receive a dividend of \$2 per share for a total dividend of $2(25) = \$50$. The dividends may be used to buy more of the asset; the price is $P_1 = 10$, so we buy $50/10 = 5$ additional shares for a total of $25 + 5 = 30$ shares at the end of time 1.

At time $t = 2$, the price of the asset is $P_2 = 12$, so those 30 shares are worth $(30)(12) = \$360$, plus we receive a dividend of \$1 per share or $1(30) = \$30$ for the 30 shares, leading to a total worth of \$390. Thus, our initial investment of \$200 is worth \$390, a net return of $390/200 - 1 = 0.95$ over the two periods, which agrees with $[(12 + 1)/10][(10 + 2)/8] - 1 = 0.95$. □

Therefore, the multiperiod return when there are dividends is based on several sources:

- The price increase of the original investment
- The dividends
- The price increase of the shares purchased by the dividends

When there are dividends, the definition of the log-return is analogous to the definition in the no-dividend case:

$$r_t = \log(1 + R_t) = \log(P_t + D_t) - \log(P_{t-1}).$$

Note, however, that the log-return is no longer directly related to the change in the log-price.

Multiperiod returns for log-returns in the presence of dividends are defined as the sum of the single-period log-returns:

$$r_t(k) = r_t + \cdots + r_{t-k+1}, \quad t = k, k + 1, \ldots, T.$$

2.3 Adjusted Prices

An alternative to including dividends explicitly in the calculation of returns is to work with *dividend-adjusted prices*, which we will refer to more simply as *adjusted prices*. Note that we expect the price of a stock to decrease after payment of a dividend.

To see why this is true, consider one share of a particular stock and suppose that a dividend D_t is paid at time t. Investors selling the stock at time $t-1$ will receive P_{t-1}; investors selling the stock at time t receive $P_t + D_t$. Under the assumption that the instrinsic value of the investment is stable from time $t-1$ to time t, we must have

$$P_{t-1} = P_t + D_t,$$

that is,

$$P_t = P_{t-1} - D_t.$$

Thus, when measuring how the value of a share of stock changes from time $t-1$ to time t, we should compare P_t to $P_{t-1} - D_t$ rather than to P_{t-1}. In a sense, the "effective price" at time $t-1$ is $P_{t-1} - D_t$.

This reasoning is the basis for defining adjusted prices. Let $P_0, P_1, \ldots, P_T$ denote a sequence of prices of an asset, let $D_1, D_2, \ldots, D_T$ denote a sequence of dividends paid by the asset, and let $\bar{P}_0, \bar{P}_1, \ldots, \bar{P}_T$ denote the corresponding sequence of adjusted prices. Define $\bar{P}_T = P_T$ and

$$\bar{P}_{T-1} = P_{T-1} - D_T = \left(1 - \frac{D_T}{P_{T-1}}\right) P_{T-1}.$$

Note that the ratio of the adjusted prices may be written

$$\frac{\bar{P}_T}{\bar{P}_{T-1}} = \frac{P_T}{P_{T-1} - D_T} \tag{2.1}$$

so that it reflects the ratio of the prices, taking into account the dividend D_T.

To define the adjusted price at time $T-2$, $\bar{P}_{T-2}$, we use the relationship between $\bar{P}_{T-1}$ and $\bar{P}_{T-2}$ implied by (2.1):

$$\frac{\bar{P}_{T-1}}{\bar{P}_{T-2}} = \frac{P_{T-1}}{P_{T-2} - D_{T-1}}.$$

Solving for $\bar{P}_{T-2}$,

$$\bar{P}_{T-2} = \frac{P_{T-2} - D_{T-1}}{P_{T-1}} \bar{P}_{T-1} = \left(1 - \frac{D_{T-1}}{P_{T-2}}\right) \frac{\bar{P}_{T-1}}{P_{T-1}} P_{T-2}.$$

Using the fact that

$$\frac{\bar{P}_{T-1}}{P_{T-1}} = 1 - \frac{D_T}{P_{T-1}},$$

it follows that

$$\bar{P}_{T-2} = \left(1 - \frac{D_T}{P_{T-1}}\right) \left(1 - \frac{D_{T-1}}{P_{T-2}}\right) P_{T-2}.$$

This relationship may be generalized to

$$\bar{P}_{T-k} = \left(1 - \frac{D_T}{P_{T-1}}\right) \left(1 - \frac{D_{T-1}}{P_{T-2}}\right) \cdots \left(1 - \frac{D_{T-k+1}}{P_{T-k}}\right) P_{T-k}, \quad k{=}1,2,\ldots,T.$$

Thus, the adjusted prices describe the changes in a stock's value, taking into account dividends.

Example 2.6 Consider an asset with prices $P_0 = 60$, $P_1 = 62.40$, and $P_2 = 63.96$ and dividends $D_1 = 2$, $D_2 = 1$. Then $\bar{P}_2 = 63.96$,

$$\bar{P}_1 = \left(1 - \frac{1}{62.40}\right) 62.40 = 61.40,$$

and

$$\bar{P}_0 = \left(1 - \frac{1}{62.40}\right)\left(1 - \frac{2}{60}\right) 60 = 57.07. \qquad \square$$

There are some unexpected properties of adjusted prices that are important to keep in mind. One is that when a dividend occurs in the current period, the entire series of adjusted prices changes.

To see this, let $\bar{P}_0, \bar{P}_1, \ldots, \bar{P}_T$ denote the adjusted prices based on observing prices and dividends for periods $0, 1, \ldots, T$. Now suppose that we observe P_{T+1} and D_{T+1}; let $\tilde{P}_0, \tilde{P}_1, \ldots, \tilde{P}_{T+1}$ denote the adjusted prices based on observing prices and dividends for periods $0, 1, \ldots, T, T+1$. Then $\tilde{P}_{T+1} = P_{T+1}$,

$$\tilde{P}_T = \left(1 - \frac{D_{T+1}}{P_T}\right) P_T = \left(1 - \frac{D_{T+1}}{P_T}\right) \bar{P}_T,$$

$$\tilde{P}_{T-1} = \left(1 - \frac{D_{T+1}}{P_T}\right)\left(1 - \frac{D_T}{P_{T-1}}\right) P_{T-1} = \left(1 - \frac{D_{T+1}}{P_T}\right) \bar{P}_{T-1},$$

and so on. In general,

$$\tilde{P}_{T-k} = \left(1 - \frac{D_{T+1}}{P_T}\right) \bar{P}_{T-k}, \quad k = 1, 2, \ldots, T.$$

Example 2.7 For the asset described in Example 2.6, suppose that we observe an additional time period, with $P_3 = 61.40$ and $D_3 = 3$. Then the updated adjusted prices are $\bar{P}_3 = 61.40$,

$$\bar{P}_2 = \left(1 - \frac{3}{63.96}\right) 63.96 = 60.96,$$

$$\bar{P}_1 = \left(1 - \frac{3}{63.96}\right)\left(1 - \frac{1}{62.40}\right) 62.40 = 58.52,$$

$$\bar{P}_0 = \left(1 - \frac{3}{63.96}\right)\left(1 - \frac{1}{62.40}\right)\left(1 - \frac{2}{60}\right) 60 = 54.39.$$

The series of adjusted prices is now 54.39, 58.52, 60.96, and 61.40, corresponding to periods $0, 1, 2, 3$, respectively. These values can be compared with the series of adjusted prices 57.07, 61.40, and 63.96 for periods $0, 1$, and 2, respectively, that were computed before observing period 3. □

This property of adjusted prices can be confusing when recording adjusted price data at different points in time.

Example 2.8 Consider the price of a share of stock in Exxon Mobil Corporation (symbol XOM); such data are available on the Yahoo Finance website, http://biz.yahoo.com/r/, by following the "Historical Quotes" link under "Research Tools."

On March 4, 2008, the adjusted price for November 30, 2005, was reported as \$57.38, on March 30, 2015, the adjusted price for November 30, 2005, was reported as \$46.77, and on January 7, 2016, the adjusted price for November 30, 2005, was reported as \$45.56.

Note that the unadjusted price for November 30, 2005, was reported as \$58.03 on each of these three dates. □

Although when there is a nonzero dividend the sequence of adjusted prices changes with the addition of a new time period of information, all adjusted prices change by the same factor; hence, the ratios of the adjusted prices are unchanged so that the returns calculated from the adjusted prices do not change.

Example 2.9 For the asset described in Example 2.6, the adjusted prices based on data from time periods $0, 1$, and 2 are 57.07, 61.40, and 63.96, respectively, while the adjusted prices based on data from time periods $0, 1, 2$, and 3 are 54.39, 58.52, 60.96, and 61.40, respectively. Using either set of adjusted prices, the return in period 1 is

$$\frac{61.40 - 57.07}{57.07} = \frac{58.52 - 54.39}{54.39} = 0.0759$$

and the return in period 2 is

$$\frac{63.96 - 61.40}{61.40} = \frac{60.96 - 58.52}{58.52} = 0.0417.$$ □

Another important property is that although adjusted prices incorporate information about dividends, returns calculated from adjusted prices are not exactly equal to returns calculated using the formula for returns in the presence of dividends. That is, although the returns based on adjusted prices do not depend on the sequence of adjusted prices used, they are not the same as the returns calculated using the formula for returns based on unadjusted prices in the presence of dividends. This is illustrated in the following example.

Example 2.10 Recall that in Example 2.6 the asset prices in periods 0 and 1 are given by $P_0 = 60$ and $P_1 = 62.40$ and the dividend in period 1 is $D_1 = 2$. Then the return in period 1 is

$$\frac{62.40 + 2}{60} - 1 = 0.0733.$$

In Example 2.9, it is shown that the return in period 1 based on adjusted prices is 0.0759, which is close to, but not exactly the same as, the value obtained here. □

In general, the return in period 1 based on prices P_0, P_1 and dividend D_1 is

$$\frac{P_1 + D_1}{P_0} - 1 = \frac{P_1}{P_0} - 1 + \frac{D_1}{P_0}. \tag{2.2}$$

The adjusted prices are $\bar{P}_1 = P_1$ and

$$\bar{P}_0 = \left(1 - \frac{D_1}{P_0}\right) P_0.$$

Therefore, the return based on adjusted prices is

$$\frac{\bar{P}_1 - \bar{P}_0}{\bar{P}_0} = \frac{1}{1 - D_1/P_0} \frac{P_1}{P_0} - 1. \tag{2.3}$$

The difference between the return based on the unadjusted prices, given in (2.2), and the return based on adjusted prices, given in (2.3), is

$$\begin{aligned}\left(\frac{P_1}{P_0} - 1 + \frac{D_1}{P_0}\right) - \left(\frac{1}{1 - D_1/P_0} \frac{P_1}{P_0} - 1\right) &= \left(1 - \frac{1}{1 - D_1/P_0}\right) \frac{P_1}{P_0} + \frac{D_1}{P_0} \\ &= -\frac{D_1/P_0}{1 - D_1/P_0} \frac{P_1}{P_0} + \frac{D_1}{P_0} \\ &= \left(1 - \frac{P_1}{P_0 - D_1}\right) \frac{D_1}{P_0}.\end{aligned}$$

Therefore, if either the dividend is a relatively small proportion of the price or the ratio

$$\frac{P_1}{P_0 - D_1}$$

is close to 1, we can expect the difference between the two calculated returns to be minor. Fortunately, in many cases, both $D_1/P_0 \doteq 0$ and $P_1 \doteq P_0 - D_1$ hold.

Example 2.11 Consider the price of a share of Target Corporation stock (symbol TGT). Let P_0 denote the price on May 15, 2015, and let P_1 denote the price on May 18, 2015, with corresponding dividend D_1. Note that May 15, 2015, was a Friday, so that May 15 and May 18 are consecutive trading days. Then $P_0 = \$78.53$, $P_1 = \$78.36$, and $D_1 = \$0.52$. The return for period 1 is

$$R_1 = \frac{78.36}{78.53} - 1 + \frac{0.52}{78.53} = 0.004457.$$

The adjusted prices are $\bar{P}_1 = P_1 = \$78.36$ and

$$\bar{P}_0 = \left(1 - \frac{0.52}{78.53}\right)(78.53) = \$78.01;$$

therefore, the return based on the adjusted prices is

$$\frac{78.36 - 78.01}{78.01} = 0.004487.$$

This is close to, but slightly different than, the actual return calculated previously, with a difference of 0.00003. Note that here $D_1/P_0 = 0.0066$ and $P_1/(P_0 - D_1) = 1.0045$.

Now consider monthly returns. Let P_0 denote the price of one share of Target stock at the end of April 2015, and let P_1 denote the price at the end of May 2015. Then $P_0 = \$78.83$, $P_1 = \$79.32$, and $D_1 = \$0.52$. The adjusted monthly prices are $\bar{P}_0 = \$78.31$ and $\bar{P}_1 = \$79.32$. Then the monthly return on Target stock is 0.01281; using the adjusted prices, the monthly return is 0.01290, again, a slight difference. □

Adjusted stock prices are generally adjusted for stock splits as well as for dividends. A stock split occurs when a company decides to proportionally increase the number of shares owned by investors. For instance, in a two-for-one stock split, the owner of each share of stock is given a second share, in a sense, splitting each share into two. Of course, the price of the shares is adjusted accordingly.

Adjusted prices better reflect changes in the asset's value over time, and they are a useful alternative to the raw prices. In the remainder of this book, the term "prices" will always refer to adjusted prices and the term "returns" will always refer to returns calculated from adjusted prices. The notations used for prices, returns, log-prices, and so on will refer to quantities based on the adjusted prices; for example, P_t will be used to denote the adjusted price of an asset at time t and R_t will be used to denote the return based on adjusted prices.

2.4 Statistical Properties of Returns

Consider the returns $R_1, R_2, \ldots$ on an asset. An important feature of such returns is that they are ordered in time. Hence, we consider the properties of a sequence of random variables $Y_1, Y_2, \ldots$ that are ordered in time.

The set of random variables $\{Y_t : t = 1, 2, \ldots\}$ is called a *stochastic process* and the sequence of observations corresponding to $Y_1, Y_2, \ldots$ is called a *time series*. When analyzing the properties of a stochastic process $\{Y_t : t = 1, 2, \ldots\}$, we consider properties of the random variable Y_t as a function of t.

Although any property of a random variable can be viewed as a function of t by computing it for each Y_t, $t = 1, 2, \ldots$, in practice, we are primarily interested in simple properties such as means and variances. For instance, let

$$\mu_t = \mathrm{E}(Y_t), \quad t = 1, 2, \ldots$$

denote the *mean function* of the process so that $\mu_3 = \mathrm{E}(Y_3)$, for example. Similarly, the *variance function* of the process is given by

$$\sigma_t^2 = \mathrm{Var}(Y_t), \quad t = 1, 2, \ldots.$$

The *covariance function* gives the covariance of two elements of $\{Y_t : t = 1, 2, \ldots\}$ as a function of their indices; it is defined as

$$\gamma_0(t, s) = \text{Cov}(Y_t, Y_s), \quad t, s = 1, 2, \ldots.$$

Hence, $\gamma_0(t,t) = \sigma_t^2$ for any t.

Note that, without further assumptions on the random variables $Y_1, Y_2, \ldots$, it is difficult, if not impossible, to obtain any information about the features of their probability distributions. For instance, if the probability distribution of Y_t is completely different for each t, and we have only one set of observations corresponding to the process, then we only have one observation from each distribution. In such a case, accurate estimation of the properties of the distribution of Y_t is not possible. Fortunately, in many cases, it is reasonable to assume that the properties of random variables Y_t and Y_s for $t \neq s$ are similar.

The strongest condition of this type is the condition that $Y_1, Y_2, \ldots$ are *independent and identically distributed*, often abbreviated to *i.i.d.*; i.i.d. random variables are independent and each has the same marginal distribution. Although this condition is appropriate in some areas of application, it is often too strong for the type of random variables used in modeling financial data.

A similar, but weaker, condition is *stationarity*. The process $\{Y_t : t = 1, 2, \ldots\}$ is said to be *stationary* if the statistical properties of the random variables in the process do not change over time. More formally, the process is stationary if for any integer m and any times $t_1, t_2, \ldots, t_m$ the joint distribution of the vector $(Y_{t_1}, Y_{t_2}, \ldots, Y_{t_m})$ is the same as the joint distribution of the vector $(Y_{t_1+h}, Y_{t_2+h}, \ldots, Y_{t_m+h})$ for any $h = 0, 1, 2, \ldots$. Thus, stationarity is a type of time invariance.

For instance, taking $m = 1$, stationarity requires that Y_t has the same distribution as Y_{t+h} for any integer h; that is, under stationarity, the marginal distribution of Y_t is the same for each t, so that $Y_1, Y_2, \ldots$ are identically distributed. Taking $m = 2$, the joint distribution of (Y_{t_1}, Y_{t_2}) is the same as the joint distribution of (Y_{t_1+h}, Y_{t_2+h}) for any time points t_1, t_2 and any $h = 0, 1, 2, \ldots$. For example, (Y_1, Y_4) must have the same distribution as (Y_2, Y_5), (Y_3, Y_6), (Y_4, Y_7), and so on. This same type of property must hold for any m-tuple of random variables. This condition holds if, in addition to being identically distributed, $Y_1, Y_2, \ldots$ are independent, but independence is not required. Although stationarity is weaker than the i.i.d. property, it is still a strong condition.

Weak Stationarity

In financial applications, the assumption of stationarity is generally stronger than is needed. Furthermore, because it refers to the entire distribution of each random variable, it is difficult to verify in practice. Hence, a weaker version of stationarity, based on means, variances, and covariances, is often used.

The process $\{Y_t : t = 1, 2, \ldots\}$ is said to be *weakly stationary* if

1. $E(Y_t) = \mu$ for all $t = 1, 2, \ldots$, for some constant μ.
2. $\text{Var}(Y_t) = \sigma^2$ for all $t = 1, 2, \ldots$, for some constant $\sigma^2 > 0$.
3. $\text{Cov}(Y_t, Y_s) = \gamma(|t - s|)$ for all $t, s = 1, 2, \ldots$, for some function $\gamma(\cdot)$.

That is, the mean and variance of Y_t do not depend on t and covariance of Y_{t+h}, Y_{s+h} does not depend on h: under Condition (3) of weak stationarity,

$$\text{Cov}(Y_{t+h}, Y_{s+h}) = \gamma(|t + h - (s + h)|) = \gamma(|t - s|).$$

Thus, weak stationarity is essentially the same as stationarity, except that it applies only to the *second-order properties* of the process, the means, variances, and covariances of the random variables.

The function $\gamma(\cdot)$ is called the *autocovariance function* of the process. Note that $\gamma(0) = \text{Cov}(Y_t, Y_t) = \sigma^2$ and $\gamma(h) = \text{Cov}(Y_{t+h}, Y_t)$, $h = 0, 1, \ldots$, for any $t = 1, 2, \ldots$. The correlation of Y_t and Y_s is given by

$$\rho(|t - s|) \equiv \frac{\text{Cov}(Y_t, Y_s)}{\sqrt{\text{Var}(Y_t)\text{Var}(Y_s)}} = \frac{\gamma(|t - s|)}{\sigma^2}.$$

The function $\rho(\cdot)$ is called the *autocorrelation function* of the process.

Example 2.12 Let $Z_0, Z_1, Z_2, \ldots$ denote i.i.d. random variables each with mean μ and standard deviation σ. Define

$$Y_t = Z_t - Z_{t-1}, \quad t = 1, 2, \ldots$$

so that $Y_1 = Z_1 - Z_0$, $Y_2 = Z_2 - Z_1$, and so on, and consider the properties of the process $\{Y_t : t = 1, 2, \ldots.\}$.

Note that

$$E(Y_t) = E(Z_t - Z_{t-1}) = E(Z_t) - E(Z_{t-1}) = \mu - \mu = 0$$

and that

$$\text{Var}(Y_t) = \text{Var}(Z_t - Z_{t-1}) = \text{Var}(Z_t) + \text{Var}(Z_{t-1}) = \sigma^2 + \sigma^2 = 2\sigma^2.$$

Hence, conditions (1) and (2) of weak stationarity are satisfied.

Consider $\text{Cov}(Y_t, Y_s)$. If $t = s - 1$, then

$$\begin{aligned}\text{Cov}(Y_t, Y_s) &= \text{Cov}(Z_{s-1} - Z_{s-2}, Z_s - Z_{s-1}) \\ &= \text{Cov}(Z_{s-1}, Z_s) - \text{Cov}(Z_{s-1}, Z_{s-1}) \\ &\quad - \text{Cov}(Z_{s-2}, Z_s) + \text{Cov}(Z_{s-2}, Z_{s-1}) \\ &= -\text{Var}(Z_{s-1}) = -\sigma^2;\end{aligned}$$

note that, because $\text{Cov}(Y_t, Y_s) = \text{Cov}(Y_s, Y_t)$, the same result holds if $s = t - 1$. Thus, $\text{Cov}(Y_t, Y_s) = -\sigma^2$ if $|t - s| = 1$.

If $|t-s|>1$, then $Y_t = Z_t - Z_{t-1}$ and $Y_s = Z_s - Z_{s-1}$ do not have any terms in common; hence, $\text{Cov}(Y_t, Y_s) = 0$. It follows that

$$\text{Cov}(Y_t, Y_s) = \begin{cases} 2\sigma^2 & \text{if } |t-s| = 0 \\ -\sigma^2 & \text{if } |t-s| = 1 \\ 0 & \text{if } |t-s| = 2, 3, \ldots \end{cases} .$$

Clearly, $\text{Cov}(Y_t, Y_s)$ is a function of $|t-s|$ and, since $\text{E}(Y_t)$ and $\text{Var}(Y_t)$ do not depend on t, $\{Y_t : t = 1, 2, \ldots\}$ is weakly stationary.

The autocovariance function of the process is given by

$$\gamma(h) = \begin{cases} 2\sigma^2 & \text{if } h = 0 \\ -\sigma^2 & \text{if } h = 1 \\ 0 & \text{if } h = 2, 3, \ldots \end{cases}$$

and the autocorrelation function is given by

$$\rho(h) = \begin{cases} 1 & \text{if } h = 0 \\ -\frac{1}{2} & \text{if } h = 1 \\ 0 & \text{if } h = 2, 3, \ldots \end{cases}$$ □

Example 2.13 Let $Z_1, Z_2, \ldots$ denote i.i.d. random variables each with mean 0 and standard deviation σ. Define

$$X_t = \frac{Z_1 + Z_2 + \cdots + Z_t}{\sqrt{t}}, \quad t = 1, 2, \ldots$$

and consider the properties of the process $\{X_t : t = 1, 2, \ldots\}$.

Note that

$$\text{E}(X_t) = \frac{1}{\sqrt{t}}\text{E}(Z_1) + \cdots + \frac{1}{\sqrt{t}}\text{E}(Z_t) = 0$$

and

$$\text{Var}(X_t) = \frac{1}{t}\text{Var}(Z_1) + \cdots + \frac{1}{t}\text{Var}(Z_t) = \frac{1}{t}t\sigma^2 = \sigma^2.$$

Now consider $\text{Cov}(X_t, X_s)$ for $t \neq s$. The calculation is simpler if we know which of t and s is smaller; note that, without loss of generality, we may assume that $t < s$. Then

$$\begin{aligned} \text{Cov}(X_t, X_s) &= \text{Cov}\left(\frac{Z_1 + Z_2 + \cdots + Z_t}{\sqrt{t}}, \frac{Z_1 + Z_2 + \cdots + Z_t}{\sqrt{s}}\right. \\ &\qquad \left. + \frac{Z_{t+1} + Z_2 + \cdots + Z_s}{\sqrt{s}}\right) \\ &= \text{Cov}\left(\frac{Z_1 + Z_2 + \cdots + Z_t}{\sqrt{t}}, \frac{Z_1 + Z_2 + \cdots + Z_t}{\sqrt{s}}\right) \\ &\qquad + \text{Cov}\left(\frac{Z_1 + Z_2 + \cdots + Z_t}{\sqrt{t}}, \frac{Z_{t+1} + Z_{t+2} + \cdots + Z_s}{\sqrt{s}}\right). \end{aligned}$$

Because, for any random variable Y, $\text{Cov}(Y, Y) = \text{Var}(Y)$,

$$\begin{aligned}
\text{Cov}&\left(\frac{Z_1+Z_2+\cdots+Z_t}{\sqrt{t}}, \frac{Z_1+Z_2+\cdots+Z_t}{\sqrt{s}}\right)\\
&= \text{Cov}\left(\frac{Z_1+Z_2+\cdots+Z_t}{\sqrt{t}}, \frac{\sqrt{t}}{\sqrt{s}}\frac{Z_1+Z_2+\cdots+Z_t}{\sqrt{t}}\right)\\
&= \frac{\sqrt{t}}{\sqrt{s}}\text{Cov}\left(\frac{Z_1+Z_2+\cdots+Z_t}{\sqrt{t}}, \frac{Z_1+Z_2+\cdots+Z_t}{\sqrt{t}}\right)\\
&= \frac{\sqrt{t}}{\sqrt{s}}\text{Cov}\left(X_t, X_t\right) = \frac{\sqrt{t}}{\sqrt{s}}\text{Var}\left(X_t\right)\\
&= \frac{\sqrt{t}}{\sqrt{s}}\sigma^2.
\end{aligned}$$

The sums $Z_1 + Z_2 + \cdots + Z_t$ and $Z_{t+1} + Z_{t+2} + \cdots + Z_s$ have no terms in common; hence, using the fact that $Z_1, Z_2, \ldots$ are independent, these sums have covariance equal to 0. It follows that

$$\text{Cov}(X_t, X_s) = \frac{\sqrt{t}}{\sqrt{s}}\sigma^2.$$

This result holds when $t < s$; if $s < t$, the same basic result holds, switching the roles of t and s:

$$\text{Cov}(X_t, X_s) = \frac{\sqrt{s}}{\sqrt{t}}\sigma^2.$$

These results may be combined by stating that, for any t, s,

$$\text{Cov}(X_t, X_s) = \sqrt{\frac{\min\{t, s\}}{\max\{t, s\}}}\,\sigma^2.$$

Therefore, although the mean and variance of X_t do not depend on t, the covariance of X_t and X_s is not a function of $|t - s|$. For instance,

$$\text{Cov}(X_1, X_5) = \frac{1}{5}\sigma^2$$

while

$$\text{Cov}(X_{11}, X_{15}) = \frac{11}{15}\sigma^2.$$

Hence, $\{X_t : t = 1, 2, \ldots\}$ is not weakly stationary. □

The property of weak stationarity greatly simplifies the statistical analysis of the process. For instance, because $\text{E}(Y_t) = \mu$ for all t, we expect that

$$\bar{Y} = \frac{1}{T}\sum_{t=1}^{T} Y_t$$

will be a reasonable estimator of μ. To estimate $\rho(1)$, the correlation of two observations one time period apart, we can use the sample correlation of pairs

$(Y_1, Y_2), (Y_2, Y_3), \ldots$, and so on. Parameter estimation using these ideas will be considered in detail in the following section.

Weak White Noise

A particularly simple example of a weakly stationary process is a sequence of random variables $Z_1, Z_2, \ldots$ such that, for each $t = 1, 2, \ldots$, $\mathrm{E}(Z_t) = \mu$ and $\mathrm{Var}(Y_t) = \sigma^2$, for some constants μ and $\sigma^2 > 0$, such that, for each $t, s = 1, 2, \ldots$, $t \neq s$, $\mathrm{Cov}(Z_t, Z_s) = 0$. A process with these properties is called *weak white noise.*

The autocovariance function of a weak white noise process $\{Z_t : t = 1, 2, \ldots\}$ is given by

$$\gamma(h) = \begin{cases} \sigma^2 & \text{if } h = 0 \\ 0 & \text{if } h = 1, 2, \ldots \end{cases}$$

and the autocorrelation function of the process is given by

$$\rho(h) = \begin{cases} 1 & \text{if } h = 0 \\ 0 & \text{if } h = 1, 2, \ldots \end{cases}.$$

Example 2.14 Let $\{Z_t : t = 1, 2, \ldots\}$ be a weak white noise process such that, for each $t = 1, 2, \ldots$, $\mathrm{E}(Z_t) = 0$ and $\mathrm{Var}(Z_t) = 1$. Let Z be a random variable with mean 0 and variance 1 such that $\mathrm{Cov}(Z, Z_t) = 0$ for $t = 1, 2, \ldots$ and let

$$Y_t = Z + Z_t, \quad t = 1, 2, \ldots.$$

Then, for all t,

$$\mathrm{E}(Y_t) = \mathrm{E}(Z) + \mathrm{E}(Z_t) = 0,$$

$$\mathrm{Var}(Y_t) = \mathrm{Var}(Z) + \mathrm{Var}(Z_t) = 1 + 1 = 2,$$

and for all t, s, $t \neq s$,

$$\begin{aligned} \mathrm{Cov}(Y_t, Y_s) &= \mathrm{Cov}(Z + Z_t, Z + Z_s) \\ &= \mathrm{Cov}(Z, Z) + \mathrm{Cov}(Z, Z_s) + \mathrm{Cov}(Z_t, Z) + \mathrm{Cov}(Z_t, Z_s) \\ &= \mathrm{Var}(Z) \\ &= 1. \end{aligned}$$

Thus, $\{Y_t : t = 1, 2, \ldots\}$ is a weakly stationary process with autocovariance function

$$\gamma(h) = \begin{cases} 2 & \text{if } h = 0 \\ 1 & \text{if } h = 1, 2, \ldots \end{cases}$$

and autocorrelation function $\rho(h) = 1/2$, $h = 1, 2, \ldots$.

Now define $X_t = Z_1 + \ldots + Z_t$, $t = 1, 2, \ldots$. Then $\mathrm{E}(X_t) = \mathrm{E}(Z_1) + \cdots + \mathrm{E}(Z_t) = 0$, $\mathrm{Var}(X_t) = \mathrm{Var}(Z_1) + \cdots + \mathrm{Var}(Z_t) = t$.

To find the covariance of X_t and X_s, we may use the same general approach used in Example 2.13. Note that, without loss of generality, we may assume that $t < s$. Then

$$\begin{aligned}\text{Cov}(X_t, X_s) &= \text{Cov}(Z_1 + \cdots + Z_t, Z_1 + \cdots + Z_s) \\ &= \text{Cov}(Z_1 + \cdots + Z_t, Z_1 + \cdots + Z_t) \\ &\quad + \text{Cov}(Z_1 + \cdots + Z_t, Z_{t+1} + \cdots + Z_s).\end{aligned}$$

Because Z_t and Z_s are uncorrelated for $t \neq s$ and each Z_t has variance 1,

$$\text{Cov}(X_t, X_s) = t;$$

thus, $\text{Cov}(X_t, X_s) = \min\{t, s\}$. Because the $\text{Var}(X_t)$ depends on t and the covariance of X_t, X_s is not a function of $|t - s|$, the process $\{X_t : t = 1, 2, \ldots\}$ is not weakly stationary. □

Application to Asset Returns

These ideas may be applied to the stochastic process $\{R_t : t = 1, 2, \ldots\}$ corresponding to the returns on an asset. Weak stationarity implies that the second-order properties of the returns do not change over time: $\text{E}(R_t)$ and $\text{Var}(R_t)$ do not depend on t and $\text{Cov}(R_t, R_s)$ is a function of $|t - s|$. Hence, under weak stationarity, we may refer to the mean return on an asset and the asset's return standard deviation, also known as the *volatility* of the asset, with the understanding that such parameters refer to all time periods under consideration.

The autocorrelation function of $\{R_t : t = 1, 2, \ldots\}$ describes the correlation structure of the returns and the relationships between returns in different time periods. In many cases, it is appropriate to model returns as a weak white noise process, simplifying the analysis; assumptions of this type are discussed in detail in Chapter 3.

2.5 Analyzing Return Data

In order to develop models for return data, it is important to understand its properties. Hence, in this section, we consider several statistical methods that are useful in describing the properties of return data as well as for investigating the appropriateness of assumptions such as weak stationarity.

Asset price data is widely available on the Internet. Here we use data taken from the finance.yahoo.com website, using the R function `get.hist.quote`, in the `tseries` package (Trapletti and Hornik 2016) that directly downloads the data into R.

The arguments of `get.hist.quote` are `instrument`, which refers to the stock symbol of interest, `start` and `end`, which specify the starting and ending dates of the time period under consideration, quote, which specifies the data to be downloaded, for which we use AdjClose for the adjusted closing price, and compression, which specifies the *sampling frequency* of the data, the time interval over which data are recorded. We may view this choice in terms of the *return interval*, the length of the time period over which each return is calculated. Typical choices are days or months, but sometimes weeks or years are used.

Daily data have the advantage that more observations are available in a given time period and that they may reflect more subtle changes in the price of the asset. On the other hand, investment decisions are often made on a monthly basis and, in many cases, monthly returns are more stable than daily returns. Hence, both daily and monthly data are commonly used. For daily, data, we use compression = "d"; for monthly data, "d" is replaced by "m."

Example 2.15 Suppose we would like to analyze data on Wal-Mart Stores, Inc., stock (symbol WMT) for the time period 2010–2014. The relevant R commands are

```
> library(tseries)
> x<-get.hist.quote(instrument="WMT", start="2009-12-31",
+ end="2014-12-31", quote="AdjClose", compression="d")
> wmt<-as.vector(x)
```

This command assigns five years of Wal-Mart price data to the variable x; the format of x is known as "zoo," an R data format for irregularly observed time series. Here, we analyze the prices as a standard vector; hence, the command wmt<-as.vector(x) converts x to a standard vector and assigns it to the variable wmt. To check the contents of wmt, we can use the command head, which displays the first few elements of the vector.

```
> head(wmt)
[1] 45.64114 46.30719 45.84608 45.74361 45.76923 45.53868
> length(wmt)
[1] 1259
```

Thus, the first adjusted price of Wal-Mart stock in the sequence is $45.64114 and there are 1259 prices in the variable wmt.

The number of significant figures displayed can be controlled by the digits argument of the options function. For instance, options(digits=5) limits the number of significant figures printed to five. Throughout this book, the number of digits will be adjusted without comment, based on the context of the example, the desire to fit the output to the page, and so on.

```
> options(digits=5)
> head(wmt)
[1] 45.641 46.307 45.846 45.744 45.769 45.539
```

Note that, when displaying the contents of a vector, the number of digits shown is chosen so that all elements of the vector have the required number of significant figures. For example,

```
> options(digits=2)
> c(11/100000, 1/3)
[1] 0.00011 0.33333
```

The corresponding returns and log-returns for Wal-Mart stock may be calculated using the commands

```
>  wmt.ret<-(wmt[-1] - wmt[-1259])/wmt[-1259]
> head(wmt.ret)
[1]  0.0145931 -0.0099576 -0.0022350  0.0005600 -0.0050372
     0.0165010
> wmt.logret<-log(wmt[-1]) - log(wmt[-1259])
> head(wmt.logret)
[1]  0.0144876 -0.0100075 -0.0022375  0.0005598 -0.0050500
     0.0163663
```

In these expressions, `wmt[-1]` returns a vector that is identical to `wmt`, except that the first element has been dropped; similarly, `wmt[-1259]` returns a vector that is identical to `wmt`, except that the 1259th or, in this case the last, element has been dropped. It follows that `wmt[-1] - wmt[-1259]` is a vector of price differences of the form $P_t - P_{t-1}$.

The function `summary` gives several summary statistics for a variable—the minimum and maximum values, the sample mean, the sample median, and the upper and lower sample quartiles; `sd` gives the sample standard deviation.

```
> summary(wmt.ret)
     Min.    1st Qu.     Median       Mean    3rd Qu.       Max.
 -0.04660   -0.00452    0.00066    0.00052    0.00558    0.04720
> sd(wmt.ret)
[1] 0.0091715
```

The function `quantile` can be used to calculate additional sample quantiles of a variable; for example,

```
> quantile(wmt.ret, probs=c(0.05, 0.10, 0.25, 0.5, 0.75, 0.90,
+ 0.95))
      5%        10%        25%        50%        75%        90%
 -0.01352   -0.00940   -0.00452    0.00066    0.00558    0.01093
    95%
 0.01445
```

Therefore, roughly 5% of the sample values are less than or equal to -0.01352; the help file for the function `quantile` gives details on the exact method of calculation of the sample quantiles. Note that the 25% and 75% quantiles correspond to the sample quartiles calculated using the function `summary` and the 50% quantile corresponds to the sample median.

The following commands give plots of the Wal-Mart stock prices and returns, given in Figures 2.1 and 2.2, respectively. Note that, if the `plot` command has only one argument, it is understood to be the y-variable, with the x-variable taken to be the corresponding index (1 to 1259 in the case of

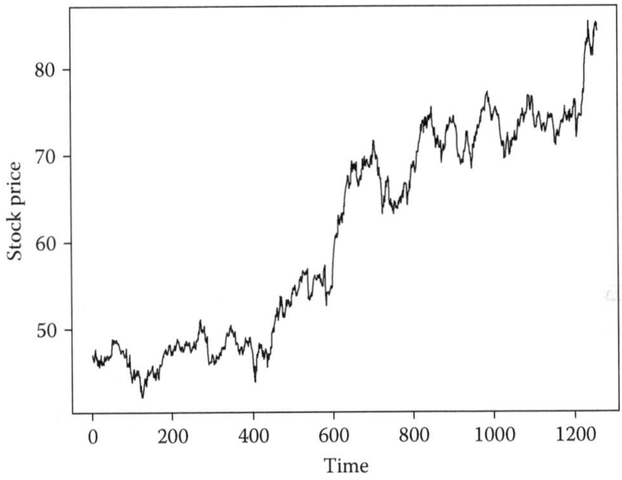

FIGURE 2.1
Time series plot of Wal-Mart daily stock prices.

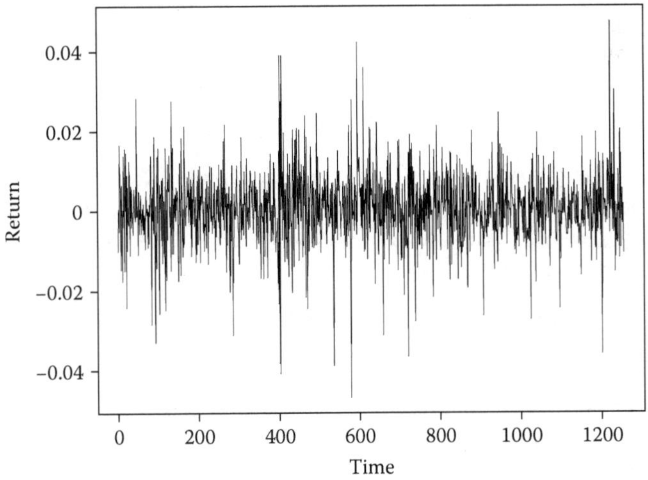

FIGURE 2.2
Time series plot of Wal-Mart daily stock returns.

wmt and 1 to 1258 in the case of wmt.ret); type="l" specifies that the plot be drawn as lines connecting the points, which are not displayed.

```
> plot(wmt, type="l", ylab="Price", xlab="Time")
> plot(wmt.ret, type="l", ylab="Return", xlab="Time")
```

□

Monthly Returns

So far in this section, we have analyzed daily prices and daily returns. In practice, models for asset returns are often based on monthly returns, which, in many cases, correspond better to the investment horizon of interest and are often more stable than daily returns. To obtain monthly returns, we use the get.hist.quote function with compression="m."

Example 2.16 The commands

```
>  x<-get.hist.quote(instrument="WMT", start="2009-12-31",
+  end="2014-12-31", quote="AdjClose", compression="m")
>  wmt.m<-as.vector(x)
```

return the prices of Wal-Mart stock for the last trading day of the month, for each month from December 2009 to December 2014, storing them in the vector wmt.m.

Thus, there are 61 monthly prices in wmt.m, which lead to 60 monthly returns.

```
> length(wmt.m)
[1] 61
> wmt.m.ret<-(wmt.m[-1]-wmt.m[-61])/wmt.m[-61]
> length(wmt.m.ret)
[1] 60
> head(wmt.m.ret)
[1] -0.000374  0.011978  0.034093 -0.035252 -0.051944 -0.049248
```

There is one possible pitfall when downloading monthly data—the last price returned corresponds to the last trading day that occurred on or before the day listed, even if that day is not the last day of the month. For example, if we use the command

```
> x<-get.hist.quote(instrument="WMT", start="2009-12-31",
+ end="2014-12-15", quote="AdjClose", compression="m")
```

the last price in x will be the price for December 15, 2014. Thus, the last monthly return will correspond to only the first half of December 2014. Therefore, when downloading monthly returns, it is important that the end date be the last day of the month under consideration.

Figure 2.3 contains a time series plot of the monthly returns on Wal-Mart stock.

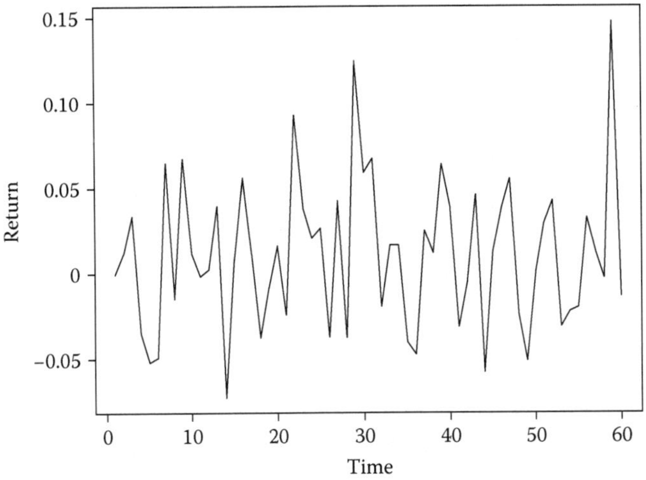

FIGURE 2.3
Time series plot of Wal-Mart stock monthly returns.

When analyzing a plot of monthly returns, sometimes it is easier to notice certain features of the sequence of returns if the plot includes the points corresponding to the return values along with the lines connecting the values. This can be achieved using the R `plot` function by using the argument `type="b,"` where `"b"` indicates both points and lines. An example of this is given in Figure 2.4. In principle, the same approach can be used when constructing a

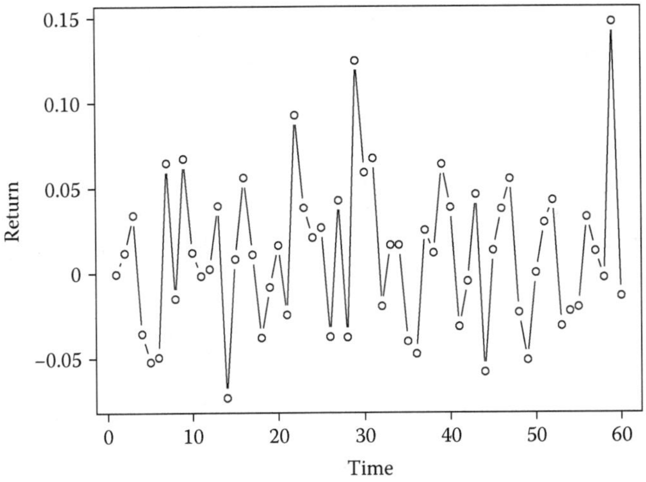

FIGURE 2.4
Alternative time series plot of Wal-Mart stock monthly returns.

plot of daily returns; however, when plotting a large number of values, the points tend to overwhelm the plot. □

Monthly returns are, of course, closely related to daily returns. If $r_1, \ldots, r_{21}$ denote the daily log-returns for a given month, then the monthly log-return for that month is simply $r_1 + \cdots + r_{21}$. Therefore, we expect the mean monthly log-return to be about 21 times as large as the mean daily return; note that there are, roughly 252 trading days in a year and $252/12 = 21$. If the daily returns are uncorrelated with each other, we expect the standard deviation of the monthly log-returns to be about 4.6 times as large as the standard deviation of the daily returns ($\sqrt{21} \doteq 4.6$). Also, a time series plot of monthly returns will be "smoother" than the corresponding plot of daily returns.

These relationships do not hold exactly for standard returns, but they hold approximately. Recall that the return at time t, R_t, and the corresponding log-return, r_t, are related by $r_t = \log(1 + R_t)$ and, using a Taylor's series approximation,

$$\log(1 + R_t) \doteq R_t.$$

Therefore, we expect that the mean monthly return on an asset will be about 21 times the mean daily return and the monthly return standard deviation will be about 4.6 times the daily return standard deviation.

Example 2.17 Let `wmt.m.ret` denote five years of monthly returns on Wal-Mart stock for the period ending December 2014. Recall that the daily returns for this stock were analyzed in Example 2.15.

The summary statistics for `wmt.m.ret` are

```
> summary(wmt.m.ret)
     Min.  1st Qu.   Median     Mean  3rd Qu.     Max.
-0.07294 -0.02235  0.01192  0.01094  0.03848  0.14780
> sd(wmt.m.ret)
[1] 0.0441572
```

The mean monthly return is about 21.6 times the mean daily return and the standard deviation of the monthly returns is about 4.8 times as large as the standard deviation of the daily returns; both of these ratios are close to what we would expect. □

Running Means and Standard Deviations

Let R_t denote the return on an asset at time t. If $\{R_t : t = 1, 2, \ldots\}$ is a weakly stationary process, then $\mu_t = \mathrm{E}(R_t)$ and $\sigma_t^2 = \mathrm{Var}(R_t)$ are constant as functions of t. A time series plot of the observed returns, like the ones in Figures 2.2 and 2.3, provides some evidence regarding the assumption of weak stationarity: If the underlying process is weakly stationary, we expect the level of the observed returns as well as their variation to be approximately constant over time. However, the variability inherent in these plots makes assessment of such properties difficult.

For instance, suppose we are interested in the mean function μ_t. If μ_t is not constant as a function of t, then we have, in general, only a single observation, R_t, with expected value μ_t; hence, it is difficult, if not impossible, to accurately estimate μ_t. Thus, it is important to attempt to determine if the observed returns are consistent with a mean function μ_t that is constant over time. One approach to assessing the extent to which μ_t varies with t is to calculate *running means.*

Suppose we observe returns $R_1, R_2, \ldots, R_T$ and consider running means based on w observations for some positive integer w. Then the first running mean is the average of the first w observations:

$$\bar{R}_{1,w} \equiv \frac{1}{w}\sum_{t=1}^{w} R_t;$$

the second running mean is the average of the next w observations, $R_2, R_3, \ldots, R_{w+1}$,

$$\bar{R}_{2,w} \equiv \frac{1}{w}\sum_{t=2}^{w+1} R_t,$$

and so on. The result is a sequence $\bar{R}_{1,w}, \bar{R}_{2,w}, \ldots, \bar{R}_{T-w+1,w}$ such that each element in the sequence is an average of w returns.

Note that

$$\mathrm{E}(\bar{R}_{t,w}) = \frac{1}{w}\sum_{j=t}^{w+t-1} \mu_j$$

so that if μ_t is constant as a function of t, $\mu_t = \mu$ for all t, then $\mathrm{E}(\bar{R}_{t,w}) = \mu$ and we expect $\bar{R}_{t,w}$ to be approximately constant as a function of t; in this sense, the information provided by the $\bar{R}_{t,w}$ is similar to the information provided by the R_t. However, because each $\bar{R}_{t,w}$ is an average of w returns, it has smaller variance than does R_t, particularly if w is fairly large. Therefore, if μ_t varies with t, we expect that variation to be reflected in the $\bar{R}_{t,w}$.

Under the assumption that the μ_t and σ_t are constant as a function of t, the standard error of $\bar{R}_{t,w}$ is given by $S/\sqrt{w}$, where S denotes the sample standard deviation of $R_1, R_2, \ldots, R_T$; this standard error provides some basis for evaluating the observed variation in $\bar{R}_{1,w}, \bar{R}_{2,w}, \ldots, \bar{R}_{T-w+1,w}$.

In R, running means may be calculated using the function `running` in the package `gtools` (Warnes et al. 2015).

Example 2.18 For the monthly returns on Wal-Mart stock, stored in the variable `wmt.m.ret`, the running means based on a width of $w = 12$ are calculated by

```
> library(gtools)
> wmt.rmean<-running(wmt.m.ret, fun=mean, width=12)
```

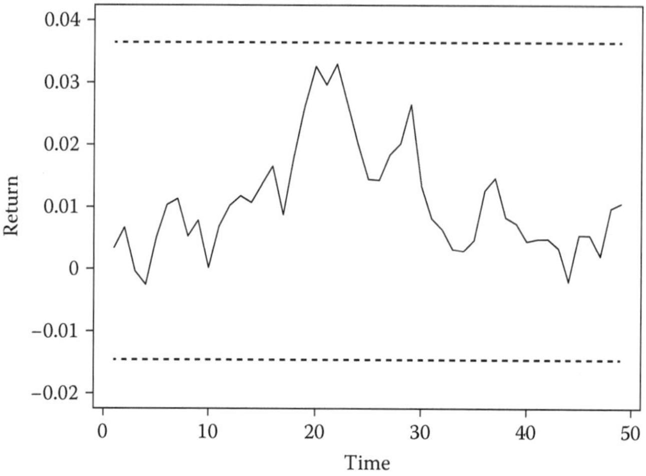

FIGURE 2.5
Time series plot of running means of monthly returns on Wal-Mart stock.

The variable `wmt.rmean` is a vector of running means, each based on 12 months of monthly returns; Figure 2.5 contains a time series plot of the running means of Wal-Mart monthly returns, together with "error bars" of the form $\bar{R} \pm 2S/\sqrt{w}$. The plot was calculated using the commands given here.

```
> plot(wmt.rmean, type="l", ylim=c(-0.02, 0.040), xlab="Time",
+ ylab="Return")
> lines(1:49, rep(mean(wmt.m.ret) + 2*sd(wmt.m.ret)/(12^.5), 49),
+ lty=2)
> lines(1:49, rep(mean(wmt.m.ret) - 2*sd(wmt.m.ret)/(12^.5), 49),
+ lty=2)
```

The function `lines` adds lines to an existing plot. In this case, each line is horizontal at $y = \bar{R} \pm 2S/\sqrt{w}$; the command `rep` constructs a vector of length 49 (based on the second argument) by repeating the first argument. The argument `lty` in `lines` sets the type of line to be used; in this case, `lty=2` specifies a dotted line. An alternative to the function `lines` is the function `abline`, which adds a line to an existing plot by specifying the slope, the argument `b`, and the y-intercept, the argument `a`, to the plot. Therefore, to add the upper error bar to the plot, we can use the command

```
> abline(a=mean(wmt.m.ret) + 2*sd(wmt.m.ret)/(12^.5), b=0, lty=2)
```

Note that the argument `ylim` is included in `plot` to set the y-limits in the plot to be large enough so that the error bars appear on the plot. □

When analyzing a plot such as Figure 2.5, there are a few things to keep in mind. For instance, the error bars are based on the standard error of a single sample mean centered at the sample mean based on all observations. It is important to realize that the plot is based on many sample means (49 in the case of Figure 2.5); the standard error does not apply to the maximum of those sample means, for example. Therefore, we would not conclude that μ_t is nonconstant simply because the plotted line crosses an error bar at some point. Also, the random variables corresponding to the plotted running means are often highly correlated; in the present case, adjacent running means are based on sets of 12 observations, 10 of which are included in both running means. Hence, if one running mean is large, it is likely that the neighboring running means are large as well. Therefore, the error bars are useful for giving a rough idea of the expected variability in the running means if the underlying process is weakly stationary; however, they should not be used for any type of formal inference.

The same approach used here for running means may be used for any other summary statistic of the returns. Most useful in this regard is the standard deviation; recall that if $\{R_t : t = 1, 2, \ldots\}$ is a weakly stationary process, then $\sigma_t^2 = \mathrm{Var}(R_t)$ does not depend on t. Running standard deviations may also be calculated using the `running` command.

Example 2.19 To calculate the running standard deviations for the Wal-Mart monthly returns in the variable `wmt.m.ret`, using a width of $w = 12$, we use the command

```
wmt.rsd<-running(wmt.m.ret, fun=sd, width=12)
```

Since the distribution of the sample standard deviation tends to be a skewed distribution (e.g., a chi-squared distribution if the returns are normally distributed), it is generally preferable to plot the log of the standard deviation rather than the standard deviation itself. It may be shown that the error bars for the logs of the running sample standard deviations based on w observations are of the form $\log(S) \pm \sqrt{2/(w-1)}$, where S is the sample standard deviation of all the values. A plot of this type for monthly returns of Wal-Mart stock is given in Figure 2.6. □

Sample Autocorrelation Function

Recall that the autocorrelation function $\rho(\cdot)$ of a weakly stationary stochastic process $\{Y_t : t = 1, 2, \ldots\}$ describes its correlation structure. For instance, $\rho(2)$ is the correlation of Y_t and Y_{t+2}; provided that the process is weakly stationary, this correlation does not depend on the value of t. The autocorrelation function may be written as $\rho(h) = \gamma(h)/\gamma(0)$, where $\gamma(\cdot)$ is the autocovariance function of the process; that is, $\gamma(h)$ is the covariance of Y_t, Y_{t+h}, $h = 0, 1, 2, \ldots$. In practice, the autocorrelation function is unknown and must be estimated from the data.

For example, suppose that we observe returns $R_1, R_2, \ldots, R_T$ and consider estimating $\rho(1)$, the correlation of two observations one time period apart.

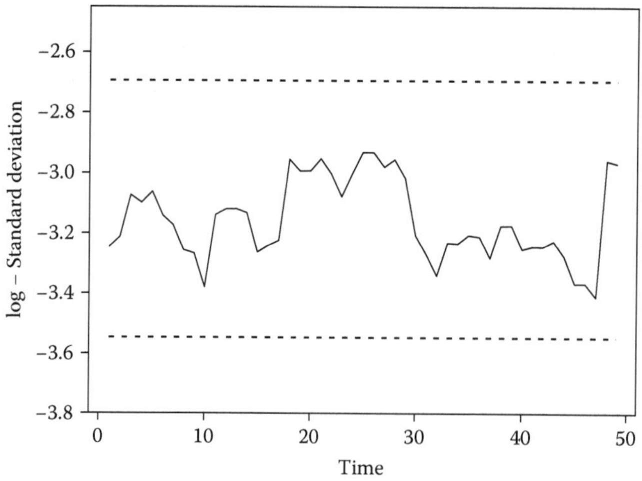

FIGURE 2.6
Time series plot of running standard deviations of monthly returns on Wal-Mart stock.

From these data we have $T - 1$ pairs of observations one time period apart:

$$(R_1, R_2), (R_2, R_3), \ldots, (R_{T-1}, R_T).$$

Then $\gamma(1)$, the covariance of two observations one time period apart, may be estimated by

$$\hat{\gamma}(1) = \frac{1}{T} \sum_{t=1}^{T-1} (R_t - \bar{R})(R_{t+1} - \bar{R}).$$

In general, we estimate $\gamma(h)$ by

$$\hat{\gamma}(h) = \frac{1}{T} \sum_{t=1}^{T-h} (R_t - \bar{R})(R_{t+h} - \bar{R}).$$

The sample autocorrelation function is given by

$$\hat{\rho}(h) = \frac{\hat{\gamma}(h)}{\hat{\gamma}(0)}, \quad h = 1, 2, \ldots.$$

Note that in $\hat{\gamma}(h)$, we use $\bar{R}$ in place of the sample means of $(R_1, R_2, \ldots, R_{T-h})$ and $(R_{h+1}, \ldots, R_T)$. Also, we take the divisor to be T, even though there are only $T - h$ terms in the sum. It may be shown that these choices lead to estimates $\hat{\rho}(h)$ with the usual properties expected of correlations. For instance, if the divisor $T - h$ is used in place of T, then it is possible for $\hat{\rho}(h)$ to take values outside the interval $[-1, 1]$. Similarly, we take $\hat{\gamma}(0)$ to be the sample variance of $Y_1, \ldots, Y_T$. An estimator of $\rho(1)$ is then given by

$$\hat{\rho}(1) = \frac{\hat{\gamma}(1)}{\hat{\gamma(0)}}.$$

To calculate the sample autocorrelation function for a set of observed returns in R, we can use the function `acf`.

Example 2.20 Consider the monthly returns on Wal-Mart stock stored in the variable `wmt.m.ret`. To calculate the sample autocorrelation of these returns, we use

```
> print(acf(wmt.m.ret, lag.max=12))
Autocorrelations of series wmt.m.ret, by lag
    0       1        2        3        4        5       6
 1.000   -0.06    -0.04    -0.156   -0.043   -0.122  -0.007
 7       8        9      10       11       12
 0.119   0.043    0.136  -0.075   -0.221    0.000
```

The argument `lag.max` sets the maximum value of h for which to calculate $\hat{\rho}(h)$; for monthly data, 12 is a reasonable maximum value. The `acf` command also produces a plot of the sample autocorrelation function; the plot for Wal-Mart monthly returns is given in Figure 2.7. Note that, without using the function `print`, only the plot is produced.

The dashed lines in the plot are at the values $\pm(2/\sqrt{T})$; if the true autocorrelation is 0, the standard error of the estimated autocorrelation is $1/\sqrt{T}$.

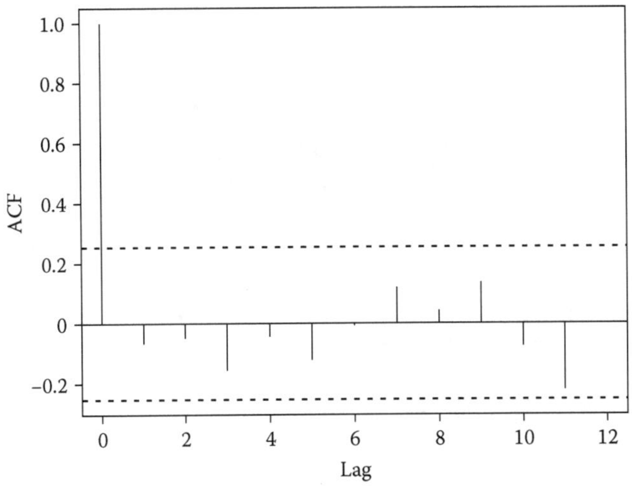

FIGURE 2.7
Sample autocorrelation function for monthly returns on Wal-Mart stock.

Thus, the dashed lines give some indication of the statistical significance of the autocorrelation estimates. However, it is important to keep in mind that this standard error applies to each individual estimate (not the maximum of a series of estimates, for example); thus, these error bars on the plot are best used as a rough guide when evaluating the magnitude of the estimates. Also note that the plot includes the value of the sample autocorrelation function at $h = 0$, which is always 1; thus, this value contains no information. □

For the Wal-Mart monthly returns, all sample autocorrelations are relatively small in magnitude and are roughly consistent with a series of uncorrelated random variables. A formal hypothesis test based on the sample autocorrelation function will be discussed in Section 3.5.

Shape of the Return Distribution

Although it is not directly related to the second-order properties of $\{R_t : t = 1, 2, \ldots\}$, the shape of the return distribution is also of interest. For instance, is the return distribution symmetric or skewed? Does it follow the familiar "bell-shaped" curve of the normal distribution or is it "long-tailed," with more very large and very small values than would be expected from normally distributed data? In order to investigate these questions, we need to assume that the returns $R_1, R_2, \ldots$ are identically distributed, an assumption we maintain through the remainder of this section.

The simplest method of assessing the shape of the distribution is to compute a histogram of the data.

Example 2.21 Consider the daily returns on Wal-Mart stock stored in the R variable `wmt.ret`. The function `hist` can be used to plot a histogram of the data

```
> hist(wmt.m.ret)
```

The plot of the histogram is given in Figure 2.8.

One drawback of histograms is that they are sensitive to the number of intervals used in their construction. The default value in R uses "Sturges' rule" in which the number of intervals is based on $\log_2(T) + 1$ where $\log_2$ denotes the base-2 logarithm and T is the sample size. An alternative method that often yields more informative histograms is the "Freedman–Diaconis rule," which is based on $2(\text{IQR})/T^{\frac{1}{3}}$, where IQR denotes the inter-quartile range of the data, the distance between the lower and the upper quartiles. Thus, Sturges' rule uses only the number of data points, while the Freedman–Diaconis rule also takes into account the variability in the data.

To use the Freedman–Diaconis rule, the argument `breaks="FD"` is added to the `hist` function.

```
> hist(wmt.ret, breaks="FD")
```

The plot of this histogram is given in Figure 2.9. □

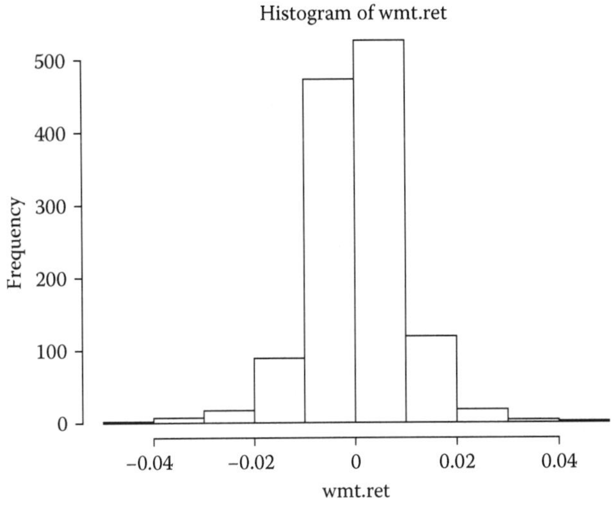

FIGURE 2.8
Histogram of daily returns on Wal-Mart stock.

It is also possible to choose the number of intervals in the histogram explicitly, by specifying the argument `breaks` as a positive integer, or to choose breakpoints between the intervals by specifying `breaks` as a vector of breakpoints.

A histogram is useful for assessing general features of the return distribution; for instance, in the example, the distribution of daily returns on Wal-Mart stock appears to be roughly symmetric. However, some features of the distribution are difficult to evaluate using a histogram. For instance, based on Figures 2.8 and 2.9, it is difficult to determine if the distribution of daily returns on Wal-Mart stock is long-tailed or not. This is particularly true when analyzing monthly returns for which the sample size is generally much smaller.

An alternative approach to assessing the shape of a distribution relative to that of a normal distribution is to use a *normal probability plot.* In a normal probability plot, the sample quantiles of a set of data are plotted versus the quantiles of the standard normal distribution; therefore, such a plot is also called a *quantile–quantile plot* or a *Q–Q plot.*

Because the quantiles of a normal distribution with mean μ and standard deviation σ are of the form

$$\mu + \sigma z_\alpha$$

where z_α is a quantile of the standard normal distribution, an approximately linear normal probability plot indicates that the shape of the sample distribution is approximately normal. Different types of nonlinearity indicate different deviations from normality.

In R, a normal probability plot can be obtained using the function `qqnorm`.

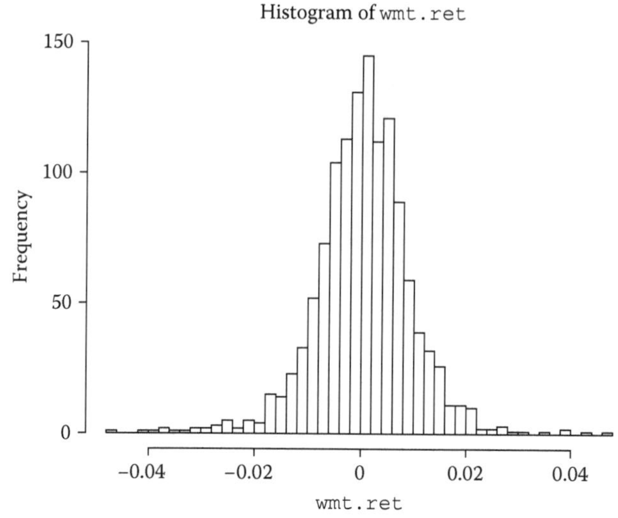

FIGURE 2.9
Histogram of daily returns on Wal-Mart stock using the Freedman–Diaconis rule.

Example 2.22 Before constructing a normal probability plot for observed return data, it is helpful to look at such plots for samples from known distributions.

Samples of 100 data values were randomly generated from four distributions: the standard normal distribution, the t-distribution with three degrees of freedom, the chi-squared distribution with two degrees of freedom, and a uniform distribution on the interval $[0, 1]$. The t-distribution is symmetric but long-tailed, the chi-squared distribution is skewed with a long right tail, and the uniform distribution is symmetric but short-tailed.

In assessing the linearity of a normal probability plot, it is helpful to include a reference line in the plot; one choice for such a line is the one with the slope given by the sample standard deviation of the data and y-intercept given by the sample mean of the data. Based on the earlier discussion, if the data are approximately normally distributed, the normal probability plot should follow such a line, at least approximately.

The R functions **rnorm**, **rt**, **rchisq**, and **runif** can be used to generate data from the four distributions described previously. The specific commands for our case are

```
> x_norm<-rnorm(100)
> x_t<-rt(100, df=3)
> x_chisq<-rchisq(100, df=5)
> x_unif<-runif(100)
```

To construct the normal probability plot for the data in **x_norm**, we use the command

```
> qqnorm(x_norm)
```

To add the reference line, we can use the function **abline**, as described previously:

```
>abline(a=mean(x_norm), b=sd(x_norm))
```

Figure 2.10 gives the normal probability plots for the four randomly generated sets of data. Note that the plotted points for the normal data generally follow the line, although there are some minor deviations. The points for the t data are below the line when the normal quantile is large and negative and are above the line when the normal quantile is large and positive. This indicates that the quantiles of the sample corresponding to probabilities close to one are more extreme than those of the normal distribution, and the quantiles of the sample corresponding to probabilities close to zero are also more extreme than those of the normal distribution; that is, the sample distribution is long-tailed.

The chi-squared distribution has a long right tail but a short left tail. Therefore, the quantiles corresponding to probabilities close to one are larger than those of the normal distribution while the reverse is true for quantiles

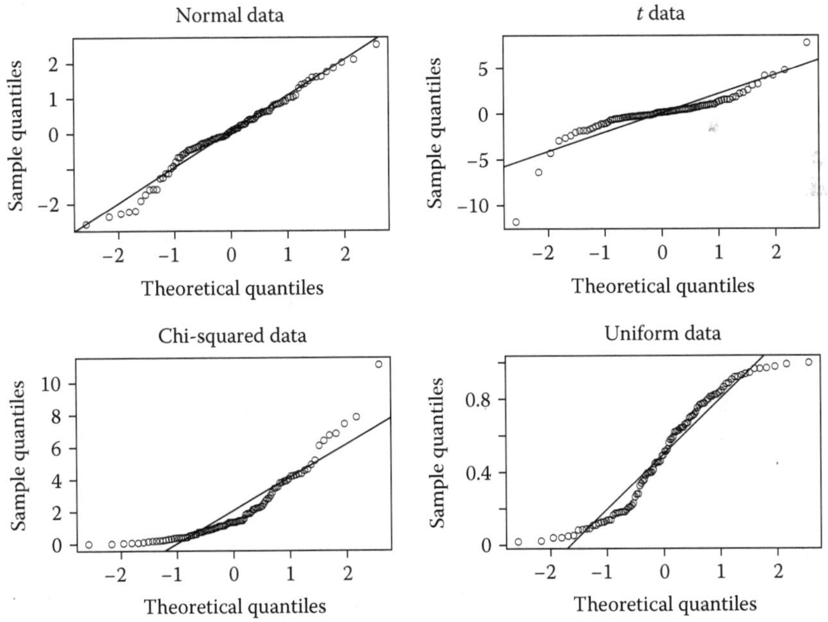

FIGURE 2.10
Normal probability plots for randomly generated data.

corresponding to probabilities close to zero. Therefore, in the plot for the chi-squared data, the right-most points are above the reference line, indicating that those sample quantiles are too large (compared to the normal quantiles); the left-most points are also above the reference line, indicating that those sample quantiles are too small. This behavior is typical of a skewed distribution.

For the uniform data, the points corresponding to probabilities close to one are below the reference line and the points corresponding to probabilities close to zero are above the reference line, indicating that both the right and the left tails of the distribution are relatively short. □

Example 2.23 The normal probability plot for the daily returns on Wal-Mart stock analyzed in Example 2.21 can be constructed using the commands

```
> qqnorm(wmt.ret)
> abline(a=mean(wmt.ret), b=sd(wmt.ret))
```

The plot is given in Figure 2.11. According to this plot, the distribution of daily returns on Wal-Mart stock is approximately symmetric but long-tailed. □

The results for the returns on Wal-Mart stock presented in the previous example are typical of stock return data—the distributions tend to be long-tailed relative to the normal distribution. For instance, a t-distribution with a small degrees of freedom, for example, six, is sometimes used as a model for return data.

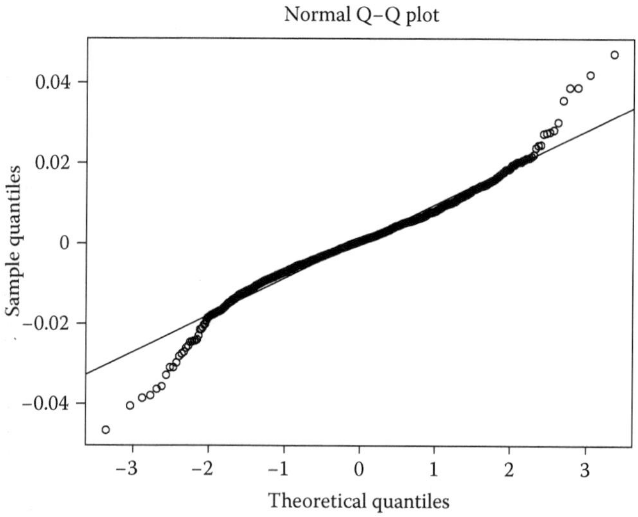

FIGURE 2.11
Normal probability plot of daily returns on Wal-Mart stock.

For the methods described in this book, an assumption of a specific distribution for returns is not needed; in particular, it is not assumed that returns are normally distributed. We do assume that the mean and standard deviation of the returns are useful summaries of the return distribution; this is generally true provided that the distribution of the data is not severely skewed or that it has very long tails so that the standard deviation does not provide useful information.

2.6 Suggestions for Further Reading

Topics such as prices, returns, and so on are covered in many books on financial modeling; hence, the following references are a sampling of those available. Benninga (2008) presents a comprehensive introduction to a wide range of topics in finance at a level suitable for beginners and with many detailed examples; see also Reilly and Brown (2009). A more rigorous treatment of these topics is available from Campbell et al. (1997).

The statistical properties of a series of observations and the analysis of such data are discussed in books on time series analysis, such as those by Montgomery et al. (2008) and Cowpertwait and Metcalfe (2009). A good introduction to time series analysis is provided by Newbold et al. (2013, Chapter 16), Ruppert (2004) and Sclove (2013) present more detailed introductions specifically geared toward financial statistics.

Stationary processes and autocovariance and autocorrelation functions are discussed by Wei (2006, Chapter 2). Chapters 1 and 2 by Montgomery et al. (2008) contain a good introduction to the analysis of time series data, including graphical methods and the numerical summaries discussed in Section 2.5. Running means are a type of "smoothing" of the data; see Chambers et al. (1983, Chapter 4) for an introduction to other types of smoothing methods. Normal probability plots and quantile–quantile plots in general are discussed by Chambers et al. (1983, Chapter 6).

2.7 Exercises

1. Consider an asset with prices $P_0 = \$10.00$, $P_1 = \$10.40$, $P_2 = \$10.20$, $P_3 = \$11.00$, and $P_4 = \$11.10$. Find the corresponding returns and log-returns.

2. Suppose that an asset has prices of the form $P_t = a\exp(bt)$, $t = 0, 1, 2, \ldots$ for some constants $a > 0$ and b. Find expressions for the return at time t, R_t, and the log-return at time t, r_t.

3. Consider an asset with return R_t and log-return r_t at time t. Using a Taylor's series approximation, find a quadratic function of R_t that approximates r_t for small values of $|R_t|$.

4. Suppose that an asset has prices $P_0 = \$4.00$, $P_1 = \$4.80$, $P_2 = \$5.00$, and $P_3 = \$5.40$ and dividends $D_1 = 0$, $D_2 = 0.40$, and $D_3 = 0$. Find the corresponding returns.

5. For the asset prices and dividends given in Exercise 4, compute the adjusted prices.

6. Let $P_0, P_1, \ldots, P_T$ denote a sequence of prices of an asset and let $D_1, D_2, \ldots, D_T$ denote the corresponding sequence of dividends. Suppose that $D_t = kP_{t-1}$, $t = 1, 2, \ldots, T$ for some constant k, so that the dividends are proportional to the price of the asset.

   a. Find an expression for the return R_t as a function of the prices P_t, P_{t-1}, and k.

   b. Find an expression for the corresponding sequence of adjusted prices.

7. Let $\{Y_t : t = 1, 2, \ldots\}$ and $\{X_t : t = 1, 2, \ldots\}$ be weakly stationary processes. Does it follow that the process $\{Y_t + X_t, t = 1, 2, \ldots\}$ is weakly stationary? Why or why not?

8. Let $\{X_t : t = 0, 1, 2, \ldots\}$ denote a weakly stationary process; let $\mu = \mathrm{E}(X_t)$, $\sigma^2 = \mathrm{Var}(X_t)$, and let $\gamma(\cdot)$ denote the autocovariance function of the process. Define

$$Y_t = X_t - X_{t-1}, \quad t = 1, 2, \ldots, .$$

   a. Find the mean and variance functions of the process $\{Y_t : t = 1, 2, \ldots\}$.

   b. Find the covariance function of $\{Y_t : t = 1, 2, \ldots\}$.

   c. Is the process $\{Y_t : t = 1, 2, \ldots\}$ weakly stationary? Why or why not?

9. Let $\{Z_t : t = 1, 2, \ldots\}$ denote a weak white noise process and let Z denote a random variable with mean 0 and variance 1 such that, for any $t = 1, 2, \ldots$, Z and Z_t are independent. Define

$$X_t = ZZ_t, \quad t = 1, 2, \ldots.$$

   a. Find the mean and variance functions of the process $\{X_t : t = 1, 2, \ldots\}$.

   b. Find the covariance function of $\{X_t : t = 1, 2, \ldots\}$.

   c. Is the process $\{X_t : t = 1, 2, \ldots\}$ weakly stationary? Is it a white noise process? Why or why not?

10. Let $r_1, r_2, \ldots$ denote the log-returns on an asset and suppose that the stochastic process $\{r_t : t = 1, 2, \ldots\}$ is weakly stationary. Let

$$\tilde{r}_1 = r_1 + r_2 + \cdots + r_{21},$$

$$\tilde{r}_2 = r_{22} + r_{23} + \cdots + r_{42},$$

and so on. For instance, the r_t might represent daily log-returns and the $\tilde{r}_t$ represent the corresponding monthly log-returns. Is $\{\tilde{r}_t : t = 1, 2, \ldots\}$ weakly stationary? Why or why not?

11. Let $X_1, X_2, \ldots, X_T$ denote independent, identically distributed random variables with mean $\mu = \mathrm{E}(X_1)$ and variance $\sigma^2 = \mathrm{Var}(X_1)$. For a given value of w, let

$$\bar{X}_{k,w} = \frac{1}{w} \sum_{t=k}^{w+k-1} X_t, \quad k = 1, 2, \ldots, T - w + 1$$

denote the running means with span w. For convenience, write

$$Y_j = \bar{X}_{k,w}, \quad j = 1, 2, \ldots, T - w + 1.$$

Is $Y_1, Y_2, \ldots, Y_{T-w+1}$ a weakly stationary process? If so, find the mean, variance, and the correlation function of the process.

12. Let $\{X_t : t = 1, 2, \ldots\}$ and $\{Y_t : t = 1, 2, \ldots\}$ denote weak white noise processes such that X_t and Y_s are independent random variables for any t, s.

    a. Is $\{X_t + Y_t : t = 1, 2, \ldots\}$ a weak white noise process? Why or why not?

    b. Is $\{X_t Y_t : t = 1, 2, \ldots\}$ a weak white noise process? Why or why not?

13. Consider stock in Papa John's International, Inc. (symbol PZZA).

    a. Using the R function `get.hist.quote`, download the adjusted prices needed to calculate three years of daily returns for the period ending December 31, 2015.

    b. Compute three years of daily returns corresponding to the prices obtained in Part (a).

    c. Use the function `summary` to compute the summary statistics for the returns computed in Part (b).

    d. Construct a time series plot of the returns calculated in Part (b).

14. Consider stock in Papa John's International, Inc. (symbol PZZA).

    a. Using the R function `get.hist.quote`, download the adjusted prices needed to calculate five years of monthly returns for the period ending December 31, 2015.

b. Compute five years of monthly returns corresponding to the prices obtained in Part (a).

c. Use the function `summary` to compute the summary statistics for the returns computed in Part (b).

d. Construct a time series plot of the returns calculated in Part (b).

15. For the five years of monthly return data for Papa John's stock calculated in Exercise 14, calculate the running means using a width of 12 months. Plot the running means versus time; include error bars in the plot; see Example 2.18. Based on this plot, what do you conclude regarding the weak stationarity of the stochastic process corresponding to monthly returns on Papa John's stock?

16. For the five years of monthly returns on Papa John's stock calculated in Exercise 14, calculate the running standard deviations using a width of 12 months. Plot the log of the running standard deviations versus time; include error bars in the plot; see Example 2.19. Based on this plot, what do you conclude regarding the weak stationarity of the stochastic process corresponding to monthly returns on Papa John's stock?

17. For the three years of daily returns on Papa John's stock calculated in Exercise 13, calculate the sample autocorrelation function and plot the results; use a maximum lag of 20. See Example 2.20. Are these results consistent with the assumption that the returns are uncorrelated random variables? Why or why not?

    Repeat these calculations using the five years of monthly return data calculated in Exercise 14 and a maximum lag of 12. Are these results consistent with the assumption that the returns are uncorrelated random variables? Why or why not?

18. For the three years of daily returns on Papa John's stock calculated in Exercise 13, construct a normal probability plot; see Example 2.23. Based on this plot, what do you conclude regarding the distribution of daily returns of Papa John's stock?

    Repeat the analysis using the five years of monthly returns on Papa John's stock calculated in Exercise 14. Do you reach the same conclusions as you did when analyzing the daily returns?

3

Random Walk Hypothesis

3.1 Introduction

Let $P_0, P_1, P_2, \ldots,$ denote a sequence of prices of an asset, such as one share of a particular stock, and let $R_1, R_2, \ldots$ denote the corresponding sequence of returns. Clearly, in making investment decisions, it would be useful to be able to use historical price and return data to predict future prices and returns. For instance, "technical analysis" is based on the belief that there are certain patterns in stock market data that tend to appear frequently and, by recognizing such patterns, it is possible to make useful predictions about the future prices and returns. A more statistical approach may look for a model that relates future prices or returns to past values and may use such a model for prediction.

An alternative viewpoint is that changes in the price of a stock are essentially unpredictable; this theory is known as the *random walk hypothesis*. Note that the random walk hypothesis does not mean that all stock prices are totally random and that one stock is as good as any other. The random walk hypothesis is a statement about the lack of useful statistical relationships between past and future prices and returns. However, one stock might still have a higher average return than another stock. Thus, past returns on a stock may provide useful information about future returns in the sense that they may be used to estimate the parameters of the return distribution.

3.2 Conditional Expectation

In discussing random walk models for asset prices, it is helpful to use the concept of conditional expectation; hence, in this section, we review some of its basic properties.

Let X and Y denote random variables. Recall that $\mathrm{E}(Y|X = x)$ is the expectation of Y computed under the assumption that X is held fixed at the value x; it is known as the *conditional expectation of* Y *given* $X = x$. Note that $\mathrm{E}(Y|X = x)$ is a function of x; that is, depending on the value at which X is held fixed, the conditional expectation of Y changes.

Define $h(x) = \mathrm{E}(Y|X = x)$; note that h is a function of the range of the random variable X. Thus, we may consider the random variable $h(X)$, known as the *conditional expectation of* Y *given* X and denoted by $\text{E}(Y|X)$. It is important to keep in mind that $\text{E}(Y|X)$ is a random variable. Conditional expectations such as $\text{E}(Y|X)$ are useful for describing certain aspects of the relationship between X and Y.

Just as the (unconditional) expected value of Y may be calculated based on knowledge of the distribution of Y, $\text{E}(Y|X)$ may be calculated based on knowledge of the joint distribution of Y and X. However, in this chapter, we are not concerned with calculating conditional expectations for specific distributions, but rather we are interested in the general properties of conditional expectation.

Some properties are straightforward extensions of the properties of unconditional expectation; for example, for random variables Y_1, Y_2, X and constants a_0, a_1, a_2,

$$\text{E}(a_0 + a_1Y_1 + a_2Y_2|X) = a_0 + a_1\text{E}(Y_1|X) + a_2\text{E}(Y_2|X).$$

However, we are more concerned with those properties that describe how $\text{E}(Y|X)$ reflects the relationship between X and Y.

The following lemma gives three such properties; see, for example, Blitzstein and Hwang (2015, Chapter 9) for proofs of these results, along with a detailed discussion of conditional expectation.

Lemma 3.1. *Let* Y *and* X *denote random variables and let* $g(\cdot)$ *be a real-valued function on the range of* X.

1. $E\{E(Y|X)\} = E(Y)$.
2. $E\{Yg(X)|X\} = g(X)E(Y|X)$.
3. *If* X *and* Y *are independent random variables, then* $E(Y|X) = E(Y)$.

Part 1 of the lemma states that the expected value of the random variable $\text{E}(Y|X)$ is the same as the expected value of Y. This result may give a convenient way to calculate $\text{E}(Y)$, if $\text{E}(Y|X)$ is easily obtained, using

$$\text{E}(Y) = \text{E}\left\{\text{E}(Y|X)\right\}.$$

Part 2 states that, when computing $\text{E}(Yg(X)|X)$, we may treat $g(X)$ as a constant and factor it out of the conditional expectation calculation. Suppose X and Y are independent, then treating X as fixed does not change the expected value of Y; it follows that $\text{E}(Y|X = x) = \text{E}(Y)$, leading to Part 3 of the lemma. Note that these three results continue to hold if X is a vector-valued random variable of the form $(X_1, X_2, \ldots, X_m)^T$.

Example 3.1 Let X and Y denote real-valued random variables and let $\hat{Y} = \text{E}(Y|X)$. Consider the covariance of $\hat{Y}$ and X, $\text{Cov}(\hat{Y}, X)$; recall that we may write

$$\text{Cov}(\hat{Y}, X) = \text{E}(\hat{Y}X) - \text{E}(\hat{Y})\text{E}(X).$$

Using Part 2 of Lemma 3.1,

$$\mathrm{E}(\hat{Y}X) = \mathrm{E}\left(\mathrm{E}(Y|X)X\right) = \mathrm{E}\left(\mathrm{E}(YX|X)\right)$$

and, using Part 1 of that lemma,

$$\mathrm{E}\left(\mathrm{E}(YX|X)\right) = \mathrm{E}(YX).$$

Since, by Part 1 of Lemma 3.1,

$$\mathrm{E}(\hat{Y}) = \mathrm{E}\left(\mathrm{E}(Y|X)\right) = \mathrm{E}(Y),$$

it follows that

$$\mathrm{Cov}(\hat{Y}, X) = \mathrm{E}(\hat{Y}X) - \mathrm{E}(\hat{Y})\mathrm{E}(X) = \mathrm{E}(YX) - \mathrm{E}(Y)\mathrm{E}(X) = \mathrm{Cov}(Y, X).$$

That is, the covariance of $\mathrm{E}(Y|X)$ and X is the same as the covariance of Y and X. □

The random variable $\mathrm{E}(Y|X)$ may be interpreted as the function of X that "best approximates" Y, and it is sometimes described as the "best predictor" of Y among functions of X. This idea is made precise by the following lemma.

Lemma 3.2. *Let Y denote a real-valued random variable and let X denote a random variable, possibly vector-valued. For any real-valued function g on the range of X,*

$$E\left\{(Y - E(Y|X))^2\right\} \leq E\left\{(Y - g(X))^2\right\}$$

with equality if and only if $g(X) = E(Y|X)$ with probability one.

Before presenting the proof of this result, consider its relationship to prediction. Suppose that our goal is to choose a function of X to approximate Y. For instance, suppose that Y is the future price of an asset and X is a vector containing past prices of the asset; in this case, approximating Y by a function of X corresponds to predicting the future price using past price data. Suppose that the quality of the approximation given by a function $g(X)$ is defined as the expected squared error of the approximation,

$$\mathrm{E}\left\{(Y - g(X))^2\right\}.$$

Then, according to Lemma 3.2, the best approximation, that is, the best predictor of Y among functions of X, is given by the conditional expectation $\mathrm{E}(Y|X)$. For many purposes, it is more useful to interpret $\mathrm{E}(Y|X)$ as the best approximation or best predictor of Y among functions of X rather than in terms of the expected value of the conditional distribution of Y given $X = x$.

Note that the lemma is a conditional version of the well-known result that the unconditional expectation $\mathrm{E}(Y)$ has the property that it minimizes $\mathrm{E}\left((Y-c)^2\right)$ over all constants c. That result may be established by writing

$$\begin{aligned}\mathrm{E}\left\{(Y-c)^2\right\} &= \mathrm{E}\left\{(Y-\mathrm{E}(Y)+\mathrm{E}(Y)-c)^2\right\}\\ &= \mathrm{E}\left\{(Y-\mathrm{E}(Y))^2\right\}+\{\mathrm{E}(Y)-c\}^2+2\mathrm{E}\left\{(Y-\mathrm{E}(Y))(\mathrm{E}(Y)-c)\right\}\end{aligned}$$

and noting that

$$\mathrm{E}\left\{(Y-\mathrm{E}(Y))(\mathrm{E}(Y)-c)\right\} = \{\mathrm{E}(Y)-c\}\,\mathrm{E}\left\{Y-\mathrm{E}(Y)\right\} = 0.$$

It follows that, for any real number c,

$$\mathrm{E}\left\{(Y-c)^2\right\} = \mathrm{E}\left\{(Y-\mathrm{E}(Y))^2\right\}+\{\mathrm{E}(Y)-c\}^2$$

and, hence,

$$\mathrm{E}\left\{(Y-c)^2\right\} \leq \mathrm{E}\left\{(Y-\mathrm{E}(Y))^2\right\}$$

with equality if and only if $c=\mathrm{E}(Y)$.

We can use a version of this idea to prove Lemma 3.2.

Proof of Lemma 3.2. The proof follows the proof of the aforementioned unconditional result. Note that

$$\begin{aligned}\mathrm{E}\left\{(Y-g(X))^2\right\} &= \mathrm{E}\left\{[Y-\mathrm{E}(Y|X)+\mathrm{E}(Y|X)-g(X)]^2\right\}\\ &= \mathrm{E}\left\{[Y-\mathrm{E}(Y|X)]^2\right\}+\{\mathrm{E}(Y|X)-g(X)\}^2\\ &\quad +2\mathrm{E}\left\{[Y-\mathrm{E}(Y|X)][\mathrm{E}(Y|X)-g(X)]\right\};\end{aligned}$$

because $\mathrm{E}(Y|X)-g(X)$ is a function of X, using Part 1 of Lemma 3.1,

$$\mathrm{E}\left\{[Y-\mathrm{E}(Y|X)][\mathrm{E}(Y|X)-g(X)]\right\}=\mathrm{E}\left\{\mathrm{E}\left[(Y-\mathrm{E}(Y|X))(\mathrm{E}(Y|X)-g(X))|X\right]\right\}$$

and, using Part 2 of Lemma 3.1,

$$\begin{aligned}&\mathrm{E}\left\{[Y-\mathrm{E}(Y|X)][\mathrm{E}(Y|X)-g(X)]|X\right\}\\ &\quad=\{\mathrm{E}(Y|X)-g(X)\}\,\mathrm{E}\left\{Y-\mathrm{E}(Y|X)|X\right\}=0.\end{aligned}$$

It follows that, for any function g,

$$\mathrm{E}\left\{(Y-g(X))^2\right\} = \mathrm{E}\left\{(Y-\mathrm{E}(Y|X))^2\right\}+\mathrm{E}\left\{[\mathrm{E}(Y|X)-g(X)]^2\right\};$$

note that $\{\mathrm{E}(Y|X)-g(X)\}^2 \geq 0$ so that

$$\mathrm{E}\left\{[\mathrm{E}(Y|X)-g(X)]^2\right\} \geq 0$$

and

$$\mathrm{E}\left\{(Y-g(X))^2\right\} \leq \mathrm{E}\left\{(Y-\mathrm{E}(Y|X))^2\right\}$$

with equality if and only if

$$\mathrm{E}\left\{[\mathrm{E}(Y|X)-g(X)]^2\right\} = 0,$$

which holds if and only if $g(X)=\mathrm{E}(Y|X)$ with probability 1. $\square$

Example 3.2 Let Y and X denote real-valued random variables such that

$$Y = \alpha + \beta X + \epsilon$$

for some constants α and β where ϵ is a mean-0 random variable such that ϵ and X are independent. Then

$$\mathrm{E}(Y|X) = \mathrm{E}(\alpha + \beta X + \epsilon|X) = \alpha + \beta \mathrm{E}(X|X) + \mathrm{E}(\epsilon|X).$$

Because ϵ and X are independent, $\mathrm{E}(\epsilon|X) = \mathrm{E}(\epsilon) = 0$ and $\mathrm{E}(X|X) = X$. Hence,

$$\mathrm{E}(Y|X) = \alpha + \beta X;$$

that is, the best predictor of Y among functions of X is $\alpha + \beta X$.

Note that this same result holds provided only that $\mathrm{E}(\epsilon|X) = 0$; independence of ϵ and X is not required. □

3.3 Efficient Markets and the Martingale Model

Why might it be difficult to use past asset returns to predict future returns? One possible reason is the efficiency of the markets that set asset prices, in the sense that those prices reflect all currently available information; hence, past return data does not include any additional useful information about future returns.

To see how efficient markets affect the properties of asset prices, we use the following simple model. Consider the price of one share of stock in a given company. Let Ω_t denote all information available at time t; that is, Ω_t consists of the values of all financial variables that have been observed up to and including time t. Because information accumulates over time,

$$\Omega_0 \subset \Omega_1 \subset \ldots \subset \Omega_t \subset \Omega_{t+1} \subset \ldots;$$

that is, the information available at time s is also available at time $s + h$ for any $h = 0, 1, 2, \ldots$.

Let V denote a random variable representing the intrinsic value of one share of stock in this company; for instance, V may include properties of the stock such as future dividends and so on. Let P_t denote the price of this share at time t, $t = 1, 2, \ldots$. Assume that

$$P_t = \mathrm{E}(V|\Omega_t);$$

that is, we assume that the price of the stock at time t is the best predictor of the value of the stock based on the information available at that time.

Suppose we are interested in predicting the price of stock at time $t + 1$, P_{t+1}, using the information available at time t. Using the interpretation of a conditional expectation value as the best predictor of a random variable, the best predictor of P_{t+1} is $\mathrm{E}(P_{t+1}|\Omega_t)$. Because $P_{t+1} = \mathrm{E}(V|\Omega_{t+1})$, the best predictor of P_{t+1} based on the information available at time t may be written

$$\mathrm{E}(P_{t+1}|\Omega_t) = \mathrm{E}(\mathrm{E}(V|\Omega_{t+1})|\Omega_t). \tag{3.1}$$

Iterated Conditional Expectations

The term on the right-hand side of 3.1 is known as an *iterated conditional expectation*, that is, the conditional expectation of a conditional expectation. Note that $\mathrm{E}(V|\Omega_{t+1})$ is the best predictor of V using the information in Ω_{t+1} and $\mathrm{E}(\mathrm{E}(V|\Omega_{t+1})|\Omega_t)$ is the best predictor of that predictor using the information in Ω_t.

Therefore, the end result, $\mathrm{E}(\mathrm{E}(V|\Omega_{t+1})|\Omega_t)$, is a predictor of V based on the information in Ω_t calculated using a two-step process. Because the information in Ω_t is also in Ω_{t+1}, this final predictor must be at least as good as $\mathrm{E}(V|\Omega_t)$. But we know that $\mathrm{E}(V|\Omega_t)$ is the best predictor of V using the information in Ω_t. Therefore, we expect that

$$\mathrm{E}(\mathrm{E}(V|\Omega_{t+1})|\Omega_t) = \mathrm{E}(V|\Omega_t).$$

Note that the same argument holds if Ω_{t+1} is replaced by any set of information that includes Ω_t.

The following proposition gives a formal statement of this result. A proof may be based on formalizing the argument described previously; the details are omitted.

Proposition 3.1. *For any random variable V and for any information sets Ω_t and Ω_{t+h} such that $\Omega_t \subset \Omega_{t+h}$,*

$$E(E(V|\Omega_{t+h})|\Omega_t) = E(V|\Omega_t)$$

with probability 1.

The Martingale Model

We now apply this result to asset prices. Recall that, according to our model, the price of a stock at time t is $P_t = \mathrm{E}(V|\Omega_t)$, where V represents the intrinsic value of the stock and Ω_t is the information available at time t. The best predictor of P_{t+1}, the price of the stock at time $t+1$, using the information available at time t is $\mathrm{E}(P_{t+1}|\Omega_t)$; using Proposition 3.1, this may be written

$$\mathrm{E}(P_{t+1}|\Omega_t) = \mathrm{E}(\mathrm{E}(V|\Omega_{t+1})|\Omega_t) = \mathrm{E}(V|\Omega_t) = P_t. \tag{3.2}$$

That is, the best predictor of tomorrow's price of the stock is today's price. Clearly, the same argument works for any price in the future: for any $h = 1, 2, \ldots$,

$$\begin{aligned} \mathrm{E}(P_{t+h}|\Omega_t) &= \mathrm{E}(\mathrm{E}(V|\Omega_{t+h})|\Omega_t) \\ &= \mathrm{E}(V|\Omega_t) = P_t. \end{aligned}$$

A sequence of random variables $P_1, P_2, \ldots$ with this property is said to be a *martingale* with respect to $\Omega_1, \Omega_2, \ldots$. Therefore, this is known as the *martingale model* for asset prices.

The martingale model for asset prices has important implications for the properties of the corresponding returns. Let $X_{t+1} = P_{t+1} - P_t$ denote the change in the price from time t to time $t+1$ and consider $\mathrm{E}(X_{t+1})$. Then

$$\begin{aligned}\mathrm{E}(X_{t+1}) &= \mathrm{E}\{\mathrm{E}(X_{t+1}|\Omega_t)\}\\ &= \mathrm{E}\{\mathrm{E}(P_{t+1} - P_t|\Omega_t)\}\\ &= \mathrm{E}\{\mathrm{E}(P_{t+1}|\Omega_t) - \mathrm{E}(P_t|\Omega)\}.\end{aligned}$$

Note that Ω_t, the information available at time t, includes P_t; that is, the random variable P_t is a function of the information in Ω_t. It follows that

$$\mathrm{E}(P_t|\Omega_t) = P_t;$$

furthermore, the result given in (3.2) shows that $\mathrm{E}(P_{t+1}|\Omega_t) = P_t$. It follows that

$$\begin{aligned}\mathrm{E}(X_{t+1}) &= \mathrm{E}\{\mathrm{E}(P_{t+1}|\Omega_t) - P_t\}\\ &= \mathrm{E}(P_t - P_t) = 0.\end{aligned}$$

Thus, price changes have an expected value of 0. Furthermore, the previous argument shows that

$$\mathrm{E}(X_{t+1}|\Omega_t) = 0$$

so that, given the information available at time t, the predicted value of the change in price from time t to time $t+1$ is 0.

Let

$$R_{t+1} = \frac{P_{t+1}}{P_t} - 1 = \frac{P_{t+1} - P_t}{P_t}, \quad t = 0, 1, 2, \ldots$$

denote the return at time $t+1$. Using the fact that P_t is a function of Ω_t,

$$\mathrm{E}(R_{t+1}|\Omega_t) = \mathrm{E}\left(\frac{P_{t+1} - P_t}{P_t}|\Omega_t\right) = \frac{1}{P_t}\mathrm{E}(P_{t+1} - P_t|\Omega_t) = 0; \tag{3.3}$$

that is, under the martingale model, the best predictor of the return in period $t+1$ using financial information in periods up to and including period t is zero. Note that this result also implies that the (unconditional) expected value of R_{t+1} is also zero:

$$\mathrm{E}(R_{t+1}) = \mathrm{E}\left\{\mathrm{E}(R_{t+1}|\Omega_t)\right\} = \mathrm{E}(0) = 0.$$

Furthermore, the following result shows that the correlation of any two returns R_t and R_s, $t \neq s$, is zero.

Proposition 3.2. *Using the framework of this chapter, define the price of an asset by*

$$P_t = E(V|\Omega_t), \quad t = 0, 1, \ldots$$

and let $R_1, R_2, \ldots,$ denote the corresponding returns.

Then, under the martingale model, for any $t, s = 1, 2, \ldots, t \neq s$,

$$Cov(R_t, R_s) = 0.$$

Proof. Without loss of generality, we may assume that $t < s$. According to 3.3, $\mathrm{E}(R_t) = \mathrm{E}(R_s) = 0$ so that

$$\mathrm{Cov}(R_t, R_s) = \mathrm{E}(R_t R_s).$$

Note that

$$\mathrm{E}(R_t R_s) = \mathrm{E}\left\{\mathrm{E}(R_t R_s | \Omega_{s-1})\right\}$$

and that $t \leq s-1$, R_t, and P_{s-1} are functions of Ω_{s-1}. It follows that

$$\begin{aligned}
\mathrm{E}(R_t R_s|\Omega_{s-1}) = R_t \mathrm{E}(R_s|\Omega_{s-1}) &= R_t \mathrm{E}\left(\frac{P_s - P_{s-1}}{P_{s-1}}|\Omega_{s-1}\right) \\
&= R_t \frac{1}{P_{s-1}} \mathrm{E}\left(P_s - P_{s-1}|\Omega_{s-1}\right) \\
&= R_t \frac{1}{P_{s-1}} (P_{s-1} - P_{s-1}) = 0,
\end{aligned}$$

using the fact that $\mathrm{E}(P_s|\Omega_{s-1}) = P_{s-1}$, establishing the result. □

3.4 Random Walk Models for Asset Prices

Under the martingale model, asset returns have zero mean and are uncorrelated. However, the framework we used to derive this model is a simple one and its assumptions are unrealistic in many respects. In particular, we assumed that the price of an asset is based only on the expectation of the asset's value.

An alternative, and more realistic, approach is based on the assumption that the current price of a stock is based on current beliefs regarding the statistical properties of future earnings of the firm; in particular, the price may be based on the expected value of the future earnings as well as their variability. Including variability in the analysis is important because investors are generally "risk averse." Under risk aversion, more risky investments are worth less than less risky investments with the same expected return.

For example, consider two investments. Stock 1 has a guaranteed return of a while Stock 2 has a return of zero with probability 1/2 and a return of b with probability 1/2. Stock 1 may be preferable, that is, it may be worth more to an investor than Stock 2, even if $a < b/2$. It follows that the argument given in the previous sections, which uses only expected values, is, at best, only a rough approximation of the process used to determine asset prices. Under a model that recognizes risk aversion, the martingale property no longer holds. Therefore, although the assumption of an efficient market suggests certain properties of asset prices, it does not necessarily follow that the martingale model holds in practice.

Hence, in this section, we consider models for asset prices that have many, but not all, of the general features of the martingale model. For instance, based on the models considered in this section, returns are statistically unrelated, but they do not necessarily have zero mean. These models are based on the concept of a *random walk*.

Consider a sequence of random variables $Y_0, Y_1, Y_2, \ldots$. A simple model for such a process is one in which the changes in process, $Y_t - Y_{t-1}$, are "random" in the sense that they have no discernible pattern and no relationships among them. Such a process is known as a random walk because it can be viewed as a model for the location on the real line of an "individual" who, at each time point, moves randomly along the line. The statistical properties of a process of this type depend on the interpretation of the term "random" used to describe the movements of the individual. Thus, there are several different technical definitions of a random walk, corresponding to different interpretations.

For a given stochastic process $\{Y_t : t = 0, 1, 2, \ldots\}$, let $Z_t = Y_t - Y_{t-1}$, $t = 1, 2, \ldots$ denote the changes in the values of Y_t; the Z_t are known as the *increments* of the process. Note that when discussing random walks, we will often include Y_0, the value at time $t = 0$, in the process; this random variable is needed to define the first increment Z_1. Thus, Y_0 represents the "starting point" of the process. The increment process $\{Z_t : t = 1, 2, \ldots\}$ will generally start at time $t = 1$.

Note that, given Y_0,

$$Y_t = Y_0 + Z_1 + \cdots + Z_t, \quad t = 1, 2, \ldots \tag{3.4}$$

so that the random variables $Y_1, Y_2, \ldots$ are equivalent to the increments $Z_1, Z_2, \ldots$ together with the starting point Y_0.

In a random walk process, the increments are "noise" in the sense that they are statistically unrelated. We have seen one example of such a noise process, weak white noise; this definition, as well as some others, are used in defining random walks.

Definitions of a Random Walk

We now consider three specific definitions of a random walk. Consider a stochastic process $\{Y_t : t = 0, 1, 2, \ldots\}$, and let $Z_t = Y_t - Y_{t-1}$, $t = 1, 2, \ldots$ denote the increments of the process.

Suppose that $Z_1, Z_2, \ldots$ are independent and identically distributed (i.i.d.) random variables each with mean μ and standard deviation σ and suppose that Y_0 is independent of $Z_1, Z_2, \ldots$. This is the strongest version of the random walk model, known as Random Walk 1 (RW1). Thus, in RW1, the changes in the position of the process, $Y_t - Y_{t-1}$, are i.i.d. random variables.

Note that, using (3.4), under RW1,

$$\begin{aligned}
\mathrm{E}(Y_t|Y_0) &= \mathrm{E}(Y_0 + Z_1 + \cdots + Z_t|Y_0) \\
&= \mathrm{E}(Y_0|Y_0) + \mathrm{E}(Z_1|Y_0) + \cdots + \mathrm{E}(Z_t|Y_0) \\
&= Y_0 + \mathrm{E}(Z_1) + \cdots + \mathrm{E}(Z_t) \\
&= Y_0 + \mu t, \quad t = 1, 2, \ldots.
\end{aligned}$$

A similar property holds for the conditional variance of Y_t given Y_0, defined as

$$\text{Var}(Y_t|Y_0) = \text{E}\left\{[Y_t - \text{E}(Y_t|Y_0)]^2 \middle| Y_0\right\} = \text{E}(Y_t^2|Y_0) - \text{E}(Y_t|Y_0)^2.$$

Note that

$$\begin{aligned} Y_t - \text{E}(Y_t|Y_0) &= (Y_0 + Z_1 + \cdots + Z_t) - (Y_0 + \mu + \cdots + \mu) \\ &= (Z_1 - \mu) + (Z_2 - \mu) + \cdots + (Z_t - \mu). \end{aligned}$$

Using this expression, together with the assumption that Y_0 and $(Z_1, Z_2, \ldots)$ are independent, it follows that

$$\begin{aligned} \text{E}\left\{[Y_t - \text{E}(Y_t|Y_0)]^2|Y_0\right\} &= \text{E}\left\{[(Z_1 - \mu) + (Z_2 - \mu) + \cdots + (Z_t - \mu)]^2|Y_0\right\} \\ &= \text{E}\left\{[(Z_1 - \mu) + (Z_2 - \mu) + \cdots + (Z_t - \mu)]^2\right\} \\ &= \text{Var}(Z_1) + \text{Var}(Z_2) + \cdots + \text{Var}(Z_t) \\ &= \sigma^2 t. \end{aligned}$$

Here, μ and σ are called the *drift* and *volatility*, respectively, of the process.

Because $Y_t = Y_0 + Z_1 + \cdots + Z_t$, it follows that Y_t and Z_{t+1} are independent; using the fact that $Y_{t+1} = Y_t + Z_{t+1}$,

$$\text{E}(Y_{t+1}|Y_t) = \text{E}(Y_t|Y_t) + \text{E}(Z_{t+1}|Y_t) = Y_t + \text{E}(Z_{t+1}) = Y_t + \mu.$$

That is, the best predictor of the position of the random walk at time $t+1$, given knowledge of the position at time t, is $Y_t + \mu$.

In fact, the same basic argument can be used to show that

$$\text{E}(Y_{t+1}|Y_0, Y_1, \ldots, Y_t) = \text{E}(Y_t + Z_{t+1}|Y_0, Y_1, \ldots, Y_t) = Y_t + \mu;$$

that is, the best predictor of Y_{t+1}, given previous knowledge of all past values of the process, is $Y_t + \mu$.

Weaker forms of the random walk model are based on weaker assumptions regarding the distribution of the increments. For instance, Random Walk 2 (RW2) assumes that the increments $Z_1, Z_2, \ldots$, together with the initial value Y_0, are independent random variables, as in RW1; $Z_1, Z_2, \ldots$ are each assumed to have mean μ and standard deviation σ, but they are not necessarily identically distributed.

Under this model, the result

$$\text{E}(Y_{t+1}|Y_0, Y_1, \ldots, Y_t) = Y_t + \mu$$

still holds. This generalization is useful because the assumption of identical distributions of the increments is difficult to verify in practice.

The weakest form of random walk that is commonly used is a weaker version of RW2 in which the independence assumption for $Z_1, Z_2, \ldots$ is replaced by the assumption that the increments are uncorrelated,

$$\text{Cov}(Z_t, Z_s) = 0 \quad \text{for any} \quad s \neq t,$$

and they are uncorrelated with Y_0,

$$\text{Cov}(Z_t, Y_0) = 0, \quad t = 1, 2, \ldots;$$

this is known as Random Walk 3 (RW3). That is, in RW3, the increment process $\{Z_t : t = 1, 2, \ldots\}$ is a weak white noise process.

In this case, we can no longer say that the best predictor of Y_{t+1} based on $Y_0, Y_1, \ldots, Y_t$ is $Y_t + \mu$. However, the *best linear predictor* of Y_{t+1} based on $Y_0, Y_1, \ldots, Y_t$ is $Y_t + \mu$.

The best linear predictor of Y_{t+1} based on $Y_0, Y_1, \ldots, Y_t$ is defined as the function of the form

$$a + b_0Y_0 + b_1Y_1 + \cdots + b_tY_t$$

that minimizes

$$\text{E}\{(Y_{t+1} - a - b_0Y_0 - \cdots - b_tY_t)^2\}. \tag{3.5}$$

Proposition 3.3. *Consider a stochastic process $\{Y_t : t = 0, 1, 2, \ldots\}$ and let $Z_t = Y_t - Y_{t-1}$, $t = 1, 2, \ldots$ denote the increments of the process. If $\{Z_t : t = 1, 2, \ldots\}$ is weak white noise and Cov$(Y_0, Z_t) = 0$ for $t = 1, 2, \ldots$, then the best linear predictor of Y_{t+1} based on $Y_0, Y_1, \ldots, Y_t$ is $Y_t + \mu$.*

Proof. Consider a linear function of $Y_0, Y_1, \ldots, Y_t$ of the form $a + b_0Y_0 + b_1Y_1 + \cdots + b_tY_t$ for some constants $a, b_0, b_1, \ldots, b_t$.

According to (3.4), each of $Y_1, Y_2, \ldots, Y_t$ is a linear function of $Y_0, Z_1, \ldots, Z_t$; hence, any function of the form

$$a + b_0Y_0 + b_1Y_1 + \cdots + b_tY_t = a + b_0Y_0 + b_1Y_1 + \cdots + (b_t - 1)Y_t + Y_t$$

may be written as

$$Y_t + c + d_0Y_0 + d_1Z_1 + \cdots + d_tZ_t \tag{3.6}$$

for some constants $c, d_0, d_1, \ldots, d_t$.

Therefore, using the fact that $Y_{t+1} - Y_t = Z_{t+1}$, the best linear predictor of Y_{t+1} is given by the constants $c, d_0, d_1, \ldots, d_t$ that minimize

$$\text{E}\{(Z_{t+1} - c - d_0Y_0 - d_1Z_1 - \cdots - d_tZ_t)^2\}, \tag{3.7}$$

which is simply 3.5 written in terms of $Y_0, Z_1, \ldots, Z_t$.

Recall that, for any random variable X, $\mathrm{E}(X^2) = \mathrm{E}(X)^2 + \mathrm{Var}(X)$. Note that $Z_{t+1} - c - d_0 Y_0 - d_1 Z_1 - \cdots - d_t Z_t$ has the expected value

$$\left(1 - \sum_{j=1}^{t} d_j\right) \mu - c - d_0 \mathrm{E}(Y_0) \tag{3.8}$$

and variance

$$\left(1 + \sum_{j=1}^{t} d_j^2\right) \sigma^2 + d_0^2 \mathrm{Var}(Y_0)$$

where here we have used the fact that any pair of $Y_0, Z_1, \ldots, Z_t, Z_{t+1}$ is uncorrelated. Clearly, the variance is minimized by $d_0 = d_1 = \cdots = d_t = 0$. Because, for these choices of $d_0, d_1, \ldots, d_t$, the expected value in (3.8) is 0 for $c = \mu$, it follows that the best linear predictor of Y_{t+1} is an expression of the form 3.6 with $d_0 = d_1 = \cdots = d_t = 0$ and $c = \mu$, yielding the expression $Y_t + \mu$, as claimed in the proposition. □

Note that the random walk models are related: RW1 implies RW2, which implies RW3. Thus, if RW3 does not hold, then neither RW2 nor RW1 holds, and if RW2 does not hold, then RW1 does not hold.

Geometric Random Walk

The same ideas can be applied after a log-transformation of the random variables; such a transformation is useful if we believe that the "random changes" in the process are multiplicative rather than additive. In this case, the original untransformed process is said to follow a *geometric random walk* model.

Let $\{U_t : t = 0, 1, 2, \ldots\}$ denote a stochastic process such that $Y_t = \log U_t$, $t = 0, 1, 2, \ldots$ follows a random walk model. For instance, if $\{Y_t : t = 0, 1, 2, \ldots\}$ follows RW1, then $\log U_t = Y_0 + Z_1 + \cdots + Z_t$, where $Z_1, Z_2, \ldots$ are i.i.d. random variables. In this case, we say that $\{U_t :, t = 0, 1, 2, \ldots\}$ follows a geometric RW1 model. Note that, under this model,

$$\begin{aligned} U_t &= \exp(Y_0) \exp(Z_1 + \cdots + Z_t) \\ &\equiv U_0 \exp(Z_1) \cdots \exp(Z_t) \end{aligned}$$

where $U_0 = \exp(Y_0)$. These ideas may also be applied to RW2 and RW3.

Application of Random Walk Models to Asset Prices

The martingale model for asset prices suggests that the stochastic process corresponding to the prices of an asset, $\{P_t : t = 0, 1, 2, \ldots\}$, might be usefully modeled as a random walk. Under this model, the increments of the process $P_t - P_{t-1}$, corresponding to changes in the price of the asset, are "noise";

for instance, if $\{P_t : t = 0, 1, 2, \ldots\}$ follows RW3, then the price changes form a weak white noise process.

However, empirical analyses suggest that changes in prices are often roughly proportional to the price, in the sense that stocks with higher prices tend to exhibit larger price changes than stocks with lower prices, generally speaking.

This behavior may be modeled by assuming that

$$P_t - P_{t-1} = W_t P_{t-1}, \quad t = 1, 2, \ldots,$$

where W_t is a random variable representing the proportional change in price. Under this assumption, the conditional expectation of $X_t = P_t - P_{t-1}$ given P_{t-1} depends on P_{t-1} in general. On the other hand,

$$p_t - p_{t-1} = \log \frac{P_t}{P_{t-1}} = \log (W_t + 1)$$

so that, letting $Z_t = \log(W_t + 1)$, $p_t - p_{t-1} = Z_t$, $t = 1, 2, \ldots$. Thus, if price changes are proportional to the price, we might expect log-prices to follow a random walk so that $\{P_t : t = 0, 1, 2, \ldots\}$ follows a geometric random walk.

Note that the increments of the process $\{p_t : t = 0, 1, 2, \ldots\}$ are simply the log-returns. However, a basic argument showing that market efficiency implies that log-returns are uncorrelated, along the lines of the one we used in Proposition 3.2 for returns, is not available.

To see why such an argument fails, consider the framework of Section 3.3, in which $P_t = \mathrm{E}(V|\Omega_t)$. Then

$$p_t = \log \mathrm{E}(V|\Omega_t)$$

so that

$$r_t = p_t - p_{t-1} = \log \frac{\mathrm{E}(V|\Omega_t)}{\mathrm{E}(V|\Omega_{t-1})}.$$

It follows that

$$\mathrm{E}(r_{t+1} r_t) = \mathrm{E}\left(\log \frac{\mathrm{E}(V|\Omega_{t+1})}{\mathrm{E}(V|\Omega_t)} \log \frac{\mathrm{E}(V|\Omega_{t+1})}{\mathrm{E}(V|\Omega_t)}\right).$$

Because of the presence of the $\log(\cdot)$ function, this expression cannot be usefully simplified.

However, suppose that the returns $R_t = (P_t - P_{t-1})/P_{t-1}$ are small. Then

$$\begin{aligned} r_t = p_t - p_{t-1} &= \log \frac{P_t}{P_{t-1}} = \log \left(1 + \frac{P_t - P_{t-1}}{P_{t-1}}\right) \\ &\doteq \frac{P_t - P_{t-1}}{P_{t-1}} = R_t. \end{aligned}$$

For example, if $R_t = 0.02$, a fairly large value for a monthly return, then $P_t/P_{t-1} = 1.02$ and $r_t = \log(1.02) = 0.01980$ and approximating r_t by R_t yields an error of 0.0002, or about 1%.

Thus, if two returns are uncorrelated, it is reasonable to expect that the corresponding log-returns will be approximately uncorrelated. Therefore, although the martingale model for log-prices does not follow directly from the efficient market assumption, the argument used in Section 3.3 suggests that it may be reasonable to model $\{p_t : t = 0, 1, 2, \ldots\}$ as a random walk.

The random walk hypothesis in finance generally refers to a geometric random walk for asset prices. Suppose $\{P_t : t = 0, 1, 2, \ldots\}$ follows a geometric random walk model; then the stochastic process corresponding to the log-prices, $\{p_t : t = 0, 1, 2 \ldots\}$ follows a random walk model.

Specifically, under RW1, $r_1, r_2, \ldots$ are i.i.d. random variables, under RW2, $r_1, r_2, \ldots$ are independent, each with mean μ and standard deviation σ, but they are not necessarily identically distributed; RW3 weakens the independence condition of RW2 to the condition that $r_1, r_2, \ldots$ are uncorrelated random variables. The key idea in the random walk hypothesis is that the past values $r_1, r_2, \ldots, r_t$ do not provide any useful information about r_{t+1}.

As noted in the introduction to this chapter, it is important to keep in mind that the random walk hypothesis does not imply that the information provided by $r_1, r_2, \ldots, r_t$ is useless in understanding the properties of r_{t+1}. For instance, under any of the random walk models we have considered, $r_1, \ldots, r_t, r_{t+1}$ each has mean μ; hence, the sample mean of $r_1, \ldots, r_t$ is an unbiased estimator of μ, which is the expected value of r_{t+1}. Thus, the past values do provide a type of indirect information about the future.

3.5 Tests of the Random Walk Hypothesis

According to the random walk hypothesis, asset prices follow a geometric random walk. Although market efficiency suggests that a geometric random walk may be appropriate for modeling price data, for the analyst, the important issue is whether or not observed prices behave like observations from a geometric random walk. Hence, a large number of statistical tests of the random walk model have been proposed.

It is worth noting that, although the term "random walk" refers to the asset log-prices, tests of the random walk model are typically based on the properties of the log-returns, which are the increments corresponding to the log-prices. Thus, these tests are designed to detect statistical relationships in the log-returns of an asset; such relationships would contradict the random walk hypothesis.

In this section, we consider four simple tests that are useful in this context; each test is designed to detect departures from some form of a random walk for log-prices; the first two tests considered test RW1 and the remaining two test RW3. Note that, because of the relationships among the three forms of the random walk model, rejection of the hypothesis that RW3 holds for log-prices implies rejection of RW2 and RW1 for log-prices.

A Test Based on the Sample Autocorrelation Function

Let $p_0, p_1, p_2, \ldots$ denote log-prices of a given asset and let $r_1, r_2, \ldots$ denote the corresponding log-returns. Under RW3 for $\{p_t : t = 0, 1, 2, \ldots\}$, $r_1, r_2, \ldots$ are uncorrelated random variables, each with mean μ and standard deviation σ.

Let $\rho(\cdot)$ denote the autocorrelation function of $\{r_t : t = 1, 2, \ldots\}$. Then, under RW3, $\rho(h) = 0$ for all $h = 1, 2, \ldots$. Therefore, a test of the RW3 version of the random walk hypothesis may be based on the sample autocorrelation function, as described in Section 2.5.

Suppose we observe T periods of return data, $r_1, r_2, \ldots, r_T$, and let $\hat{\rho}(h)$, $h = 1, 2, \ldots$, denote the sample autocorrelation function. Although a test of the random walk hypothesis can be based on any one sample autocorrelation, a better approach is to construct a test statistic based on several sample autocorrelations, such as the first m sample autocorrelations, for some given value of m. A test statistic of this type is given by

$$B = T(T+2)\sum_{h=1}^{m} \frac{\hat{\rho}(h)^2}{T-h}.$$

Note that B tends to be large when the sample autocorrelations are far from 0. Under the null hypothesis that RW3 holds for log-prices, so that $\rho(1) = \rho(2) = \cdots = \rho(m) = 0$, B has a chi-squared distribution with m degrees of freedom. This is known as the *Box–Ljung test.*

In order to carry out the test, m, the number of lags used to compute B, must be selected. When the data consist of five years of monthly returns, a relatively small value of m should be used; for example, $m = 12$ is a reasonable choice. For longer series of daily returns, a larger value of m could be used.

Example 3.3 Consider the monthly log-returns for Wal-Mart stock, stored in the variable `wmt.m.logret`. To compute the test statistic B and the corresponding p-value, we may use the `Box.test` function.

```
> Box.test(wmt.m.logret, lag=12, type="L")

        Box-Ljung test

data:  wmt.m.logret
X-squared = 9.9011, df = 12, p-value = 0.6246
```

The argument `lag` in the `Box.test` function specifies m, the number of lags to be used and the `type="L"` argument specifies that the Box–Ljung test be used.

These results indicate that, for $m = 12$, $B = 9.9011$ and the p-value is 0.6246. Therefore, based on this test, there is no evidence to reject the hypothesis that the data do not exhibit autocorrelation (at least up to lag 12), confirming our informal conclusion in Section 2.5. □

Variance-Ratio Test

The variance-ratio test is based on the following observation. Suppose that RW3 holds for the log-prices, so that $r_1, r_2, \ldots, r_T$ each has mean μ and standard deviation σ and r_t, r_s are uncorrelated for all $t \neq s$. Then

$$\mathrm{E}(r_t + r_{t-1}) = \mathrm{E}(r_t) + \mathrm{E}(r_{t-1}) = \mu + \mu = 2\mu$$

and

$$\mathrm{Var}(r_t + r_{t-1}) = \mathrm{Var}(r_t) + \mathrm{Var}(r_{t-1}) = \sigma^2 + \sigma^2 = 2\sigma^2.$$

More generally, $r_t + r_{t-1} + \cdots + r_{t-q+1}$ has mean $q\mu$ and variance $q\sigma^2$. Note that $r_t + r_{t-1} + \cdots + r_{t-q+1}$ is simply the q-period log-return at time t. Therefore, if RW3 holds for log-prices, then there is a simple relationship between the variance of multiperiod log-returns and the variance of single-period log-returns.

This fact can be used to test RW3 by comparing an estimate of the variance of

$$r_t + r_{t-1} + \cdots + r_{t-q+1}, \quad t = q, \ldots, T$$

to an estimate of the variance of $r_1, r_2, \ldots, r_T$; if RW3 holds, the ratio of these estimates should be roughly q.

For a given value of q, let

$$S_q^2 = \frac{\sum_{t=q}^{T}(r_t + r_{t-1} + \cdots + r_{t-q+1} - q\bar{r})^2}{T-q},$$

which is essentially the sample variance of $r_t + r_{t-1} + \cdots + r_{t-q+1}$, $t = q$, $q+1, \ldots, T$, with the divisor equal to the sample size minus one, except that instead of subtracting the sample mean of these values we subtract $q\bar{r}$, where

$$\bar{r} = \frac{1}{T}\sum_{t=1}^{T} r_t.$$

Let S^2 denote the usual sample variance of $r_1, r_2, \ldots, r_T$. The variance-ratio statistic is given by

$$V_q = \frac{T}{T-q+1}\frac{1}{q}\frac{S_q^2}{S^2}.$$

If RW3 holds, we expect that

$$\frac{1}{q}\frac{S_q^2}{S^2} \doteq 1;$$

the factor $T/(T-q+1)$ is an adjustment term designed to improve the accuracy of the normal approximation to the distribution of V_q in small samples. Note that, like the Box–Ljung test, the variance-ratio test is a test of the correlation structure of the log-returns.

Under the null hypothesis that RW3 holds for the log-returns, $\sqrt{T}(V_q - 1)$ is approximately normally distributed with mean 0 and variance given by $2(2q-1)(q-1)/(3q)$. Therefore, the standardized test statistic is

$$\bar{V}_q = \sqrt{T}\sqrt{\frac{3q}{2(2q-1)(q-1)}}(V_q - 1)$$

and the null hypothesis is rejected for large values of $|\bar{V}_q|$. The p-value of the test is given by

$$\mathrm{P}(|Z| > |\bar{V}_{q,0}|)$$

where Z has a standard normal distribution and $\bar{V}_{q,0}$ is the observed value of $\bar{V}_q$; hence,

$$\mathrm{P}(|Z| > |\bar{V}_{q,0}|) = 2\left(1 - \Phi(|\bar{V}_{q,0}|)\right)$$

where Φ denotes the standard normal distribution function.

Example 3.4 Consider the statistic V_3 applied to the log-returns on Wal-Mart stock. This statistic may be calculated using the following commands:

```
> x<-wmt.m.logret - mean(wmt.m.logret)
> x3<-x[3:60] + x[2:59] + x[1:58]
> (60/58)*(1/3)*(sum(x3^2)/57)/var(x)
[1] 0.90135
```

Here, `x` is the vector of mean-corrected log-returns for Wal-Mart stock and `x3` is the vector of three-month mean-corrected log-returns,

$$r_t - \bar{r} + r_{t-1} - \bar{r} + r_{t-2} - \bar{r}, \quad t = 3, 4, \ldots, T.$$

The observed value of the statistic V_3 is 0.90135.

To compute a p-value, we use the fact that, under the null hypothesis, V_3 has mean 1 and variance

$$\frac{2(2q-1)(q-1)}{3q}\frac{1}{T} = \frac{2(5)(2)}{9}\frac{1}{60} = \frac{1}{27}.$$

Therefore, the observed value of the standardized test statistic is given by

$$\bar{V}_{3,0} = \sqrt{27}(0.90135 - 1) = -0.51260;$$

this corresponds to a two-tailed p-value of 0.6082, calculated by

```
> 2*(1-pnorm(0.51260))
[1] 0.6082
```

Therefore, there is no evidence to reject the null hypothesis that Wal-Mart log-prices follow RW3. Note that here the function `pnorm` is the standard normal distribution function.

A similar conclusion is obtained using V_6. The observed value of V_6 is 0.610, corresponding to a p-value of 0.222. □

Runs Test

Not all types of dependence are reflected in correlation. Another approach to detecting relationships in a series of returns is to look at patterns of above-average and below-average returns.

More formally, let $r_1, r_2, \ldots, r_T$ denote a sequence of log-returns and let $\text{med}(r_1, r_2, \ldots, r_T)$ denote the sample median of $r_1, r_2, \ldots, r_T$. For $t = 1, 2, \ldots, T$, define

$$G_t = \begin{cases} 1 & \text{if } r_t > \text{med}(r_1, \ldots, r_T) \\ 0 & \text{if } r_t \leq \text{med}(r_1, \ldots, r_T) \end{cases}$$

Then $G_1, G_2, \ldots, G_T$ is a sequence of indicator variables showing if the return in a given period exceeds the median ($G_t = 1$) or not ($G_t = 0$).

Suppose that $r_1, r_2, \ldots r_T$ are i.i.d. random variables; that is, suppose that RW1 holds. Then $G_1, G_2, \ldots, G_T$ should exhibit a random pattern of zeros and ones. On the other hand, if $r_1, r_2, \ldots, r_T$ have some type of dependence structure, or if certain features of the distribution of r_t depend on t, then there may be patterns of zeros and ones; for instance, if r_t, r_{t+1} are dependent, then $G_{t+1} = 1$ may be more likely if $G_t = 1$ than if $G_t = 0$.

Therefore, we can test the hypothesis that $p_0, p_1, p_2, \ldots, p_T$ follow RW1 by counting the number of "runs" in the sequence $G_1, G_2, \ldots, G_T$, where a run is defined as a sequence of one symbol. For example, if the sequence of indicator variables is 0 0 0 1 1 1 0 0 1 1, there are four runs, while if the sequence is 0 1 0 0 1 1 0 1 1 0, there are seven runs.

For convenience, assume that there are $T/2$ zeros in sequence $G_1, G_2, \ldots, G_T$ and $T/2$ ones. This holds if T is even and $r_1, r_2, \ldots, r_T$ are unique. In general, the number of zeros and the number of ones are both approximately equal to $T/2$ with high probability and the results described as follows continue to hold.

Let M_0 denote the observed number of runs and let M denote the number of runs in a random sequence of length T with $T/2$ ones and $T/2$ zeros; to compute a p-value for the test, we can compare M_0 to the distribution of M. Although the exact distribution of M is complicated, it may be shown that M is approximately distributed as a binomial random variable with parameters T and $1/2$. To see why this might hold, consider building a sequence of length T by adding randomly selected ones and zeros, one step at a time; at each stage, there is a 50% chance of increasing the number of runs by one. This fact may be used to calculate a p-value for the test.

Example 3.5 Consider the calculation of M_0 and the corresponding p-value for log-returns on Wal-Mart stock, stored in the variable `wmt.m.logret`.

These calculations can be performed using the function **runs.test**, which is available in the **randtests** package (Caeiro and Mateus 2014).

```
> library(randtests)
Warning message:
package randtests was built under R version 3.2.3
> runs.test(wmt.m.logret)

        Runs Test

data:  wmt.m.logret
statistic = 1.0417, runs = 35, n1 = 30, n2 = 30, n = 60, p-value =
0.2976
alternative hypothesis: nonrandomness
```

Therefore, the p-value of the test is 0.2976 so that there is no evidence to reject the null hypothesis that RW1 holds for the log-returns. □

Rescaled Range Test

The Box–Ljung, variance-ratio, and runs tests are useful for detecting association among log-returns from nearby time periods; however, another way in which the random walk hypothesis may fail is if the log-returns are related over a long period of time. For instance, there may be multiyear periods during which the monthly log-returns are generally (but not always) large. The *rescaled range test* is designed to detect this type of long-range dependence.

The test statistic is given by

$$H = \frac{\max_{1\leq k\leq T}\sum_{t=1}^{k}(r_t-\bar{r}) - \min_{1\leq l\leq T}\sum_{t=1}^{l}(r_t-\bar{r})}{S\sqrt{T}}$$

where S is the sample standard deviation of $r_1, r_2, \ldots, r_T$; that is, H is the range of the variables

$$\sum_{t=1}^{k}(r_t-\bar{r}), \quad k = 1, 2, \ldots, T.$$

Large values of H are evidence against the null hypothesis that RW1 holds for the log-prices.

A large value of H indicates that there are times t_0, t_1 such that

$$\sum_{t=1}^{t_1}(r_t-\bar{r})$$

is a large positive value and

$$\sum_{t=1}^{t_0}(r_t-\bar{r})$$

TABLE 3.1
Critical Values for the Rescaled Range Test

Significance Level	Critical Value
0.10	1.620
0.05	1.747
0.025	1.862
0.005	2.098

is a large negative value; note that the values $r_t - \bar{r}$, $t = 1, 2, \ldots, T$ must sum to 0. That is, there is a time period over which the log-returns differ greatly from their sample mean.

To determine if the observed value of H is statistically significant, we compare it to the critical values in Table 3.1.

Example 3.6 Consider calculation of H for the Wal-Mart log-returns in `wmt.m.logret`. To calculate H in R, we can use the `cumsum` function, which returns the cumulative sums of the values in a vector:

```
> x<-c(1, 3, -2, -4, 5)
> cumsum(x)
[1]  1  4  2 -2  3
```

Let

$$H_1 = \max_{1 \leq k \leq T} \sum_{j=1}^{k} (r_j - \bar{r})$$

and

$$H_2 = \min_{1 \leq l \leq T} \sum_{j=1}^{l} (r_j - \bar{r})$$

so that $H = (H_1 - H_2)/(S\sqrt{T})$.

For the Wal-Mart monthly log-returns, H_1 and H_2 may be calculated by

```
> H1<-max(cumsum(wmt.m.logret-mean(wmt.m.logret)))
> H1
[1] 0.085685
> H2<-min(cumsum(wmt.m.logret-mean(wmt.m.logret)))
> H2
[1] -0.19496
```

and H is given by

```
> H<-(H1 - H2)/(sd(wmt.m.logret)*(60^.5))
> H
[1] 0.83819
```

To compute the p-value for a test of RW2, we compare the observed value of H to critical values in Table 3.1. It follows that the p-value is greater than 0.10.

Therefore, according to the rescaled range test, there is no evidence to reject the hypothesis that the RW2 model holds for Wal-Mart stock. □

3.6 Do Stock Returns Follow the Random Walk Model?

Although assumptions regarding market efficiency suggest that asset log-prices are approximately uncorrelated and, hence, some form of a geometric random walk may be reasonable for asset prices, the issue is essentially an empirical one. For an analyst, the important issue is whether or not observed prices behave like observations from a random walk model. In the previous section, several tests of the random walk model were presented. These tests show that, when applied to log-returns on Wal-Mart stock, there is no evidence to reject the hypothesis that the log-returns form a random sequence in the sense that the log-returns in different time periods are unrelated statistically. In this section, we extend those results by applying the tests to a wide range of stock returns.

Stocks for firms represented in the Standard & Poor's (S&P) 100 index were considered. The S&P 100 is a stock market index based on stocks from 100 large U.S. firms, spread across several different industries; stock market indices will be discussed in detail in Chapter 8. For present purposes, we may view the S&P 100 firms as a cross section of large U.S. companies. For each stock, five years of monthly returns were analyzed for the period ending December 31, 2014; of the 100 stocks represented in the S&P 100, four stocks did not have five years of monthly returns available, leading to a set of 96 stocks for analysis.

For each of the 96 stocks, the four tests of the random walk model discussed in the previous section were performed; the Box–Ljung test uses $m = 12$ and the variance-ratio test is based on the statistic V_3. Figure 3.1 contains the results of these tests in the form of histograms of the results for the 96 stocks. For the Box–Ljung test, the variance-ratio test, and the runs test, the histograms contain the p-values of the tests; for the rescaled range test, the histograms are based on values of the test statistics.

If the random walk model holds in general for stocks in the S&P 100, p-values should be approximately distributed according to a uniform distribution on the interval $(0, 1)$; this result is a more general version of the result that, if the random walk model holds for all stocks, then about 5% of the stocks should have a p-value less than 0.05. Conversely, if the random walk model does not hold for some stocks, we expect that there should be more than 10 (roughly 10% of 96) p-values in the range from 0 to 0.10 (the first interval on the histograms).

The histograms in Figure 3.1 suggest that neither the Box–Ljung test, the runs test, nor the variance-ratio test contradicts the hypothesis that the random walk model holds. In fact, the Box–Ljung test has a large number

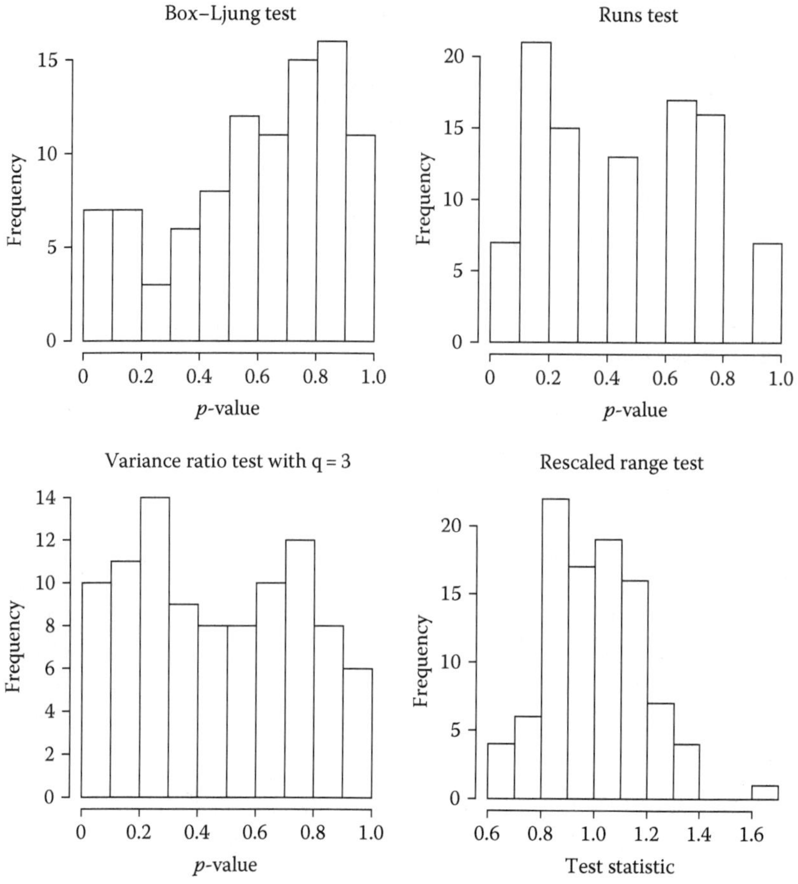

FIGURE 3.1
Results of tests of the random walk model for stocks in the S&P 100.

of relatively large p-values, indicating that the autocorrelations of the stock log-returns tend to be smaller than might be expected from a sequence of uncorrelated random variables.

For the rescaled range test, we have a table of critical values for the test but not a simple method for computing p-values; thus, the histogram presented in Figure 3.1 is of the observed test statistics. In interpreting this histogram, we may use the fact that the critical value for a test with level 10% is 1.620 and the critical value of a test with level 5% is 1.747; see Table 3.1. Therefore, only 1 of the 96 stocks yields a significant result at the 10% level in the rescaled range test.

According to these results, it is reasonable to conclude that, based on the tests used, the properties of the log-returns of stocks for firms in the S&P 100 are not inconsistent with the random walk model. However, these results should not be considered to be strong evidence in favor of the random walk hypothesis. To reach such a conclusion, we would need to consider the *power* of the tests, the probability of rejecting the null hypothesis when it does not hold. In the present context, the power is the probability of rejecting the random walk model when the data, in fact, do not follow that model. Therefore, evaluating the power of a test requires consideration of models for a time series of log-returns that may be appropriate if the random walk does not hold. If the power of a test is low against reasonable alternative models, it may be the case that the random walk model does not hold, but the tests used are unable to recognize this fact; this might be the case, for example, if the sample sizes of the data used in the tests are relatively small.

However, even if we cannot conclude with certainty that the random walk hypothesis holds, the results of this section clearly show that there is not a strong relationship between past returns on an asset and future returns; in particular, the results suggest that it will be difficult to use past returns to accurately predict future returns.

Hence, the remainder of the book focuses on methods that can be applied when the returns on a given asset form a series of uncorrelated random variables. These methods use the fact that the returns on different assets in the same time period are correlated in order to construct a combination of assets yielding a large expected return and small risk. That is, rather than trying to predict the future returns of different assets, we use the statistical properties of a set of asset returns to guide investment decisions.

3.7 Suggestions for Further Reading

The random walk hypothesis is one of the most widely discussed topics in financial statistics. The bestselling book, *A Random Walk Down Wall Street*, by Princeton economist Burton G. Malkiel (Malkiel 2003), contains a detailed, nontechnical, discussion of the random walk hypothesis and its implications; in particular, this book discusses many of the criticisms of the random walk hypothesis that have been raised and gives arguments against them. An alternative view is provided by *A Nonrandom Walk Down Wall Street*, by econometricians Andrew Lo and A. Craig MacKinlay (Lo and MacKinlay 2002), in which it is shown that there is an element of predictability in stock prices; a key part of this argument is the variance-ratio test statistic, presented in Section 3.5. Although the details of the analyses presented in this book are highly technical, the basic ideas are not; an accessible summary of many of the issues raised by Lo and MacKinlay (2002) is given by Lo (1997).

Conditional expectation and its relationship to optimal prediction is discussed by Rice (2007, Section 4.4) and Ross (2006, Section 7.6). The discussion of the relationship between efficient markets and the martingale model for asset prices given in Section 3.3 is based on the work of Campbell et al. (1997, Section 1.5); see also Fama (1965, 1970) and Samuelson (1965). A proof of

Proposition 3.1 on iterated conditional expectations is given by Severini (2005, Chapter 2).

The three versions of a random walk process discussed in Section 3.4 are based on Chapter 2 from the work of Campbell et al. (1997, Section 2.1); see Fabozzi et al. (2006, Chapter 10) for further discussion. Statistical tests of the random walk model, including the Box–Ljung test, the variance-ratio test, and the rescaled-range test, are discussed in detail in Chapter 2 of Campbell et al. (1997). The Box–Ljung statistic is discussed by Montgomery et al. (2008, Section 2.6; where it is called the Ljung–Box statistic)). A good introduction to the runs test is presented by Newbold et al. (2013, Section 14.7); see Granger (1963) for a more detailed discussion, including its application to economic data. Properties of the rescaled-range test are given by Lo (1991); in particular, this paper discusses the derivation of the critical values listed in Table 3.1.

Hypothesis tests and p-values are a central topic in statistics. See Newbold et al. (2013, Chapters 9, 10) or Woolridge (2013, Appendix C) for an introduction and Hogg and Tanis (2006, Chapters 8 and 9) for a more advanced discussion, including the important issue of the power of a test.

3.8 Exercises

1. Let X and Y denote random variables, each with mean 0, such that

$$\text{E}(Y|X) = X + a$$

for some constant a. Find a.

2. Let X and Y denote random variables and let

$$\hat{Y} = \text{E}(Y|X).$$

Show that

$$\text{Cov}(h(X), \hat{Y}) = \text{Cov}(h(X), Y)$$

for any function h.

Does it follow that the correlation of $h(X)$ and $\hat{Y}$ is equal to the correlation of $h(X)$ and Y? Why or why not?

3. Let Y and X denote random variables such that

$$Y = \beta_0 + \beta_1 X + \epsilon$$

where β_0 and β_1 are constants and ϵ is a random variable with zero mean and standard deviation σ_ϵ; assume that X and ϵ are independent. Find the function of X that is the best predictor of Y^2 and compare it to the square of the best predictor of Y.

4. Let $\Omega_1 \subset \Omega_2 \subset \cdots$ denote information sets and let $P_1, P_2, \ldots$ denote a corresponding sequence of asset prices such that $P_t \in \Omega_t$. Suppose that

$$\mathrm{E}(P_{t+h}|\Omega_t) = (1+\eta)^h P_t, \quad h = 0, 1, \ldots$$

for some constant $\eta > -1$.

a. Let $R_{t+1} = P_{t+1}/P_t - 1$ denote the return in period $t+1$. Find $\mathrm{E}(R_{t+1})$.

b. For $t \neq s$, find $\mathrm{Cov}(R_t, R_s)$.

5. Let V denote the instrinsic value of a share of stock in a given firm and let Ω_t denote the set of financial information available at time t, $t = 1, 2, \ldots$. Let P_t denote the price of one share of stock in the firm. However, instead of $P_t = \mathrm{E}(V|\Omega_t)$, we have

$$P_t = \mathrm{E}(V|\Omega_t) + Z_t$$

where Z_t is a random variable such that $\mathrm{E}(Z_t|\Omega_t) = 0$. That is, the price P_t is approximately, but not exactly, equal to $\mathrm{E}(V|\Omega_t)$.

Does the martingale property

$$\mathrm{E}(P_{t+h}|\Omega_t) = P_t, \; h = 1, 2, \ldots$$

hold in this case? Why or why not?

6. Suppose that $\{Y_t : t = 0, 1, \ldots\}$ follows RW3 and, for some given integer $m > 1$, define a sequence of random variables $X_1, X_2, \ldots$ by $X_0 = Y_0$,

$$X_1 = Y_m, \; X_2 = Y_{2m}, \; X_3 = Y_{3m},$$

and so on, so that $X_t = Y_{tm}$.

Does $\{X_t : t = 0, 1, 2, \ldots\}$ follow RW3? Why or why not?

7. Suppose $\{Y_t : t = 0, 1, \ldots\}$ follows RW3 and, for each $t = 1, 2, \ldots$, define $X_t = aY_t + b$ for constants a, b.

Does $\{X_t : t = 0, 1, \ldots\}$ follow RW3? Why or why not?

8. Let $\{Y_t : t = 0, 1, \ldots\}$ and $\{X_t : t = 0, 1, \ldots\}$ denote stochastic processes that both follow RW1. Assuming that $\{Y_t : t = 1, 2, \ldots\}$ and $\{X_t : t = 1, 2, \ldots\}$ are independent, does $\{X_t + Y_t : t = 0, 1, \ldots\}$ follow RW1? Why or why not?

9. Let $\{Y_t : t = 0, 1, \ldots\}$ denote a random walk process. For a given positive integer h, define

$$X_t = Y_{t+h} - Y_h, \; t = 0, 1, \ldots.$$

Is $\{X_t : t = 0, 1, \ldots\}$ a random walk process? Why or why not? Consider each of the three definitions of a random walk, RW1, RW1, and RW3.

10. Let $\{Y_t : t = 1, 2, \ldots\}$ denote a weakly stationary process; let σ^2 denote the variance of Y_t and let $\rho(\cdot)$ denote the autocorrelation function of the process.

    For a given positive integer q, find an expression for

$$\frac{\text{Var}(Y_1 + \cdots + Y_q)}{\text{Var}(Y_1)}$$

    as a function of $\rho(\cdot)$.

    Interpret this result in terms of the properties of the variance-ratio test.

11. Consider stock in Best Buy Company, Inc. (symbol BBY).

    a. Using the R function `get.hist.quote`, download the adjusted prices needed to calculate five years of monthly returns for the period ending December 31, 2015, and compute the log-prices.
    b. Compute five years of monthly log-returns corresponding to the log-prices obtained in Part (a).
    c. Using the function `summary`, calculate the summary statistics for the returns computed in Part (b).
    d. Using the function `acf`, calculate the sample autocorrelation function for the log-returns computed in Part (b); use a maximum lag of 12.

12. For the data calculated in Exercise 11, use the Box–Ljung test to test the hypothesis that

$$H_0 : \rho(1) = \rho(2) = \cdots = \rho(12) = 0$$

    where $\rho(\cdot)$ denotes the autocorrelation of the stochastic process corresponding to the log-returns on Best Buy stock; see Example 3.3.

    Is this result consistent with the hypothesis that the log-prices of Best Buy stock follow RW3? Why or why not?

13. Use the variance ratio test to test the hypothesis that the log-prices of Best Buy stock, calculated in Exercise 11, follow RW3; see Example 3.4.

    a. Using the log-returns on Best Buy stock, compute the variance ratio test statistic V_6.
    b. Find the mean and variance of V_6 under the null hypothesis that the stochastic process corresponding to the log-prices of Best Buy stock follows RW3.
    c. Determine the p-value of the test.
    d. Based on these results, are the observed log-prices of Best Buy stock consistent with the RW3 model? Why or why not?

14. For the log-prices of Best Buy stock calculated in Exercise 11, calculate the p-value of the runs test; see Example 3.5.

    Based on your results, what do you conclude regarding the random walk hypothesis as applied to the log-prices of Best Buy stock?

15. For the log-returns on Best Buy stock calculated in Exercise 11, calculate the rescaled-range statistic H; see Example 3.6.

    Based on this result, what do you conclude regarding random walk hypothesis as applied to the log-prices of Best Buy stock?

4

Portfolios

4.1 Introduction

Suppose we have a given amount of capital to invest and a number of possible assets in which to invest. How should we place our investment in the various assets? This is known as the *portfolio selection problem* and the mathematical and statistical methods developed to solve it are known as *portfolio theory.*

If it were possible to accurately forecast the future returns of the assets we could simply invest in the asset or assets with the largest predicted returns over the future time period of interest. However, empirical evidence, such as that presented in the previous chapter, suggests that such forecasting is difficult at best.

Hence, in portfolio theory, we attempt to choose the combination of assets that yields a portfolio return with desirable statistical properties; specifically, we seek a large expected return, while minimizing the "risk" of the portfolio, defined here as the standard deviation of the return. Thus, in an ideal case, the return on the portfolio will have a large expected value and a small standard deviation so that the portfolio realizes a large return with high probability. However, complicating the situation is the fact that the two goals are typically in conflict: Riskier assets generally have a higher expected return, as a reward for assuming the risk.

4.2 Basic Concepts

Suppose there are N assets in the market and let $R_1, R_2, \ldots, R_N$ denote random variables representing their respective returns in a given time period. Consider a portfolio in which we place a proportion w_j of our total investment in asset j, $j = 1, 2, \ldots, N$. The quantities $w_1, w_2, \ldots, w_N$ are known as the *portfolio weights.* Because w_j represents the proportion invested in asset j, we must have

$$w_1 + w_2 + \cdots + w_N = \sum_{j=1}^{N} w_j = 1.$$

The return on the portfolio, R_p, can be expressed in terms of the returns on the individual assets and the portfolio weights. For each $j = 1, 2, \ldots, N$, let $P_{j,0}$ and $P_{j,1}$ denote the prices of asset j in periods 0 and 1, respectively, so that

$$R_j = \frac{P_{j,1}}{P_{j,0}} - 1.$$

Suppose we wish to invest capital C in the N assets, according to the weights $w_1, w_2, \ldots, w_N$. Then, in period 0, we buy

$$\frac{w_j C}{P_{j,0}}$$

shares of asset j, $j = 1, 2, \ldots, N$.

In period 1, the shares in asset j are worth

$$\frac{w_j C}{P_{j,0}} P_{j,1}.$$

Thus, the total worth of the portfolio in period 1 is

$$\sum_{j=1}^{N} \frac{w_j C}{P_{j,0}} P_{j,1}$$

and the return on the portfolio is

$$\frac{\sum_{j=1}^{N} \frac{w_j C}{P_{j,0}} P_{j,1}}{C} - 1.$$

Writing $P_{j,1}/P_{j,0}$ as $R_j + 1$, the return may be written

$$\sum_{j=1}^{N} w_j(R_j + 1) - 1 = \sum_{j=1}^{N} w_j R_j.$$

That is, the return on the portfolio is a linear function of the individual asset returns, with the coefficients given by the portfolio weights:

$$R_p = w_1 R_1 + w_2 R_2 + \cdots + w_N R_N = \sum_{j=1}^{N} w_j R_j.$$

The goal in the statistical approach to portfolio theory is to select the portfolio weights so that $\mathrm{E}(R_p)$ is large and the standard deviation of R_p, or equivalently $\mathrm{Var}(R_p)$, is small.

In this chapter, a number of basic concepts and results are presented, focusing on the case in which $N = 2$; the general case will be considered in the following chapter.

Thus, consider two assets, with returns R_1 and R_2. For $j = 1, 2$, let

$$\mu_j = \mathrm{E}(R_j) \quad \text{and} \quad \sigma_j^2 = \mathrm{Var}(R_j)$$

and let ρ_{12} denote the correlation of R_1 and R_2. Let w_1 and w_2 denote the portfolio weights; because $w_1 + w_2 = 1$, we can take $w_1 = w$ and $w_2 = 1 - w$ for some real number w.

The portfolio placing weight w on asset 1 and weight $1-w$ on asset 2 has return

$$R_p = wR_1 + (1-w)R_2.$$

Then, using results on the mean and variance of a sum of random variables,

$$\begin{aligned}\mu_p \equiv \mu_p(w) &= \mathrm{E}(R_p) = w\mathrm{E}(R_1) + (1-w)\mathrm{E}(R_2)\\ &= w\mu_1 + (1-w)\mu_2\end{aligned}$$

and

$$\sigma_p^2 \equiv \sigma_p^2(w) = \mathrm{Var}(R_p) = w^2\sigma_1^2 + (1-w)^2\sigma_2^2 + 2w(1-w)\rho_{12}\sigma_1\sigma_2.$$

The portfolio problem for the case of two assets is concerned with choosing w so that $\mu_p(w)$ is large and $\sigma_p(w)$ is small.

Example 4.1 Consider two assets. Suppose that $\mu_1 = 0.10$, $\mu_2 = 0.20$, $\sigma_1 = 0.20$, $\sigma_2 = 0.25$, and $\rho_{12} = 0.5$. Then the portfolio that places weight w on asset 1 and weight $1-w$ on asset 2 has mean return

$$\mu_p(w) = w(0.10) + (1-w)(0.20) = 0.20 - 0.10w$$

and return variance

$$\begin{aligned}\sigma_p^2(w) &= w^2(0.20)^2 + (1-w)^2(0.25)^2 + 2w(1-w)(0.50)(0.20)(0.25)\\ &= 0.0625 - 0.075w + 0.0525w^2.\end{aligned}$$

For instance, the portfolio placing half its weight on asset 1 and half its weight on asset 2 has expected return

$$\mu_p(0.5) = 0.20 - 0.10(0.5) = 0.15$$

and return standard deviation

$$\sigma_p(0.5) = \left(0.0625 - 0.075(0.5) + 0.0525(0.5)^2\right)^{\frac{1}{2}} \doteq 0.195. \qquad \square$$

Diversification

Note that, although the mean return on the portfolio depends only on the mean returns on the individual assets, the risk of the portfolio depends on the relationship between the asset returns, as measured by their correlation. This is a fundamental idea in portfolio theory; in particular, it plays an important role in the concept of diversification, which refers to reducing the risk of a portfolio by investing in many assets, a central idea in portfolio theory. We illustrate this concept by considering two examples.

Example 4.2 Consider the case in which we have two assets with returns R_1, R_2 with the same mean and variance: $\mathrm{E}(R_1) = \mathrm{E}(R_2) = \mu$ and $\mathrm{Var}(R_1) = \mathrm{Var}(R_2) = \sigma^2$. Furthermore, assume that R_1 and R_2 are uncorrelated.

Investing entirely in either asset 1 or asset 2 yields the same expected return and the same risk. Now consider a portfolio consisting of both asset 1 and asset 2. Let w be the proportion of our investment in asset 1 so that the portfolio weights are $w_1 = w$ and $w_2 = 1 - w$. Then the expected return on the portfolio is

$$\mu_p(w) = \mathrm{E}(R_p) = w\mathrm{E}(R_1) + (1-w)\mathrm{E}(R_2) = w\mu + (1-w)\mu = \mu.$$

Therefore, the expected return on the portfolio does not depend on the value of w; in particular, investing in the portfolio yields the same expected return as investing in either asset 1 or asset 2. However including both assets in the portfolio reduces risk.

Because R_1 and R_2 are uncorrelated,

$$\sigma_p^2(w) = \mathrm{Var}(R_p) = w^2\mathrm{Var}(R_1) + (1-w)^2\mathrm{Var}(R_2) = (w^2 + (1-w)^2)\sigma^2.$$

For $w = 0$ or 1, $\sigma_p = \sigma$. However, for any other value of w, $0 < w < 1$, $\sigma_p < \sigma$. Choosing $w = 1/2$ minimizes σ_p yielding a value of $\sigma/\sqrt{2}$. □

The scenario in this example is a special one since the assets have the same mean and variance, and they are uncorrelated; the following example generalizes this by assuming the asset returns are correlated.

Example 4.3 Consider the framework of Example 4.2 but now assume that the correlation of R_1 and R_2 is ρ_{12}, $-1 < \rho_{12} < 1$. Consider the portfolio placing equal weight on the two assets so that $w_1 = w_2 = 1/2$. Then $R_p = (1/2)R_1 + (1/2)R_2$; hence, $\mu_p = \mathrm{E}(R_p) = \mu$ as in Example 4.2. The variance of R_p is given by

$$\begin{aligned}\sigma_p^2 = \mathrm{Var}(R_p) &= \frac{1}{4}\mathrm{Var}R_1 + \frac{1}{4}\mathrm{Var}R_2 + 2\frac{1}{2}\frac{1}{2}\mathrm{Cov}(R_1, R_2)\\ &= \frac{1}{4}\sigma^2 + \frac{1}{4}\sigma^2 + \frac{1}{2}\rho_{12}\sigma^2\\ &= \frac{1}{2}(1+\rho_{12})\sigma^2.\end{aligned}$$

Therefore, for any value of ρ_{12}, $-1 < \rho_{12} < 1$, $\sigma_p < \sigma$.

When ρ_{12} is close to 1, then σ_p is close to σ, the standard deviation of R_1 and R_2; therefore, there is little benefit to including both assets in the portfolio. This is not surprising because, when $\rho_{12} \doteq 1$, R_1 and R_2 have a linear relationship; hence, there is little diversity in including both assets in the portfolio. On the other hand, the standard deviation of the portfolio return is the smallest when the asset returns are *negatively* correlated so that one asset return increases when the other decreases and vice versa. □

4.3 Negative Portfolio Weights: Short Sales

Although it may be more natural to consider portfolio weights that are constrained to be in the interval $[0, 1]$, often it is desirable to allow weights to be negative. This is illustrated in the following example.

Example 4.4 Let R_1, R_2 denote returns on two assets and suppose that $\mathrm{E}(R_1) = \mathrm{E}(R_2) = \mu$, $\mathrm{Var}(R_1) = 0.3$, $\mathrm{Var}(R_2) = 0.1$, and suppose that the correlation of R_1, R_2 is $\sqrt{3}/2$. If we construct a portfolio by investing a proportion w of our investment in asset 1 and $1 - w$ of our investment in asset 2, the return on our portfolio is $R_p = wR_1 + (1 - w)R_2$. It is straightforward to show that R_p has mean $\mu_p(w) = \mu$ and variance $\sigma_p^2(w)$, given by

$$\begin{aligned}\sigma_p^2(w) &= (0.3)w^2 + (0.1)(1-w)^2 + \frac{\sqrt{3}}{2}\sqrt{(0.3)(0.1)}w(1-w)\\ &= (0.1)(w^2 + w + 1).\end{aligned}$$

Because all portfolios have the same expected return, we should choose w to minimize risk. Note that $\sigma_p^2(w)$ is a quadratic function of w with a positive coefficient of w^2; hence, it can be minimized by taking the derivative, setting it equal to 0, and solving for w, yielding $w = -1/2$. That is, the minimum variance portfolio places weight $-1/2$ on asset 1 and weight $3/2$ on asset 2. □

Such a negative weight in a portfolio represents a "short sale" in which the investor sells, rather than buys, the asset. Because the investor does not own the asset in question, first she must borrow it before selling it; when the asset is sold, the investor receives the current price for the asset. At the end of the investment period, to pay back the loan, the investor must buy the asset at its current price. Therefore, the entire procedure is the opposite of what takes place when buying an asset.

Example 4.5 Consider the portfolio constructed in Example 4.4. Suppose the price of asset 1 is \$10 per share in period 0 and \$12 per share in period 1 so that the realized value of R_1 is $(12 - 10)/10 = 0.2$. For asset 2, suppose the price is \$15 per share in period 0 and \$21 per share in period 1 so that R_2 is observed to be 0.4.

Suppose we have \$100 to invest. Because $w_1 = -1/2$, we borrow \$50 (i.e., $1/2 \times 100$) worth of asset 1, 5 shares at \$10 per share, and immediately sell them, yielding \$50. We now have \$150 to invest in asset 2, which allows us to buy 10 shares at \$15 per share.

At the end of period 1, our 10 shares of asset 2 are worth \$21 per share, or \$210 total. However, we must still repay our loan of 5 shares of asset 1. Each share of asset 1 is now worth \$12, so we must use \$60 (i.e., 5×12) of our \$210 to buy 5 shares of asset 1. Therefore, our net worth at the end of period 1 is 210\$ − 60\$ = \$150. This corresponds to a return of

$$\frac{150 - 100}{100} = 0.5.$$

In terms of w_1, w_2, R_1, and R_2 the return on our portfolio is

$$w_1 R_1 + w_2 R_2 = -0.5R_1 + 1.5R_2 = -0.5(0.2) + 1.5(0.4) = 0.5$$

as determined previously. □

A short sale of an asset may be viewed as a loan, with the interest rate of the loan equal to the return on that asset. Therefore, if the price of the asset decreases over the loan period, so that the return is negative, then the interest rate on the loan is effectively negative and the investor makes money on the loan as well. Because of this, short sales are often described as an appropriate investment for an asset expected to have a negative return. Although this may be true, a short sale can be useful for controlling risk even in cases in which the asset price is expected to increase. This illustrated in Example 4.4.

Although portfolios with negative weights are convenient from a mathematical point of view, and we will use them here, there are some important practical considerations. Note that there is a fundamental difference between a short sale and a more typical "long" position. If we purchase \$100 of an asset, the most we can lose is \$100, which occurs if the share price decreases to 0. However, if we borrow \$100 worth of an asset, our potential losses are unbounded because the price of the asset can, in principle, increase indefinitely. Therefore, short sales are tightly regulated.

For instance, when shares are borrowed, a certain amount of collateral must be placed in a "margin account," effectively increasing the cost of the loan. If the price of the stock increases, then more collateral may be needed since the cost of buying back the stock is greater; in this case, the investor receives a "margin call" demanding that additional funds be added to the margin account. Also, short positions may be subject to a "forced buy-in," in which the lender of the asset requires that the loan be repaid immediately, even if such a repayment requires the investor who borrowed the shares to lose money.

Because of such considerations, analysts sometimes only consider portfolios in which the weights are restricted to lie in the interval $[0, 1]$, although such constraints complicate the details of the analysis. Constraints of this type will be considered in Section 5.8.

4.4 Optimal Portfolios of Two Assets

Consider two assets, with returns R_1 and R_2, respectively; for $j = 1, 2$, let R_j have mean and standard deviation μ_j and σ_j, respectively, and let ρ_{12}

denote the correlation of R_1 and R_2. The portfolio based on portfolio weights $w_1 = w$ and $w_2 = 1 - w$ has return $R_p = wR_1 + (1 - w)R_2$. We summarize the properties of the portfolio with return R_p by the mean and standard deviation of the return, $\mu_p(w)$ and $\sigma_p(w)$, respectively, viewed as functions of w, $-\infty < w < \infty$. Recall that

$$\mu_p(w) = w\mu_1 + (1 - w)\mu_2$$

and

$$\sigma_p^2(w) = w^2\sigma_1^2 + (1 - w)^2\sigma_2^2 + 2w(1 - w)\rho_{12}\sigma_1\sigma_2.$$

In this section, we consider the problem of choosing the value of w.

As w varies, $\mu_p(w)$ and $\sigma_p(w)$ vary; we may view these values as points $(\sigma_p(w), \mu_p(w))$ in the "risk-return space." A plot of $(\sigma_p(w), \mu_p(w))$ as w varies is a useful way to understand the relationship between the expected return and risk of a portfolio.

Example 4.6 Suppose that $\mu_1 = 0.2$, $\sigma_1 = 0.1$, $\mu_2 = 0.1$, $\sigma_2 = 0.05$, and $\rho_{12} = 0.25$. Then $\mu_p(w) = w(0.2) + (1 - w)(0.1) = 0.1 + 0.1w$ and

$$\begin{aligned}\sigma_p^2(w) &= w^2(0.1)^2 + (1 - w)^2(0.05)^2 + 2w(1 - w)(0.25)(0.1)(0.05)\\ &= 0.01w^2 - 0.0025w + 0.0025.\end{aligned}$$

Figure 4.1 contains a plot of $(\sigma_p(w), \mu_p(w))$ as w varies. □

The curve in Figure 4.1 represents the set of all $(\sigma_p(w), \mu_p(w))$ pairs that are available to the investor; it is known as the *opportunity set*. This term

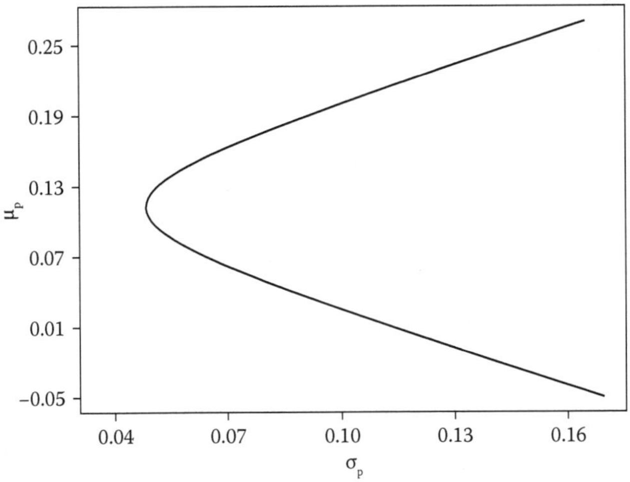

FIGURE 4.1
Expected return and risk for different portfolios in Example 4.6.

will also be used to describe the corresponding portfolios; for instance, in the previous example, a portfolio in the opportunity set has a value of (σ_p, μ_p) on the curve in Figure 4.1.

Note that, unless $\mu_1 = \mu_2$, for a given value m there is exactly one value of w such that $\mu_p(w) = m$. On the other hand, for a given value $s > 0$, there may be zero, one, or two values of w such that $\sigma_p(w) = s$, depending on the number of solutions to the quadratic equation $\sigma_p^2(w) - s^2 = 0$.

Example 4.7 Consider the assets described in Example 4.6. Note that, for a given value of μ_p, say m, the portfolio with that expected return may be found by solving

$$\mu_p(w) = 0.1 + 0.1w = m$$

for w, yielding $w = 10m - 1$. On the other hand, for a given value of σ_p, say s, there may be zero, one, or two portfolios with that standard deviation, depending on the solutions to the quadratic equation

$$\sigma_p^2(w) = 0.01w^2 - 0.0025w + 0.0025 = s^2.$$

For instance, there are no portfolios with $\sigma_p(w) = 0.0375$ because there are no (real) solutions to the equation

$$0.01w^2 - 0.0025w + 0.0025 = (0.0375)^2;$$

there are two portfolios with $\sigma_p(w) = 0.0625$, corresponding to the two solutions to the equation

$$0.01w^2 - 0.0025w + 0.0025 = (0.0625)^2,$$

$w = (1 + \sqrt{10})/8 \doteq 0.52$ and $w = (1 - \sqrt{10})/8 \doteq -0.27$; there is one portfolio with $\sigma_p(w) = \sqrt{15}/80$, corresponding to the one solution to the equation

$$0.01w^2 - 0.0025w + 0.0025 = \left(\frac{\sqrt{15}}{80}\right)^2 = \frac{15}{6400},$$

$w = 1/8$. These three possibilities are illustrated in Figure 4.2. □

Efficient Portfolios

For the case in which two portfolios have the same return standard deviation, only the one with the larger return mean is of interest. For example, in Figure 4.1, each portfolio with a $(\sigma_p(w),\ \mu_p(w))$ pair on the lower half of the curve is dominated by a portfolio with a $(\sigma_p(w),\ \mu_p(w))$ pair on the upper half of the curve. Thus, the portfolios corresponding to the upper half of the curve are known as *efficient* portfolios.

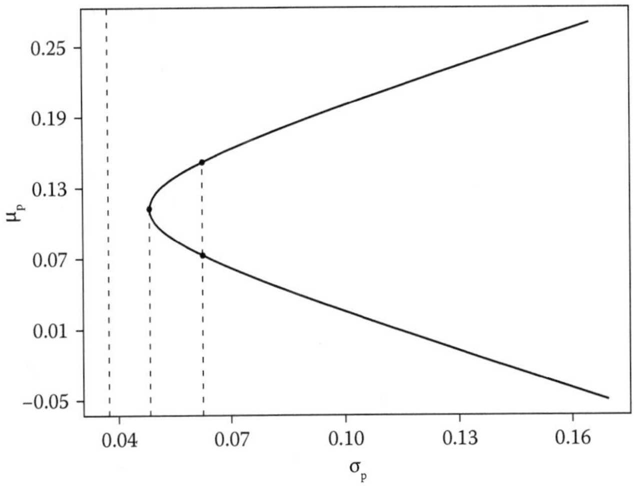

FIGURE 4.2
Three possibilities for the solutions to $\sigma_p^2(w) = s$ in Example 4.7.

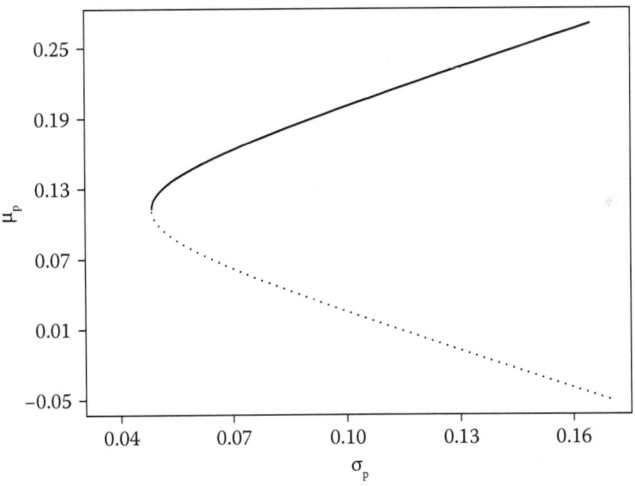

FIGURE 4.3
Efficient frontier in Example 4.7.

That is, for an efficient portfolio, it is not possible to have a larger expected return without increasing the risk or, conversely, it is not possible to have lower risk without decreasing the expected return. This upper half of the opportunity set is known as the *efficient frontier*; the efficient frontier for the assets described in Example 4.7 is given in Figure 4.3. The term "efficient frontier" will also be used to be describe the portfolios with a $(\sigma_p(w),\ \mu_p(w))$ pair on the efficient frontier.

Each portfolio on the efficient frontier has the largest possible expected return for a given level of risk and the lowest possible risk for a given expected return. Therefore, there is no objective way to choose from among the efficient portfolios; such a choice depends on an investor's view of the relative importance of a portfolio's expected return and risk.

The Minimum-Variance Portfolio

Suppose that our goal is simply to minimize risk, without consideration of the expected return of the portfolio. Then we choose w to minimize the return standard deviation $\sigma_p(w)$ or, equivalently, to minimize the return variance $\sigma_p^2(w)$. The resulting portfolio is known as the *minimum-variance portfolio.*

To find the minimum-variance portfolio, we need to find the value of w that minimizes $\sigma_p^2(w)$. To solve this minimization problem, we may use the approach used in calculus. Note that

$$\frac{d\sigma_p^2(w)}{dw} = 2w\sigma_1^2 - 2(1-w)\sigma_2^2 + 2(1-2w)\rho_{12}\sigma_1\sigma_2$$

and

$$\frac{d^2\sigma_p^2(w)}{dw^2} = 2\sigma_1^2 + 2\sigma_2^2 - 4\rho_{12}\sigma_1\sigma_2.$$

Clearly,

$$\begin{aligned}\frac{d^2\sigma_p^2(w)}{dw^2} &\geq 2\sigma_1^2 + 2\sigma_2^2 - 4\sigma_1\sigma_2 \\ &= 2(\sigma_1 - \sigma_2)^2 \geq 0\end{aligned}$$

and, provided that $\rho_{12} < 1$,

$$\frac{d^2\sigma_p^2(w)}{dw^2} > 0.$$

Hence, $\sigma_p^2(w)$ can be minimized by solving

$$w\sigma_1^2 - (1-w)\sigma_2^2 + (1-2w)\rho_{12}\sigma_1\sigma_2 = 0$$

for w, yielding the solution

$$w = w_{mv} \equiv \frac{\sigma_2^2 - \rho_{12}\sigma_1\sigma_2}{\sigma_1^2 + \sigma_2^2 - 2\rho_{12}\sigma_1\sigma_2}$$

and

$$1 - w_{mv} = \frac{\sigma_1^2 - \rho_{12}\sigma_1\sigma_2}{\sigma_1^2 + \sigma_2^2 - 2\rho_{12}\sigma_1\sigma_2}.$$

Note that $\sigma_1^2 + \sigma_2^2 - 2\rho_{12}\sigma_1\sigma_2 = \text{Var}(R_1 - R_2)$. Using the fact that

$$\text{Var}(R_1 - R_2) = \sigma_1^2 + \sigma_2^2 - 2\rho_{12}\sigma_1\sigma_2 = (\sigma_1 - \sigma_2)^2 + 2(1 - \rho_{12})\sigma_1\sigma_2,$$

it follows that if $\rho_{12} < 1$, then $\text{Var}(R_1 - R_2) > 0$. Therefore, provided that $\rho_{12} < 1$, the denominator in the expression for w_{mv} is nonzero.

Example 4.8 Consider Example 4.6 in which $\sigma_1 = 0.1$, $\sigma_2 = 0.05$, and $\rho_{12} = 0.25$. Then

$$w_{mv} = \frac{(0.05)^2 - (0.25)(0.1)(0.05)}{(0.1)^2 + (0.05)^2 - 2(0.25)(0.1)(0.05)} = \frac{1}{8}.$$

Recall that in Example 4.6 we saw that the quadratic equation $\sigma_p^2(w) - 15/6400 = 0$ has a single root, at $w = 1/8$. Such a single root always occurs at the point of minimum risk; see Figure 4.2.

Therefore, risk is minimized by placing 1/8th of our investment in asset 1. Here, $\mu_1 = 0.2$ and $\mu_2 = 0.1$, so that the minimum-variance portfolio has expected return

$$(1/8)\mu_1 + (7/8)\mu_2 = 0.1125$$

and, using the result in Example 4.6, the standard deviation of the return is

$$\left.(0.01w^2 - 0.0025w + 0.0025)^{\frac{1}{2}}\right|_{w=1/8} \doteq 0.0484.$$

This gives the portfolio with minimum risk. However, it is only the optimal choice if our goal is to minimize risk. For example, suppose that we are willing to increase the standard deviation of the portfolio to 0.05; the solutions to $0.01w^2 - 0.0025w + 0.0025 = (0.05)^2$ are $w = 0.25$ and $w = 0$. Only the first of these corresponds to a portfolio on the efficient frontier (why?), and its expected return is $0.1 + 0.1(0.25) = 0.125$. Hence, a 3.3% increase in risk (i.e., 0.0484 to 0.05) yields a 11% increase in expected return (i.e., 0.1125 to 0.125). □

The Risk-Aversion Criterion

The minimum-variance portfolio completely ignores the expected return of the portfolio. An alternative, and usually preferable, approach is to use a criterion that takes into account both the risk and the expected return of the portfolio.

Because of risk aversion, investors are generally willing to accept a lower expected return if that lower expected return corresponds to lower risk as well. Conversely, high portfolio risk may be tolerable if the portfolio has a high expected return.

Consider the portfolio placing weight w on asset 1; let $\mu_p(w)$ and $\sigma_p(w)$ denote the mean and standard deviation, respectively, of the return on that portfolio. We might consider evaluating this portfolio by

$$f_\lambda(w) = \mu_p(w) - \frac{\lambda}{2}\sigma_p^2(w)$$

where $\lambda > 0$ is a given parameter, known as the *risk aversion parameter*.

The function f_λ is a type of penalized mean return, with a penalty based on the return variance; the magnitude of the penalty is controlled by the parameter λ. Thus, larger values of $f_\lambda(\cdot)$ correspond to portfolios with a greater expected return relative to the portfolio risk, with the tradeoff between expected return and risk controlled by λ. For a given value of λ, let w_λ denote the value of w that maximizes $f_\lambda(w)$.

To find w_λ, we may use an approach similar to the one used to find the weights of the minimum-variance portfolio. Note that

$$\begin{aligned} f'_\lambda(w) &= \mu'_p(w) - \lambda\sigma_p(w)\sigma'_p(w) \\ &= (\mu_1 - \mu_2) - \lambda(w\sigma_1^2 - (1-w)\sigma_2^2 + (1-2w)\rho_{12}\sigma_1\sigma_2) \end{aligned}$$

and

$$\begin{aligned} f''_\lambda(w) &= -\lambda(2\sigma_1^2 + 2\sigma_2^2 - 4\rho_{12}\sigma_1\sigma_2) \\ &= -2\lambda \text{Var}(R_1 - R_2) < 0 \end{aligned}$$

provided that $\rho_{12} < 1$. Hence, we can maximize $f_\lambda(w)$ by solving $f'_\lambda(w) = 0$ for w, yielding the solution

$$w_\lambda = \frac{\sigma_2^2 - \rho_{12}\sigma_1\sigma_2 + (\mu_1 - \mu_2)/\lambda}{\sigma_1^2 + \sigma_2^2 - 2\rho_{12}\sigma_1\sigma_2}; \tag{4.1}$$

it follows that

$$1 - w_\lambda = \frac{\sigma_1^2 - \rho_{12}\sigma_1\sigma_2 - (\mu_1 - \mu_2)/\lambda}{\sigma_1^2 + \sigma_2^2 - 2\rho_{12}\sigma_1\sigma_2}.$$

Example 4.9 Consider two assets such that $\mu_1 = 0.04$, $\mu_2 = 0.02$, $\sigma_1 = 0.2$, $\sigma_2 = 0.1$, and $\rho_{12} = 0.25$. Then, using (4.1),

$$w = w_\lambda \equiv \frac{1}{8} + \frac{1}{2\lambda}, \quad \lambda > 0.$$

Figure 4.4 contains plots of $\mu_p(w_\lambda)$ and $\sigma_p(w_\lambda)$ as λ varies. Note that, for large values of λ, there is a large penalty on the variance of the return; hence, the optimal portfolio has small risk. To achieve this, the portfolio must also have small expected return. Conversely, for a small λ, the variance of the return is largely irrelevant; hence, the optimal portfolio has large risk. As a reward for the large risk incurred, the portfolio also has a large expected return.

When λ is small, the weight on asset 1 is large; hence, the weight on asset 2 is negative. That is, in order to achieve a large expected return, we must borrow asset 2 (which has a low expected return) in order to buy asset 1 (which has a large expected return). For instance, if $\lambda = 0.25$, the optimal portfolio places weight 2.125 on asset 1 and weight -1.125 on asset 2. □

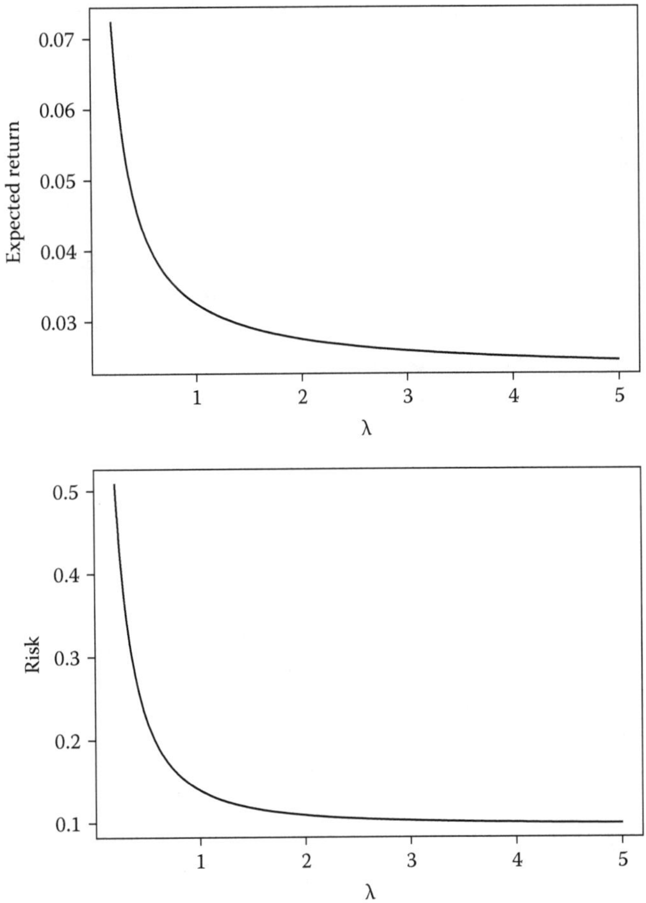

FIGURE 4.4
Properties of the optimal portfolio as a function of λ in Example 4.9.

4.5 Risk-Free Assets

In constructing a portfolio, we might consider the possibility of *not investing* some of our capital. That is, if, as in the previous section, there are two assets and w_i denotes the proportion of our investment in asset $i, i = 1, 2$, then we might have $w_1 + w_2 < 1$. The proportion not invested, $1 - w_1 - w_2$, does not contribute to the expected return on the portfolio, but it reduces the proportion of the investment contributing to the risk of the portfolio.

A better approach is to invest in a *risk-free asset*, an asset that realizes a small return but does so with complete certainty. Let R_f denote the return on a risk-free asset. Then $\text{E}(R_f) = \mu_f$, the risk-free rate of return, and $\text{Var}(R_f) = 0$, that is, $\Pr(R_f = \mu_f) = 1$.

Investment in a risk-free asset might contribute only a small return to the portfolio but it reduces the risk of our investment, giving the investor a convenient way to control risk; note that, since R_f has zero variance, it also has zero covariance with any other asset return. Although we might consider a simple "savings account" as a risk-free asset, the usual risk-free asset used in portfolio analysis is a three-month U.S. Treasury Bill.

For instance, consider a portfolio consisting of a risk-free asset, with return R_f, and a standard, "risky" asset, with return R. Let $\mu = \text{E}(R)$ and $\sigma^2 = \text{Var}(R)$; assume that $\mu > \mu_f$. Suppose we invest a proportion w of our investment in the risky asset, with the remainder invested in the risk-free asset; assume that $w > 0$ so that we do not borrow the risky asset to buy the risk-free asset. Then the return on our portfolio is $wR + (1-w)R_f$.

This portfolio has expected return

$$\mu_0(w) = w\mu + (1-w)\mu_f \tag{4.2}$$

and return standard deviation

$$\sigma_0(w) = w\sigma. \tag{4.3}$$

Note that, although R_f and μ_f are equal with probability one, we will use R_f when referring to returns, for example, the return on the portfolio is $wR + (1-w)R_f$, and use μ_f when referring to properties of the distribution of returns, for example, the expected return on the portfolio is $w\mu + (1-w)\mu_f$.

Hence, the risk of the portfolio can be made as small as we like, by choosing a value of w close to 0. Of course, reducing the risk of the portfolio also reduces the expected return. Note that solving (4.3) for w yields

$$w = \frac{\sigma_0(w)}{\sigma}$$

and using this result in (4.2) yields the following relationship between the expected return on such a portfolio and its risk:

$$\mu_0(w) = \mu_f + \frac{\mu - \mu_f}{\sigma}\sigma_0(w).$$

Example 4.10 Consider a risky asset with expected return $\mu = 0.02$ and return standard deviation $\sigma = 0.1$; suppose that the risk-free return is $\mu_f = 0.001$. Then

$$\frac{\mu - \mu_f}{\sigma} = \frac{0.02 - 0.001}{0.1} = 0.19$$

so that the mean return $\mu_0(w)$ and return standard deviation $\sigma_0(w)$ of the portfolio placing weight w on the risky asset and weight $1-w$ on the risk-free asset are related by

$$\mu_0(w) = 0.001 + 0.19\sigma_0(w).$$ □

Because of the simple relationship between the expected return and return standard deviation for portfolios formed from a risky asset and a risk-free asset, the set of risk, expected-return pairs available to the investor takes a particularly simple form.

Example 4.11 Consider the asset and risk-free asset described in Example 4.10. Let $\mu_0(w)$ and $\sigma_0(w)$ denote the mean and standard deviation, respectively, of the portfolio placing weight w on the risky asset. Figure 4.5 contains a plot of the efficient frontier (the solid line) together with the opportunity set (the solid line together with the dotted line). The minimum risk portfolio in this context is the one with $w = 0$ corresponding to the vertex in the plot, which occurs at $(0, \mu_f)$; note that the portfolio with $w = 0$ is the one in which the investor neither invests in nor borrows the risky asset.

The efficient frontier in this case is that part of the opportunity set corresponding to $w \geq 0$, that is, the line segment with positive slope. The slope of this line segment is $(\mu - \mu_f)/\sigma$, which may be viewed as a measure of the mean excess return on the asset per unit of risk. Note that if we have two possible risky assets, the one with the larger slope leads to portfolios with a higher expected return for a given level of risk. This property plays an important role in constructing a portfolio from two risky assets together with the risk-free asset, and it will be discussed in detail in Section 4.6. □

The risk-free return provides a convenient baseline for measuring the return on an asset; for a given asset with return R, the quantity $R - R_f$ is known as the *excess return* of the asset. Therefore, for the portfolio discussed earlier, the expected excess return on the portfolio is proportional to the portfolio risk:

$$\mu_0(w) - \mu_f = \frac{\mu - \mu_f}{\sigma}\sigma_0(w).$$

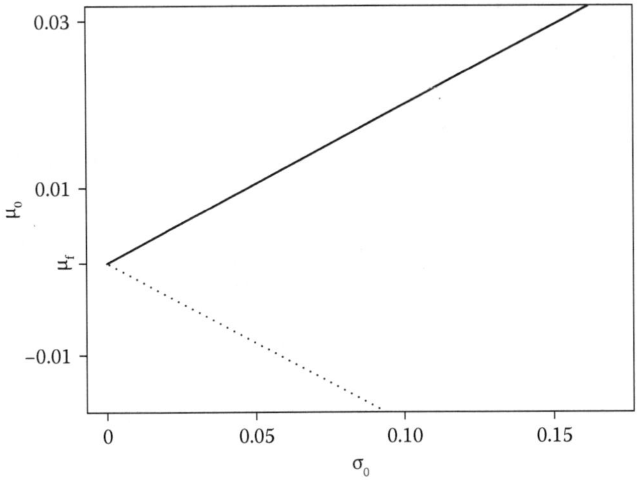

FIGURE 4.5
Efficient frontier and opportunity set in Example 4.11.

Note that a short sale of a risk-free asset corresponds to a loan at the risk-free rate. Therefore, we are assuming that investors can borrow and lend at the same rate, an assumption that is not true in practice for the vast majority of investors. However, it may be considered to be a reasonable approximation for an institutional investor, such as a pension fund or an insurance company.

Thus, in some cases, in order to achieve a high expected return, the investor must borrow at the risk-free rate and invest the proceeds from that loan in the risky asset.

Example 4.12 Consider the assets discussed in Example 4.10 in which the risky asset has mean return 0.02 and the risk-free return is 0.001. To construct a portfolio with an expected return of 0.0295, we must use portfolio weights $w = 1.5$ and $1 - w = -0.5$ (note that $(1.5)(0.02) + (-0.5)(0.001) = 0.0295$); that is, we must take a loan for an amount equal to half of our available capital and invest our capital, together with the proceeds from the loan, in asset 1. □

4.6 Portfolios of Two Risky Assets and a Risk-Free Asset

Now suppose that there are two risky assets available for investment, as in Section 4.4, plus a risk-free asset, as discussed in the previous section. Consider risky assets with returns R_1 and R_2, respectively. For $j = 1, 2$, let $\mu_j = \mathrm{E}(R_j)$ and $\sigma_j^2 = \mathrm{Var}(R_j)$ and let ρ_{12} denote the correlation of R_1, R_2; assume that $|\rho_{12}| < 1$. Let R_f denote the return on the risk-free asset, and let $\mu_f = \mathrm{E}(R_f)$; recall that, by definition, $\mathrm{Var}(R_f) = 0$.

A portfolio consisting of the two risky assets together with the risk-free asset has a return of the form

$$w_1 R_1 + w_2 R_2 + w_f R_f \tag{4.4}$$

where $w_1 + w_2 + w_f = 1$; assume that $w_f \neq 1$ so that the portfolio contains one or both of the risky assets. Note that we can write (4.4) as

$$(1 - w_f)(\bar{w}_1 R_1 + \bar{w}_2 R_2) + w_f R_f$$

where $\bar{w}_j = w_j/(1 - w_f)$, $j = 1, 2$. Hence, $\bar{w}_1 + \bar{w}_2 = 1$.

That is, it is convenient to view the portfolio selection problem as having two stages. In the first stage, we construct a portfolio of the two risky assets, with weights $\bar{w}_1$ and $\bar{w}_2$; in the second stage, that portfolio is combined with the risk-free asset by choosing the value of w_f. However, when choosing $\bar{w}_1$ and $\bar{w}_2$, it is important to keep in mind that the portfolio of the risky assets will be combined with a risk-free asset.

Including the possibility of investing in a risk-free asset has a large effect on the portfolio selection problem. Because we may always decrease risk by investing in the risk-free asset, there is a sense in which the risk in the portfolio of risky assets becomes less important.

Let R_p denote the return on the portfolio of risky assets; once the portfolio of risky assets has been selected, we are effectively back to the case of one risky asset together with the risk-free asset.

The return on the entire portfolio may be written

$$(1 - w_f)R_p + w_f R_f.$$

This portfolio has expected return $(1 - w_f)\mu_p + w_f \mu_f$, where $\mu_p = \mathrm{E}(R_p)$ and return variance

$$(1 - w_f)^2 \sigma_p^2.$$

For $w_f \geq 0$, a plot of the expected return of the portfolio versus its risk is simply a line segment starting at $(0, \mu_f)$ and passing through (σ_p, μ_p), similar to the efficient frontier in Example 4.11.

Example 4.13 Suppose two risky assets have returns with means $\mu_1 = 0.05$ and $\mu_2 = 0.15$, respectively, standard deviations $\sigma_1 = \sigma_2 = 0.25$, and correlation $\rho_{12} = 0.125$ and suppose that there is a risk-free asset with expected return $\mu_f = 0.025$. Figure 4.6 gives a plot of the efficient frontier (the solid line) for portfolios of the two risky assets. Thus, in the first stage of the portfolio selection problem, we choose a portfolio from among those corresponding to the points on the efficient frontier.

In choosing such a portfolio, the fact that the portfolio will be combined with the risk-free asset plays an important role. Figure 4.7 contains a plot of the risk-expected-return pairs of the portfolio, consisting of the two risky assets plus the risk-free asset, for a particular choice of a portfolio of risky assets.

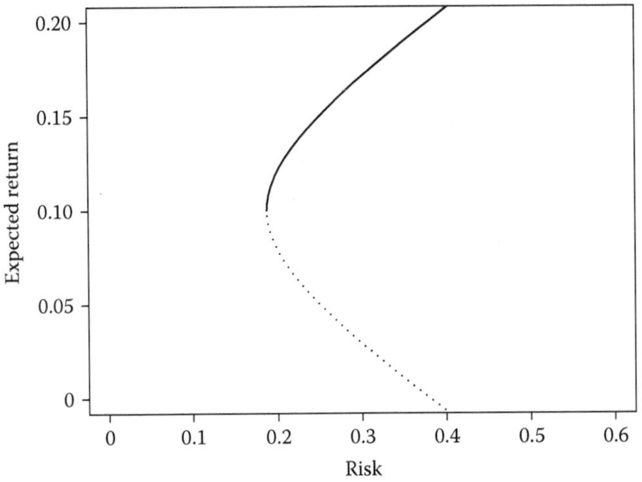

FIGURE 4.6
Efficient frontier for portfolios of the two risky assets in Example 4.13.

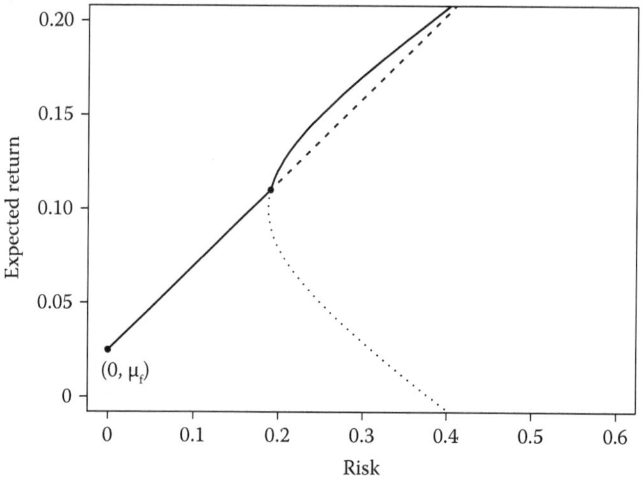

FIGURE 4.7
Risk, mean–return pairs corresponding to a particular portfolio of risky assets in Example 4.13.

This plot illustrates a number of points regarding combining a portfolio of risky assets with a risk-free asset.

- The line segment connecting $(0, \mu_f)$ and the point on the curve corresponds to portfolios placing weight w_f, $0 \leq w_f \leq 1$, on the risk-free asset and weight $1 - w_f$ on the portfolio of risky assets corresponding to the point on the curve.
- The dashed line segment extending beyond the curve corresponds to portfolios with $w_f < 0$, for which the investor borrows at the risk-free rate in order to purchase the portfolio of risky assets. Note that such a portfolio has a smaller expected return than does a portfolio of risky assets alone with the same risk; such portfolios correspond to the points on the solid curve that lies above the dotted line segment.
- Thus, the portfolio problem consists of choosing a point on the curve (i.e., choosing a portfolio of risky assets) together with choosing a point on the half-line that starts at $(0, \mu_f)$ and passes through the point on the curve (i.e., choosing w_f). □

Let (σ_p, μ_p) denote the risk, mean–return pair for a portfolio of risky assets; for example, in Figure 4.7, (σ_p, μ_p) is a point on the curve. Because all half-lines starting at $(0, \mu_f)$ and passing through (σ_p, μ_p) have the same starting point, the different possible half-lines may be described by their slopes,

$$\frac{\mu_p - \mu_f}{\sigma_p}.$$

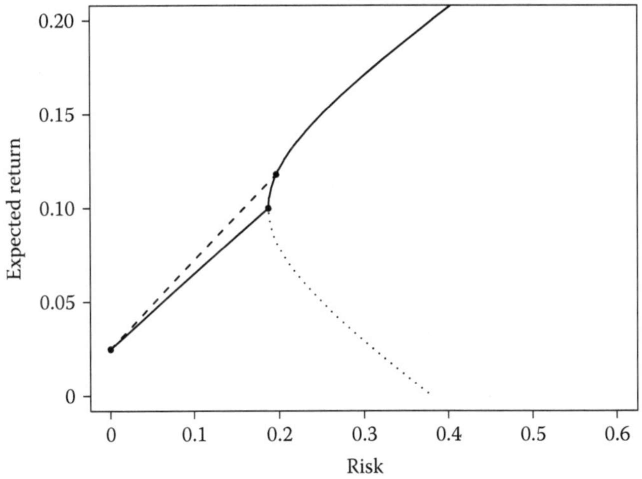

FIGURE 4.8
Comparison of two portfolios of risky assets in Example 4.13.

Therefore, choosing between two risky portfolios is essentially choosing between two possible slopes for such lines; see Figure 4.8 for such a comparison in the context of Example 4.13.

Note that the half-line with the larger slope (the dashed line) is preferred for any level of risk; that is, for any desired level of risk, the portfolio corresponding to the larger slope has a greater expected return. This suggests that the portfolio of risky assets with the value of (σ_p, μ_p) that yields the largest possible slope for the line connecting $(0, \mu_f)$ and (σ_p, μ_p) is optimal.

Sharpe Ratio

Consider a portfolio of risky assets with expected return μ_p and risk σ_p. The slope of the line connecting $(0, \mu_f)$ and (σ_p, μ_p) is known as the *Sharpe ratio* of the portfolio; hence, the Sharpe ratio is given by

$$\text{SR} = \frac{\mu_p - \mu_f}{\sigma_p}.$$

The Sharpe ratio has a useful interpretation as the expected excess return on the portfolio per unit of risk.

Note that the portfolio giving weight w_f to the risk-free asset and weight $1 - w_f$ to the portfolio of risky assets has expected return

$$(1 - w_f)\mu_p + w_f \mu_f$$

and return standard deviation $|1 - w_f|\sigma_p$.

Suppose that an investor would like to choose w_f so that the expected return on the portfolio consisting of the two risky assets together with the risk-free asset is m for some value $m > \mu_f$. This may be achieved by choosing w_f to solve

$$(1 - w_f)\mu_p + w_f \mu_f = m$$

$$w_f = \frac{\mu_p - m}{\mu_p - \mu_f};$$

note here we may assume that $\mu_p > \mu_f$ because, otherwise, there would be no reason to invest in the portfolio of risky assets and that $0 \le w_f < 1$ provided that $\mu_f < m \le \mu_p$. If $m > \mu_p$, then $w_f < 0$, indicating that the investor will need to borrow capital at the risk-free rate in order to attain an expected return of m.

The resulting portfolio has risk

$$\left|1 - \frac{\mu_p - m}{\mu_p - \mu_f}\right| \sigma_p = \frac{m - \mu_f}{\mu_p - \mu_f}\sigma_p = \frac{m - \mu_f}{\text{SR}},$$

where SR denotes the Sharpe ratio of the portfolio of risky assets. Hence, the risk of the portfolio with expected return m is inversely proportional to the Sharpe ratio of the portfolio of risky assets. That is, we should choose the portfolio of risky assets to have the largest Sharpe ratio possible.

Conversely, if w_f is chosen to achieve a given level of risk, s, then either $w_f = 1 + s/\sigma_p$ or $w_f = 1 - s/\sigma_p$. It is easy to show that the second of these yields the larger expected return,

$$\left(1 - \frac{s}{\sigma_p}\right)\mu_f + \frac{s}{\sigma_p}\mu_p = \mu_f + s(\text{SR}).$$

Hence, the expected excess return of the porfolio with risk s is proportional to the Sharpe ratio of the portfolio of risky assets, showing again that we should choose the portfolio of risky assets to have the largest Sharpe ratio possible.

Thus, we have proven the following important result. Note the previous analysis is not limited to the case of two risky assets; it applies to any portfolio of risky assets.

Proposition 4.1. *Consider a portfolio consisting of the risk-free asset and a portfolio of risky assets. Then the optimal portfolio of risky assets is the one with the largest Sharpe ratio.*

Tangency Portfolio

Thus, to construct the optimal portfolio of risky assets with returns R_1 and R_2, we find $\bar{w}_1, \bar{w}_2$, $\bar{w}_1 + \bar{w}_2 = 1$, so that the portfolio with return

$$\bar{w}_1 R_1 + \bar{w}_2 R_2$$

has the maximum possible Sharpe ratio.

Write $\bar{w}_1 = w$ and $\bar{w}_2 = 1 - w$. Our goal is to find the value of w that maximizes

$$f(w) = (\mu_p(w) - \mu_f)/\sigma_p(w)$$

where

$$\mu_p(w) = w\mu_1 + (1-w)\mu_2$$

and

$$\sigma_p^2(w) = w^2\sigma_1^2 + (1-w)^2\sigma_2^2 + 2w(1-w)\rho_{12}\sigma_1\sigma_2.$$

We may maximize $f(w)$ using standard results from calculus. Using the rule for taking the derivative of a ratio, it follows that

$$f'(w) = \frac{\mu_p'(w) - f(w)\sigma_p'(w)}{\sigma_p(w)};$$

hence, the solution to $f'(w) = 0$ solves

$$\frac{\mu_p'(w)}{\sigma_p'(w)} = f(w). \tag{4.5}$$

It may be shown that $f(w)$ is maximized by

$$w = w_{\mathcal{T}} \equiv \frac{(\mu_1 - \mu_f)\sigma_2^2 - (\mu_2 - \mu_f)\rho_{12}\sigma_1\sigma_2}{(\mu_2 - \mu_f)\sigma_1^2 + (\mu_1 - \mu_f)\sigma_2^2 - [(\mu_1 - \mu_f) + (\mu_2 - \mu_f)]\rho_{12}\sigma_1\sigma_2}; \tag{4.6}$$

see Proposition 5.7 for a proof of this result in a more general setting.

Hence, the Sharpe ratio of the portfolio of the two risky assets is maximized by choosing $w = w_{\mathcal{T}}$, as given by (4.6). The portfolio corresponding to $w = w_{\mathcal{T}}$ is known as the *tangency portfolio*, a term that is based on an important property of the line connecting $(0, \mu_f)$ and $(\sigma_p(w_{\mathcal{T}}), \mu_p(w_{\mathcal{T}}))$.

Consider a curve $(x(z), y(z))$ parameterized by a real number z. The tangent vector to the curve at z is given by $(x'(z), y'(z))$ and the slope of the tangent vector is $y'(z)/x'(z)$. Thus, $\mu_p'(w)/\sigma_p'(w)$ is the slope of the tangent vector to the curve $(\sigma_p(w), \mu_p(w))$ at w and the fact that the first-order condition (4.5) is satisfied at $w = w_{\mathcal{T}}$ may be interpreted as the condition that the slope of the line connecting $(0, \mu_f)$ and $(\sigma_p(w_{\mathcal{T}}), \mu_p(w_{\mathcal{T}}))$ is equal to the slope of the tangent vector to the efficient frontier at $w = w_{\mathcal{T}}$. Since $(\sigma_p(w_{\mathcal{T}}),\ \mu_p(w_{\mathcal{T}}))$ is on the efficient frontier, it follows that the line connecting $(0, \mu_f)$ and $(\sigma_p(w_{\mathcal{T}}),\ \mu_p(w_{\mathcal{T}}))$ is tangent to the efficient frontier.

Example 4.14 As in Example 4.13, consider two risky assets with expected returns $\mu_1 = 0.05$ and $\mu_2 = 0.15$, respectively, return standard deviations $\sigma_1 = \sigma_2 = 0.25$, and correlation of the returns given by $\rho_{12} = 0.125$, and suppose that there is a risk-free asset with expected return $\mu_f = 0.025$. Then, using the formula (4.6), the weight given to asset 1 in the tangency portfolio is $w_{\mathcal{T}} = 1/14 \doteq 0.071$. Therefore, the optimal portfolio of the risky assets is obtained by placing weight $1/14$ on asset 1 and weight $13/14$ on asset 2.

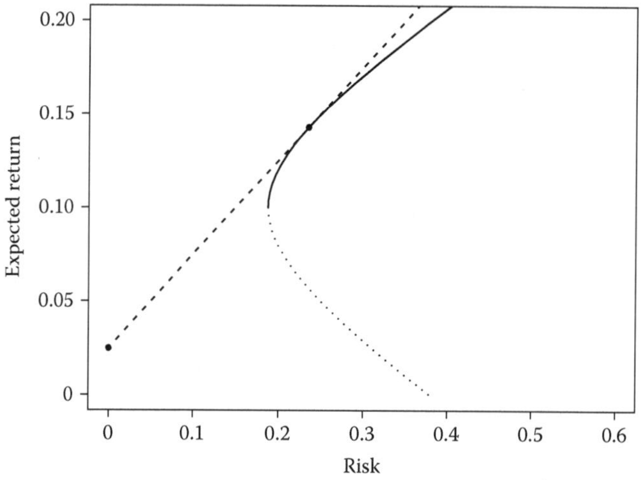

FIGURE 4.9
Tangency portfolio in Example 4.14.

Figure 4.9 shows the efficient frontier in this example with the location of the tangency portfolio (the point shown on the efficient frontier) and a dashed line representing the risk-expected return pairs for portfolios constructed from the tangency portfolio and the risk-free asset.

Note that, since the efficient frontier of the two risky assets lies below the dashed line, for any desired level of risk, a portfolio based on the tangency portfolio and the risk-free asset will have an expected return at least as large (and usually larger) than that of any portfolio on the efficient frontier with that level of risk. □

Consider the problem of finding the best combination of risky assets, with returns R_1, and R_2, and risk-free asset, with return R_f; that is, consider the problem of finding the optimal portfolio return of the form

$$w_f R_f + (1 - w_f)[w R_1 + (1 - w) R_2].$$

The previous results show that the optimal solution is to first take $w = w_{\mathcal{T}}$, corresponding to the tangency portfolio; this gives the optimal combination of risky assets. Let $\mu_{\mathcal{T}}$ and $\sigma_{\mathcal{T}}$ denote the mean and standard deviation of the return on the tangency portfolio.

Given a desired level of risk s we then choose w_f so that $w_f R_f + (1 - w_f)$ $R_{\mathcal{T}}$ has a standard deviation equal to s, that is, take $w_f = s/\sigma_{\mathcal{T}}$; alternatively, given a desired value for the expected return, m, we find w_f so that

$$w_f \mu_f + (1 - \mu_f)\mu_{\mathcal{T}} = m.$$

Note that, according to this theory, all investors should use the same combination of risky assets; only the proportion of the tangency portfolio versus the risk-free asset depends on the investor's goals.

4.7 Suggestions for Further Reading

The statistical approach to portfolio theory, as described in this book, originated with Markowitz (1952); hence, it is often referred to as *Markowitz portfolio theory.* See Benninga (2008, Chapter 8), Elton et al. (2007, Chapters 4 and 5), Francis and Kim (2013, Chapter 6), and Fabozzi et al. (2006, Chapter 2) for further discussion of the results of this chapter, along with a number of extensions, from the financial modeling perspective. Ruppert (2004, Chapter 5) and Sclove (2013, Section 6.2) consider these topics from more of a statistical point of view. Bernstein (2001) discusses the mean-variance approach to portfolio selection, focusing on the main ideas, rather than on technical results; Francis and Kim (2013, Chapter 5) solve a number of portfolio problems using graphical techniques, which some readers may find to be a useful alternative to the more mathematical approach used here.

4.8 Exercises

1. Suppose that there are N assets, with log-returns $r_1, r_2, \ldots, r_N$. Consider a portfolio placing weight w_j on asset j, $j = 1, 2, \ldots, N$ and let r_p denote the log-return on the portfolio.

   Is it true that

   $$r_p = w_1 r_1 + w_2 r_2 + \cdots + w_N r_N?$$

   Why or why not?

2. Consider two assets. Suppose that the return on asset 1 has expected value 0.05 and standard deviation 0.1 and suppose that the return on asset 2 has expected value 0.02 and standard deviation 0.05. Suppose that the asset returns have correlation 0.4.

   Consider a portfolio placing weight w on asset 1 and weight $1 - w$ on asset 2; let R_p denote the return on the portfolio.

   Find the mean and variance of R_p as a function of w.

3. For the assets described in Exercise 2, plot the opportunity set of possible (σ_p, μ_p) pairs.

4. Consider two assets. Suppose that the return on asset 1 has expected value 0.02 and standard deviation 0.05 and suppose that the return on asset 2 has expected value 0.04 and standard deviation 0.06.

Consider an equally weighted portfolio in which each asset receives weight 1/2 and let R_p denote the return on the portfolio.

Find the expected value of R_p and the variance of R_p as functions of ρ_{12}, the correlation of the returns on the two assets.

5. Consider two assets. Suppose that the return on asset 1 has expected value 0.004 and standard deviation 0.05 and suppose that the return on asset 2 has expected value 0.002 and standard deviation 0.06. Suppose that the asset returns have correlation 0.2.

   Find the portfolios based on these assets that have a return standard deviation of 0.045.

6. Consider two assets. Suppose that the return on asset 1 has expected value 0.08 and standard deviation 0.1 and suppose that the return on asset 2 has expected value 0.02 and standard deviation 0.05. Suppose that the asset returns have correlation 0.25.

   Consider the opportunity set corresponding to these assets. For each of the following pairs of portfolio return variances (s^2) and portfolio return means (m) determine whether or not the pair is an element of the opportunity set.

   a. $s^2 = 0.00375$ and $m = 0.05$
   b. $s^2 = 0.01$ and $m = 0.1$
   c. $s^2 = 0.00625$ and $m = 0.065$

7. Consider two assets. Suppose that the return on asset 1 has expected value 0.02 and standard deviation 0.1 and suppose that the return on asset 2 has expected value 0.005 and standard deviation 0.1. Suppose that the asset returns have correlation 0.5.

   Find the weight given to asset 1 in the efficient portfolio with return standard deviation 0.09.

8. Consider two assets with returns R_1 and R_2. Suppose that

$$\mathrm{E}(R_1) = \mathrm{E}(R_2).$$

   Describe the opportunity set and the efficient frontier for these assets.

9. For the assets described in Exercise 2 find the minimum-variance portfolio.

10. Let w_{mv} denote the weight given to asset 1 in the minimum-variance portfolio. Find conditions on ρ_{12}, in terms of σ_1, σ_2, so that $0 < w_{mv} < 1$.

11. For the assets described in Exercise 5, find the weight of asset 1 in the risk-averse portfolio with parameter λ.

12. Consider the risk-averse portfolio based on the risk-aversion parameter λ. Describe the properties of the portfolio in the limiting case in which $\lambda \to \infty$ and the limiting case in which $\lambda \to 0$. Interpret the results.

13. Let w_λ denote the weight given to asset 1 in the risk-averse portfolio with parameter λ. Suppose we wish to enforce the restriction that the weight given to asset 1 is in the interval $[0, 1]$. Let $\bar{w}_\lambda$ denote the weight given to asset 1 that maximizes the risk-aversion criterion function under this restriction.

    Give an expression for $\bar{w}_\lambda$ in terms of w_λ.

14. For the assets described in Exercise 2, find the tangency portfolio. Suppose that $\mu_f = 0.005$.

15. Consider two risky assets with returns R_1 and R_2, respectively, and a risk-free asset with return R_f. For $j = 1, 2$, let $\mu_j = \mathrm{E}(R_j)$ and $\sigma_j^2 = \mathrm{Var}(R_j)$ and let ρ_{12} denote the correlation of R_1 and R_2. Let $w_\mathcal{T}$ denote the weight given to asset 1 in the tangency portfolio.

    Suppose that the values of σ_1 and σ_2 used in determining the tangency portfolio are incorrect and that the correct values are $\alpha\sigma_1$ and $\alpha\sigma_2$, respectively, for some $\alpha > 0$. Compare the tangency portfolio based on the correct values to the tangency portfolio based on the incorrect values.

16. Consider two risky assets with returns R_1 and R_2, respectively, and a risk-free asset with return R_f. For $j = 1, 2$, let $\mu_j = \mathrm{E}(R_j)$ and $\sigma_j^2 = \mathrm{Var}(R_j)$ and let $\mu_f = \mathrm{E}(R_f)$.

    Suppose that the Sharpe ratios for the two assets are equal:

    $$\frac{\mu_1 - \mu_f}{\sigma_1} = \frac{\mu_2 - \mu_f}{\sigma_2}.$$

    Show that $w_\mathcal{T}$, the weight given to asset 1 in the tangency portfolio, depends only on σ_1 and σ_2 and give an expression for $w_\mathcal{T}$.

17. Consider two assets with returns R_1 and R_2, respectively, and a risk-free asset with return R_f. For $j = 1, 2$, let $\mu_j = \mathrm{E}(R_j)$ and $\sigma_j^2 = \mathrm{Var}(R_j)$ and let $\mu_f = \mathrm{E}(R_f)$.

    Suppose that $\mu_1 = \mu_f$ and that $\sigma_1 > \sigma_2$. Assume that $\mu_2 > \mu_f$. Give conditions under which $0 < w_\mathcal{T}$. Interpret your results.

18. Consider two assets with returns R_1 and R_2, respectively, and a risk-free asset with return R_f. For $j = 1, 2$, let $\mu_j = \mathrm{E}(R_j)$, $\sigma_j^2 = \mathrm{Var}(R_j)$, and let ρ_{12} denote the correlation of R_1 and R_2. Let $\mu_f = \mathrm{E}(R_f)$.

    Let $\mathrm{SR}_j = (\mu_j - \mu_f)/\sigma_j$ denote the Sharpe ratio of asset j, $j = 1, 2$. Find conditions on $\mathrm{SR}_1, \mathrm{SR}_2$, and ρ_{12} under which the tangency portfolio consists entirely of asset 1.

19. Let R_1 and R_2 denote the returns on two assets. Suppose that $R_2 = R_1 + \epsilon$ where $\mathrm{E}(\epsilon) = 0$ and R_1 and ϵ are uncorrelated. Assume that $\mathrm{Var}(\epsilon) > 0$. Thus, the return on asset 2 is equal to the return on asset 1 plus "noise."

    a. Find the minimum variance portfolio of R_1, R_2.

    b. Find the tangency portfolio of R_1, R_2.

5

Efficient Portfolio Theory

5.1 Introduction

In Chapter 4, we considered the problem of constructing a portfolio of two risky assets, possibly together with a risk-free asset. The focus of this chapter is the extension of those results to the general case of N risky assets, for an integer $N \geq 2$.

Although the basic approaches we will consider are the same as the ones used in the $N = 2$ case, there are a number of important details in which the general N case is different. The most obvious difference is in the scale of the problem. When $N = 2$, the portfolio is described by a single weight, w, representing the investment in asset 1, with $1 - w$ invested in asset 2; the mean and variance of the portfolio return are linear and quadratic functions, respectively, of w. The statistical properties of the portfolio returns depend on five parameters: two mean returns, two return standard deviations, and the correlation of the returns.

In the general N case, the mean and variance of the portfolio return are functions of the weights $w_1, w_2, \ldots, w_N$; because the weights must sum to 1, there are effectively $N - 1$ weights that must be selected. The mean and standard deviation of the portfolio return depend on N asset expected returns, N return standard deviations, and $N(N-1)/2$ correlations between the returns on different assets, representing the ways in which the asset returns are interrelated.

Perhaps the most important mathematical difference between the two cases is that, in the two-asset case, there is only one portfolio with a given mean return. That is, if we require a portfolio mean return of 0.02, for example, this specifies the value of w that must be used. Furthermore, this value of w determines the value of the return standard deviation σ_p, so that there is only a single possible value of σ_p for a given value of μ_p. That is, σ_p is a function of μ_p. When $N > 2$, typically there are infinitely many portfolios with a given mean return.

5.2 Portfolios of N Assets

Consider a set of N assets with returns $R_1, R_2, \ldots, R_N$ and consider a portfolio placing weight w_j on asset j; then $w_1 + w_2 + \cdots + w_N = 1$. The return on the portfolio with weights $w_1, w_2, \ldots, w_N$ is a function of the random variables $R_1, R_2, \ldots, R_N$, $R_p = \sum_{j=1}^N w_j R_j$, and hence, the properties of R_p depend on the properties of $R_1, R_2, \ldots, R_N$.

For instance, the expected return of the portfolio, $\mathrm{E}(R_p)$, is a simple function of the expected returns on the individual assets,

$$\mathrm{E}(R_p) = \sum_{j=1}^{N} w_j \mathrm{E}(R_j) = \sum_{j=1}^{N} w_j \mu_j$$

where $\mu_j = \mathrm{E}(R_j)$, $j = 1, 2, \ldots, N$.

The standard deviation of R_p depends on the standard deviations of R_1, $R_2, \ldots, R_N$, but it also depends on the relationships among $R_1, R_2, \ldots, R_N$, as measured by their covariances or correlations. Specifically,

$$\mathrm{Var}(R_p) = \sum_{j=1}^{N} w_j^2 \mathrm{Var}(R_j) + 2 \sum_{i<j} w_i w_j \mathrm{Cov}(R_i, R_j) \tag{5.1}$$

where the second summation in this expression is the sum over all i, j from 1 to N for which i is less than j.

Let σ_j denote the standard deviation of R_j, $j = 1, 2, \ldots, N$, let σ_{ij} denote the covariance of R_i, R_j for $i, j = 1, 2, \ldots, N$, $i \neq j$, and let σ_p denote the standard deviation of the portfolio with weights $w_1, w_2, \ldots, w_N$. Then (5.1) may be written

$$\sigma_p^2 = \sum_{j=1}^{N} w_j^2 \sigma_j^2 + 2 \sum_{i<j} w_i w_j \sigma_{ij}. \tag{5.2}$$

Alternatively, σ_p may be expressed in terms of the asset return standard deviations together with their correlations, $\rho_{ij} = \sigma_{ij}/(\sigma_i \sigma_j)$:

$$\sigma_p^2 = \sum_{j=1}^{N} w_j^2 \sigma_j^2 + 2 \sum_{i<j} w_i w_j \rho_{ij} \sigma_i \sigma_j.$$

Matrix Notation

In the case of N assets, the expressions for the expected return and risk of a portfolio may be conveniently expressed using matrix notation. Let

$$\boldsymbol{R} = \begin{pmatrix} R_1 \\ R_2 \\ \vdots \\ R_N \end{pmatrix}$$

or, equivalently, $\boldsymbol{R} = (R_1, R_2, \ldots, R_N)^T$, denote the vector of asset returns; note that all vectors used here will be column vectors. Then $\boldsymbol{R}$ is a *random vector*;

it has mean vector

$$\boldsymbol{\mu} = \mathrm{E}(\boldsymbol{R}) = \begin{pmatrix} \mu_1 \\ \mu_2 \\ \vdots \\ \mu_N \end{pmatrix}.$$

Similarly, the portfolio weights $w_1, w_2, \ldots, w_N$ describing a portfolio can be represented by a weight vector $\boldsymbol{w} = (w_1, w_2, \ldots, w_N)^T$. Then $R_p = \sum_{j=1}^N w_j R_j$, which may be written $R_p = \boldsymbol{w}^T \boldsymbol{R}$.

Using this notation, the mean return on the portfolio is given by

$$\mu_p = \sum_{j=1}^{N} w_j \mu_j = \boldsymbol{w}^T \boldsymbol{\mu},$$

which is simply the "dot product," or *inner product*, of the mean and weight vectors. The requirement that $\sum_{j=1}^N w_j = 1$ may be written $\mathbf{1}_N^T \boldsymbol{w} = 1$. Here $\mathbf{1}_N$ denotes the N-dimensional vector of all ones; when the dimension is clear from the context, we will write it simply as $\mathbf{1}$. Because inner products are symmetric in their arguments, we may also write $\mu_p = \boldsymbol{\mu}^T \boldsymbol{w}$ and $\boldsymbol{w}^T \mathbf{1} = 1$.

The simplification achieved by matrix notation is most apparent when considering the variance of a portfolio return. Let $\boldsymbol{\Sigma}$ denote the *covariance matrix* of $\boldsymbol{R}$, the $N \times N$ matrix given by

$$\boldsymbol{\Sigma} = \begin{pmatrix} \sigma_1^2 & \sigma_{12} & \cdots & \sigma_{1N} \\ \sigma_{21} & \sigma_2^2 & \cdots & \sigma_{2N} \\ \vdots & \vdots & \ddots & \vdots \\ \sigma_{N1} & \sigma_{N2} & \cdots & \sigma_N^2 \end{pmatrix}.$$

Thus, $\boldsymbol{\Sigma}_{ij}$, the (i, j)th element of $\boldsymbol{\Sigma}$, is $\mathrm{Cov}(R_i, R_j)$ for $i \neq j$ and is $\mathrm{Var}(R_i)$ for $i = j$. Because $\mathrm{Cov}(R_i, R_j) = \mathrm{Cov}(R_j, R_i)$, so that $\sigma_{ij} = \sigma_{ji}$, $\boldsymbol{\Sigma}$ is a symmetric matrix.

A covariance matrix gives a particularly simple way of expressing the variance of a linear function of a random vector. Let $\boldsymbol{a}$ denote an N-dimensional vector, that is, an element of $\Re^N$; then $\boldsymbol{a}^T \boldsymbol{R} = \sum_{j=1}^N a_j R_j$ and

$$\mathrm{Var}(\boldsymbol{a}^T \boldsymbol{R}) = \boldsymbol{a}^T \boldsymbol{\Sigma} \boldsymbol{a}.$$

To see why such a result holds, write $\boldsymbol{a} = (a_1, a_2, \ldots, a_N)^T$; then

$$\boldsymbol{\Sigma a} = \begin{pmatrix} a_1\sigma_1^2 + a_2\sigma_{12} + \cdots + a_N\sigma_{1N} \\ a_1\sigma_{21} + a_2\sigma_2^2 + a_3\sigma_{23} + \cdots + a_N\sigma_{2N} \\ \vdots \\ a_1\sigma_{N1} + a_2\sigma_{N2} + \cdots + a_{N-1}\sigma_{N,N-1} + a_N\sigma_N^2 \end{pmatrix}$$

and

$$\boldsymbol{a}^T\boldsymbol{\Sigma}\boldsymbol{a} = \begin{pmatrix} a_1 & a_2 & \cdots & a_N \end{pmatrix} \begin{pmatrix} a_1\sigma_1^2 + a_2\sigma_{12} + \cdots + a_N\sigma_{1N} \\ a_1\sigma_{21} + a_2\sigma_2^2 + a_3\sigma_{23} + \cdots + a_N\sigma_{2N} \\ \vdots \\ a_1\sigma_{N1} + a_2\sigma_{N2} + \cdots + a_{N-1}\sigma_{N,N-1} + a_N\sigma_N^2 \end{pmatrix}$$

$$= (a_1^2\sigma_1^2 + a_1a_2\sigma_{12} + \cdots + a_1a_N\sigma_{1N})$$
$$+ (a_2a_1\sigma_{21} + a_2^2\sigma_2^2 + a_2a_3\sigma_{23} + \cdots + a_2a_N\sigma_{1N})$$
$$+ \cdots + (a_Na_1\sigma_{N1} + \cdots + a_Na_{N-1}\sigma_{N,N-1} + a_N^2\sigma_N^2).$$

Note that, in this sum, each term of the form $a_j^2\sigma_j^2$ occurs once and each term of the form $a_ja_k\sigma_{jk}$, $j < k$, occurs twice; hence, the sum is equal to

$$\sum_{j=1}^{N} a_j^2\sigma_j^2 + 2\sum_{j<k} a_ja_k\sigma_{jk}.$$

In particular, (5.1) for the variance of the return on the portfolio based on the weight vector $\boldsymbol{w}$ may be written using matrix notation as

$$\text{Var}(R_p) = \boldsymbol{w}^T\boldsymbol{\Sigma}\boldsymbol{w}.$$

Furthermore, the covariance matrix may be used to calculate the covariance of two linear functions of a random vector. Let $\boldsymbol{a}, \boldsymbol{b}$ be elements of $\Re^N$. Then

$$\text{Cov}(\boldsymbol{a}^T\boldsymbol{R}, \boldsymbol{b}^T\boldsymbol{R}) = \boldsymbol{a}^T\boldsymbol{\Sigma}\boldsymbol{b}. \tag{5.3}$$

Conversely, if for a given $N \times N$ matrix $\boldsymbol{A}$,

$$\text{Cov}(\boldsymbol{a}^T\boldsymbol{R}, \boldsymbol{b}^T\boldsymbol{R}) = \boldsymbol{a}^T\boldsymbol{A}\boldsymbol{b}$$

for any $\boldsymbol{a}, \boldsymbol{b} \in \Re^N$, then $\boldsymbol{A}$ must be the covariance matrix of $\boldsymbol{R}$.

Example 5.1 Consider a set of four assets and suppose that the return vector $\boldsymbol{R}$ has mean vector $(0.10, 0.20, 0.05, 0.10)^T$ and covariance matrix

$$\boldsymbol{\Sigma} = \begin{pmatrix} 0.05 & 0.01 & 0.02 & 0 \\ 0.01 & 0.10 & 0.05 & 0.02 \\ 0.02 & 0.05 & 0.20 & 0.10 \\ 0 & 0.02 & 0.10 & 0.20 \end{pmatrix}.$$

Consider the portfolio with the weight vector $(0.20, 0.30, 0.10, 0.40)^T$. To calculate the mean and variance of the portfolio return in R, we may use the following commands.

```
> mu<-c(0.10, 0.20, 0.05, 0.10)
> Sigma<-matrix(c(0.05, 0.01, 0.02, 0, 0.01, 0.10, 0.05, 0.02,
```

```
+ 0.02, 0.05, 0.20, 0.10, 0, 0.02, 0.10, 0.20), 4, 4)
> Sigma
     [,1] [,2] [,3] [,4]
[1,] 0.05 0.01 0.02 0.00
[2,] 0.01 0.10 0.05 0.02
[3,] 0.02 0.05 0.20 0.10
[4,] 0.00 0.02 0.10 0.20
> w<-c(0.2, 0.3, 0.1, 0.4)
> sum(w*mu)
[1] 0.125
> w%*%Sigma%*%w
       [,1]
[1,] 0.0628
```

Therefore, the return on the portfolio has mean 0.125 and variance 0.0628; it follows that the return standard deviation is given by

```
> (0.0628)^.5
[1] 0.251
```

In these calculations, the product `w*mu` of vectors `w` and `mu` creates a new vector with the jth element given by the product of the jth elements of `w` and `mu`; hence, `sum(w*mu)` returns the mean return corresponding to the weight vector `w` and mean vector `mu`. The function `matrix` is used to construct a matrix from a vector; note that the entries in the matrix are populated by column, unless `byrow=T` is specified. The matrix multiplication operator in R is `%*%`; in the expression `w%*%Sigma%*%w`, the first `w` is automatically interpreted as a row vector and the second `w` is interpreted as a column vector.

Consider a second portfolio, with the weight vector $(0.50, 0.10, 0.10, 0.30)^T$. Then the covariance of the returns on the two portfolios may be calculated by

```
> w1<-c(0.5, 0.1, 0.1, 0.3)
> w1%*%Sigma%*%w
       [,1]
[1,] 0.0487.
```

Thus, the covariance of the returns on the two portfolios is 0.0487. □

Because the variance of $\boldsymbol{a}^T\boldsymbol{R}$ must be nonnegative for any $\boldsymbol{a} \in \Re^N$, we must have

$$\boldsymbol{a}^T\boldsymbol{\Sigma}\boldsymbol{a} \geq 0 \quad \text{for all} \quad \boldsymbol{a} \in \Re^N;$$

a matrix satisfying this property is said to be *nonnegative definite*; hence, all covariance matrices are nonnegative definite.

If, in addition,

$$\boldsymbol{a}^T\boldsymbol{\Sigma}\boldsymbol{a} = 0 \quad \text{if and only if} \quad \boldsymbol{a} = \mathbf{0}_N,$$

where $\mathbf{0}_N$ denotes the zero vector in $\Re^N$, then $\boldsymbol{\Sigma}$ is said to be *positive definite*.

In the present context, $\mathbf{\Sigma}$ is positive definite provided that there is no nontrivial linear function of $R_1, R_2, \ldots, R_N$ that has zero variance (a "trivial" linear function being one with all coefficients equal to zero). In particular, since portfolio weights must sum to 1, if the covariance matrix of the returns is positive definite, any portfolio has positive variance. Unless stated otherwise, we will always assume that the covariance matrix of any return vector is positive definite.

Example 5.2 Consider the covariance matrix `Sigma` described in Example 5.1. One way to determine if a symmetric matrix is positive definite is to compute its *eigenvalues*; if the eigenvalues are all positive, then the matrix is positive definite; if the eigenvalues are all nonnegative, then the matrix is nonnegative definite.

Therefore, to determine if `Sigma` is positive definite, we can compute its eigenvalues using the `eigen`function:

```
> eigen(Sigma)$values
[1] 0.3127 0.1188 0.0731 0.0455
```

All eigenvalues are positive; therefore, the matrix is positive definite. □

Note that we may write $\mathbf{\Sigma}$ in terms of the asset return standard deviations $\sigma_1, \sigma_2, \ldots, \sigma_N$ and their correlations, ρ_{ij}, $i, j = 1, \ldots, N$, $i \neq j$. Using the fact that $\sigma_{ij} = \rho_{ij}\sigma_i\sigma_j$, we may write

$$\mathbf{\Sigma} = \begin{pmatrix} \sigma_1 & 0 & \ldots & 0 \\ 0 & \sigma_2 & \ddots & \vdots \\ \vdots & \ddots & \ddots & 0 \\ 0 & \ldots & 0 & \sigma_N \end{pmatrix} \begin{pmatrix} 1 & \rho_{12} & \ldots & \rho_{1N} \\ \rho_{21} & 1 & \ldots & \rho_{2N} \\ \vdots & \vdots & \ddots & \vdots \\ \rho_{N1} & \rho_{N2} & \ldots & 1 \end{pmatrix}$$

$$\times \begin{pmatrix} \sigma_1 & 0 & \ldots & 0 \\ 0 & \sigma_2 & \ddots & \vdots \\ \vdots & \ddots & \ddots & 0 \\ 0 & \ldots & 0 & \sigma_N \end{pmatrix}.$$

The matrix

$$\boldsymbol{C} = \begin{pmatrix} 1 & \rho_{12} & \ldots & \rho_{1N} \\ \rho_{21} & 1 & \ldots & \rho_{2N} \\ \vdots & \vdots & \ddots & \vdots \\ \rho_{N1} & \rho_{N2} & \ldots & 1 \end{pmatrix}$$

is known as the *correlation matrix* of $\boldsymbol{R}$ and the matrix

$$\boldsymbol{V} = \begin{pmatrix} \sigma_1^2 & 0 & \ldots & 0 \\ 0 & \sigma_2^2 & \ddots & \vdots \\ \vdots & \ddots & \ddots & 0 \\ 0 & \ldots & 0 & \sigma_N^2 \end{pmatrix}$$

is known as the *variance matrix* of $\boldsymbol{R}$. Then

$$\boldsymbol{\Sigma} = \boldsymbol{V}^{\frac{1}{2}}\boldsymbol{C}\boldsymbol{V}^{\frac{1}{2}}$$

where $\boldsymbol{V}^{\frac{1}{2}}$ denotes the matrix square root of $\boldsymbol{V}$, the diagonal matrix with the asset standard deviations on the diagonal.

The requirement that $\boldsymbol{\Sigma}$ be nonnegative definite can be expressed in terms of $\boldsymbol{C}$: using the fact that all σ_j are nonnegative, $\boldsymbol{\Sigma}$ is nonnegative definite if and only if $\boldsymbol{C}$ is nonnegative definite. This condition on $\boldsymbol{C}$ is the N-asset analogue of the requirement that $-1 \leq \rho_{12} \leq 1$ in the two-asset case.

Similarly, $\boldsymbol{\Sigma}$ is positive definite if and only if all σ_j are positive and $\boldsymbol{C}$ is positive definite, the analogue of the condition that $-1 < \rho_{12} < 1$ in the two-asset case. In particular, the condition that $-1 < \rho_{ij} < 1$ for all $i \neq j$ is not sufficient for the return covariance matrix $\boldsymbol{\Sigma}$ to be positive definite.

Diversification

In Section 4.2, we saw the benefits of a diversified portfolio of two assets. When there are N assets available, the analysis is more complicated but the potential benefits of diversification are potentially even greater.

Example 5.3 Consider a set of N assets and suppose that the returns on the assets all have mean μ, standard deviation σ, and that they are uncorrelated. Therefore, $\boldsymbol{\mu}$, the mean vector of the returns, may be written $\mu\mathbf{1}$, and $\boldsymbol{\Sigma}$, the covariance matrix of the returns, is $\sigma^2\boldsymbol{I}_N$ where $\boldsymbol{I}_N$ denotes the $N \times N$ identity matrix; when the dimension of the identity matrix is clear from the context, we will write it simply as $\boldsymbol{I}$.

Consider an equally-weighted portfolio with weights $w_1 = w_2 \cdots = w_N = 1/N$, which we may write as $\boldsymbol{w} = (1/N)\mathbf{1}$. Then the expected return on the portfolio is

$$\boldsymbol{w}^T\boldsymbol{\mu} = \frac{1}{N}\mathbf{1}^T(\mu\mathbf{1}) = \frac{\mu}{N}\mathbf{1}^T\mathbf{1} = \frac{\mu}{N}N = \mu$$

and the variance of the portfolio return is

$$\boldsymbol{w}^T\boldsymbol{\Sigma}\boldsymbol{w} = \left(\frac{1}{N}\mathbf{1}\right)^T(\sigma^2\boldsymbol{I})\left(\frac{1}{N}\mathbf{1}\right) = \frac{\sigma^2}{N^2}\mathbf{1}^T\boldsymbol{I}\mathbf{1} = \frac{\sigma^2}{N};$$

note that, for any matrix $\boldsymbol{A}$, $\mathbf{1}^T\boldsymbol{A}\mathbf{1}$ is the sum of all elements of $\boldsymbol{A}$. Thus, the larger the number of assets under consideration, the smaller is the variance of the equally-weighted portfolio; that is, the larger the number of assets, the greater is the benefit of diversification, at least in this simple setting. Furthermore, when the asset returns are uncorrelated, as we have assumed, the portfolio variance approaches zero as N increases. □

The previous example shows that when there are a large number of assets with uncorrelated returns, then it is possible to construct a portfolio with a small standard deviation. The following example shows that having the returns be uncorrelated is important for this result to hold.

Example 5.4 Consider the same scenario as in Example 5.3, except that now assume that the correlation of any two asset returns is ρ, where $0 < \rho < 1$. Then,

$$\boldsymbol{\Sigma} = \sigma^2 \begin{pmatrix} 1 & \rho & \cdots & \cdots & \rho \\ \rho & 1 & \rho & \cdots & \rho \\ \vdots & \ddots & \ddots & \ddots & \vdots \\ \rho & \cdots & \rho & 1 & \rho \\ \rho & \cdots & \cdots & \rho & 1 \end{pmatrix}. \tag{5.4}$$

Recall that $\boldsymbol{\Sigma}$ must be a positive-definite matrix; thus, all the eigenvalues of $\boldsymbol{\Sigma}$ must be positive. Note that all the rows of $\boldsymbol{\Sigma}$ sum to $1+(N-1)\rho$; hence, the vector $\mathbf{1}$ is an eigenvector of $\boldsymbol{\Sigma}$ and, because $\boldsymbol{\Sigma}\mathbf{1} = (1+(N-1)\rho)\mathbf{1}$, it follows that $1+(N-1)\rho$ is an eigenvalue of $\boldsymbol{\Sigma}$. Hence, we must have $\rho > -1/(N-1)$ and, for large N, this lower bound is close to 0. It may be shown that the remaining eigenvalues of $\boldsymbol{\Sigma}$ are all $1-\rho$ so that ρ must satisfy $-1/(N-1) < \rho < 1$.

Because

$$\mathbf{1}^T \ \boldsymbol{\Sigma}\mathbf{1} = N\sigma^2 + (N^2 - N)\rho\sigma^2,$$

the variance of the equally-weighted portfolio is given by

$$\sigma_p^2 = \boldsymbol{w}^T \boldsymbol{\Sigma} \boldsymbol{w} = \frac{\sigma^2}{N} + \left(1 - \frac{1}{N}\right)\rho\sigma^2.$$

Therefore, for any $0 < \rho < 1$, $\sigma_p < \sigma$. However, unlike the case of uncorrelated assets, if $\rho > 0$, the standard deviation of the portfolio return cannot be made arbitrarily small by including more assets.

If ρ is negative, it is possible for σ_p to be close to zero. However, a negative value of ρ is generally unrealistic—try to imagine a large number of assets such that, for any pair, a greater than average return on one corresponds to less than average returns on the others.

It may be shown that, for $\boldsymbol{\Sigma}$ of the form (5.4), the minimum possible variance of a portfolio return is

$$\frac{\sigma^2}{N} + \left(1 - \frac{1}{N}\right)\rho\sigma^2$$

so that the equally-weighted portfolio has the smallest possible return variance; this result will be discussed in Example 5.7. □

Although the specific results given here depend on the form of $\boldsymbol{\Sigma}$, the basic conclusion holds more generally. That is, when asset returns are positively correlated, as is usually the case, diversification generally reduces the risk of a portfolio, but there are limits to its benefits.

5.3 Minimum-Risk Frontier

We begin by describing the set of mean returns and return standard deviations that are available to the investor based on a given set of assets. Let $\boldsymbol{R}$ denote an $N \times 1$ vector of asset returns, with mean vector $\boldsymbol{\mu}$ and covariance matrix $\boldsymbol{\Sigma}$. Therefore, the return on the portfolio based on a weight vector $\boldsymbol{w}$ is given by $\boldsymbol{w}^T\boldsymbol{R}$, which has mean $\mu_p = \boldsymbol{w}^T\boldsymbol{\mu}$ and standard deviation σ_p, where $\sigma_p^2 = \boldsymbol{w}^T\boldsymbol{\Sigma}\boldsymbol{w}$.

Consider portfolios with an expected return of m, where m is a given real number. This is achieved by a portfolio corresponding to a weight vector $\boldsymbol{w}$ satisfying

$$\boldsymbol{w}^T\boldsymbol{\mu} = m; \tag{5.5}$$

of course, the weights must sum to 1 so that the requirement (5.5) is in addition to the requirement that $\boldsymbol{w}^T\boldsymbol{1} = 1$. It follows that, for $N > 2$, assuming there are at least three distinct values in the vector $\boldsymbol{\mu}$, there are infinitely many weight vectors leading to a portfolio with expected return m.

Therefore, when $N > 2$, the set of (σ_p, μ_p) pairs for all possible portfolios is a region in $\Re^2$, known as the *opportunity set*; recall that in the $N = 2$ case, the opportunity set is a curve. Figure 5.1 shows the general form of the *opportunity set*, that is, the set of possible risk-mean pairs, for a typical $N > 2$ case.

Let R_{p1} and R_{p2} denote the returns on two portfolios, both with mean return m, for some real number m. Because $\mathrm{E}(R_{p1}) = \mathrm{E}(R_{p2})$, if $\mathrm{Var}(R_{p1}) < \mathrm{Var}(R_{p2})$, then portfolio 1 is preferred to portfolio 2. That is, although there

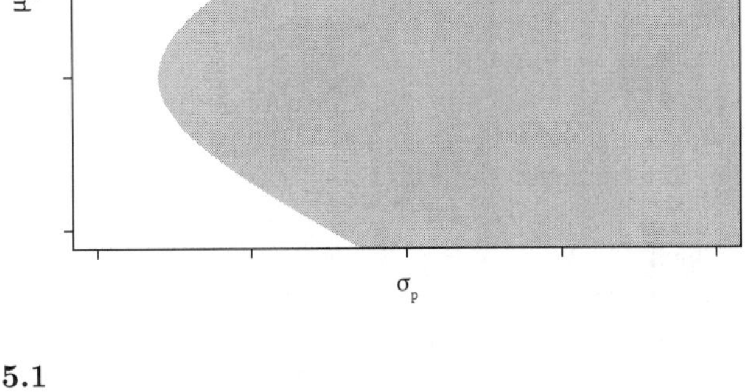

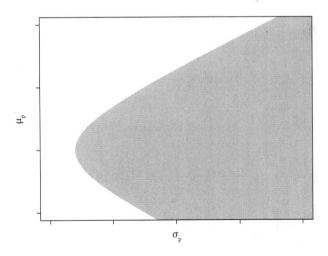

FIGURE 5.1
Opportunity set for an $N > 2$ case.

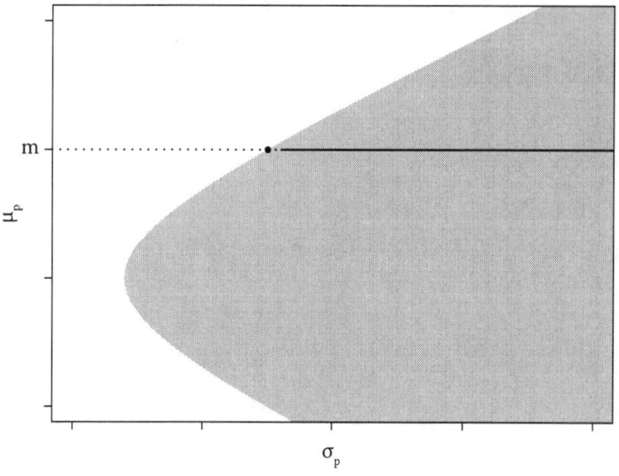

FIGURE 5.2
Minimum risk corresponding to an expected return of m.

are infinitely many portfolios with a given mean return, we are only interested in the one with the smallest return standard deviation.

Graphically, the portfolio with a given mean return m that has the smallest standard deviation corresponds to the leftmost value on the line segment of points (σ_p, m) in the opportunity set; see Figure 5.2, where the value of (σ_p, μ_p) for a minimum risk portfolio with mean m is indicated with a dot. The solid horizontal line represents the possible values of σ_p corresponding to $\mu_p = m$.

The weight vector for this portfolio may be obtained by minimizing $\text{Var}(R_p)$ subject to the restriction that $\text{E}(R_p) = m$. That is, we solve a constrained minimization problem:

$$\begin{aligned} \underset{\boldsymbol{w} \in \Re^N}{\text{minimize}} \quad & \boldsymbol{w}^T \boldsymbol{\Sigma} \boldsymbol{w} \\ \text{subject to} \quad & \boldsymbol{w}^T \boldsymbol{\mu} = m \\ & \boldsymbol{w}^T \mathbf{1} = 1. \end{aligned} \tag{5.6}$$

The weight vector for the minimum-risk portfolio for any given mean return m is the solution to a minimization problem of this type. In terms of the opportunity set of possible pairs (σ_p, μ_p), as shown in Figure 5.1, solving (5.6) for each value of m finds the left boundary of the set; see Figure 5.3. This boundary is known as the *minimum-risk frontier.*

For convenience, we will use the term "minimum-risk frontier" to refer to risk-expected return pairs of the form (σ_p, μ_p) as well as to refer to the corresponding portfolios and their weight functions. For instance, the statement that a weight vector is on the minimum-risk frontier means that the

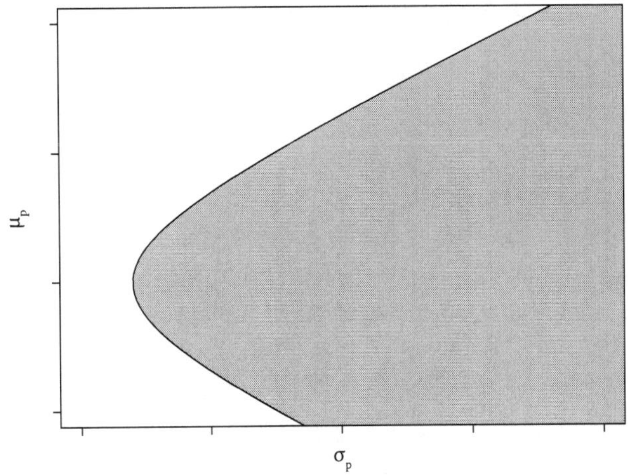

FIGURE 5.3
An example of a minimum-risk frontier.

value of (σ_p, μ_p) for the portfolio corresponding to that weight vector is on the minimum-risk frontier.

The minimum-risk frontier corresponds to the portfolios with the smallest risk for a given expected return. Note that the minimum-risk frontier for the N-asset case is similar in many respects to the opportunity set in the two-asset case. For instance, it includes portfolios for which there is a portfolio with the same risk but a larger expected return.

The *efficient frontier* consists of those portfolios with the smallest risk for a given expected return as well as the largest expected return for a given risk; recall that this is the same definition used in the $N = 2$ case. Efficient frontiers for an arbitrary number of assets will be discussed in detail in Section 5.5.

The remainder of this section is devoted to describing the minimum-risk frontier. Although the results are somewhat technical, they all rely on a simple result from linear algebra, the Cauchy–Schwarz inequality, together with some basic properties of quadratic functions of a single variable.

The following lemma gives a statement of the Cauchy–Schwarz inequality; the proof, which is available in many books on linear algebra, is omitted.

Lemma 5.1 (Cauchy–Schwarz Inequality). *Let $\boldsymbol{x}$ and $\boldsymbol{y}$ be elements of $\Re^n$, for some $n = 1, 2, \ldots$. Then*

$$(\boldsymbol{x}^T \boldsymbol{y})^2 \leq (\boldsymbol{x}^T \boldsymbol{x})(\boldsymbol{y}^T \boldsymbol{y})$$

with equality if and only if either one of $\boldsymbol{x}$ and $\boldsymbol{y}$ is the zero vector or $\boldsymbol{x} = c\boldsymbol{y}$ for some scalar c.

Because $\boldsymbol{x}^T\boldsymbol{y} \leq |\boldsymbol{x}^T\boldsymbol{y}|$, the Cauchy–Schwarz inequality also states that

$$\boldsymbol{x}^T\boldsymbol{y} \leq (\boldsymbol{x}^T\boldsymbol{x})^{1/2}(\boldsymbol{y}^T\boldsymbol{y})^{1/2}$$

with equality if either one of $\boldsymbol{x}, \boldsymbol{y}$ is the zero vector or $\boldsymbol{x} = c\boldsymbol{y}$ for some $c > 0$.

Zero-Investment Portfolios

Consider a vector $\boldsymbol{v} \in \Re^N$ such that $\boldsymbol{v}^T\mathbf{1} = 0$. Then the portfolio based on the vector $\boldsymbol{v}$, that is, with return $\boldsymbol{v}^T\boldsymbol{R}$, has no net investment—the coordinates of $\boldsymbol{v}$ that are greater than zero are offset by one or more coordinates that take negative values. A portfolio based on weights given by such a vector is said to be a *zero-investment portfolio*; the vector $\boldsymbol{v}$ will be called a *zero-investment weight vector*, to distinguish such vectors from standard weight vectors that sum to 1.

Let $\boldsymbol{w}_1, \boldsymbol{w}_2$ be two portfolio weight vectors. Then $\boldsymbol{v} = \boldsymbol{w}_1 - \boldsymbol{w}_2$ satisfies $\boldsymbol{v}^T\mathbf{1} = 0$ so that $\boldsymbol{w}_1 - \boldsymbol{w}_2$ is a zero-investment weight vector. Conversely, if $\boldsymbol{w}$ is a portfolio weight vector and $\boldsymbol{v}$ is a zero-investment weight vector, then $\boldsymbol{w} + \boldsymbol{v}$ is also a portfolio weight vector.

Define a set of N-dimensional vectors by

$$\mathcal{V}_0 = \{\boldsymbol{v} \in \Re^N : \boldsymbol{v}^T\mathbf{1} = 0,\ \boldsymbol{v}^T\boldsymbol{\mu} = 0\}. \tag{5.7}$$

If $\boldsymbol{\delta} \in \mathcal{V}_0$, then $\boldsymbol{\delta}^T\boldsymbol{R}$ is the return on a zero-investment portfolio that has zero expected return; in this chapter, $\boldsymbol{\delta}$ generally will be used to denote an element of $\mathcal{V}_0$.

Elements of $\mathcal{V}_0$ play a central role in describing the minimum-risk frontier because if $\boldsymbol{w}$ satisfies the constraints $\boldsymbol{w}^T\mathbf{1} = 1$ and $\boldsymbol{w}^T\boldsymbol{\mu} = m$ and $\boldsymbol{v} \in \mathcal{V}_0$, then $\boldsymbol{w} + \boldsymbol{v}$ also satisfies these constraints:

$$(\boldsymbol{w} + \boldsymbol{v})^T\mathbf{1} = \boldsymbol{w}^T\mathbf{1} + \boldsymbol{v}^T\mathbf{1} = 1 + 0 = 1$$

and

$$(\boldsymbol{w} + \boldsymbol{v})^T\boldsymbol{\mu} = \boldsymbol{w}^T\boldsymbol{\mu} + \boldsymbol{v}^T\boldsymbol{\mu} = m + 0 = m.$$

Conversely, if $\boldsymbol{w}_1$ and $\boldsymbol{w}_2$ are weight vectors satisfying the constraint that $\boldsymbol{w}_j^T\boldsymbol{\mu} = m$ for $j = 1, 2$, then $\boldsymbol{w}_1 - \boldsymbol{w}_2 \in \mathcal{V}_0$. Note that $\mathcal{V}_0$ is a linear subspace of $\Re^N$, with dimension $N - 2$.

Characterizing the Minimum-Risk Frontier

Note that the constrained minimization problem (5.6) is an N-dimensional problem; that is, the decision variable in the minimization problem is N-dimensional. However, we can describe its solutions by looking at all "one-dimensional subproblems" of this N-dimensional problem. This is a useful technique that will appear several times in this and later chapters. One result of this type is the following proposition, which gives a simple characterization of the minimum-risk frontier.

Proposition 5.1. *Consider a vector of portfolio returns* $\boldsymbol{R}$ *with mean vector* $\boldsymbol{\mu}$ *and covariance matrix* $\boldsymbol{\Sigma}$. *The portfolio with weight function* $\hat{\boldsymbol{w}}$ *is on the minimum-risk frontier if and only if*

$$\hat{\boldsymbol{w}}^T \boldsymbol{\Sigma} \boldsymbol{\delta} = 0 \quad \textit{for all} \quad \boldsymbol{\delta} \in \mathcal{V}_0.$$

Proof. Suppose that the portfolio corresponding to $\hat{\boldsymbol{w}}$ is on the minimum-risk frontier. Let $\hat{m} = \hat{\boldsymbol{w}}^T \boldsymbol{\mu}$ so that $\hat{\boldsymbol{w}}$ solves the constrained minimization problem (5.6) with $m = \hat{m}$. Note that this implies that $\hat{\boldsymbol{w}}^T \mathbf{1} = 1$. Let $\boldsymbol{\delta}$ be an element of $\mathcal{V}_0$; we want to show that $\hat{\boldsymbol{w}}^T \boldsymbol{\Sigma} \boldsymbol{\delta} = 0$.

Note that for any real number z, $\hat{\boldsymbol{w}} + z\boldsymbol{\delta}$ satisfies

$$(\hat{\boldsymbol{w}} + z\boldsymbol{\delta})^T \mathbf{1} = \hat{\boldsymbol{w}}^T \mathbf{1} + z\boldsymbol{\delta}^T \mathbf{1} = 1 + z(0) = 1$$

and

$$(\hat{\boldsymbol{w}} + z\boldsymbol{\delta})^T \boldsymbol{\mu} = \hat{\boldsymbol{w}}^T \boldsymbol{\mu} + z\boldsymbol{\delta}^T \boldsymbol{\mu} = \hat{m} + z(0) = \hat{m}.$$

That is, weight vectors of the form $\hat{\boldsymbol{w}} + z\boldsymbol{\delta}$ satisfy the constraints in the minimization problem (5.6).

Define

$$f(z) = (\hat{\boldsymbol{w}} + z\boldsymbol{\delta})^T \boldsymbol{\Sigma} (\hat{\boldsymbol{w}} + z\boldsymbol{\delta}), \quad z \in \Re.$$

Because $\hat{\boldsymbol{w}}$ is the weight vector of the minimum-risk portfolio with mean return $\hat{m}$, and for any z, $\hat{\boldsymbol{w}} + z\boldsymbol{\delta}$ is the weight vector of a portfolio with mean return $\hat{m}$, it follows that $f(z)$ must be minimized at $z = 0$. Note that the function $f(z)$ is a quadratic function of the real-valued variable z. Therefore, we may use the properties of the function $f(\cdot)$ of a single variable to describe the properties of a portfolio on the minimum-risk frontier.

Because f is a quadratic function in z, we must have $f'(0) = 0$. Note that

$$f(z) = \hat{\boldsymbol{w}}^T \boldsymbol{\Sigma} \hat{\boldsymbol{w}} + 2z\hat{\boldsymbol{w}}^T \boldsymbol{\Sigma} \boldsymbol{\delta} + z^2 \boldsymbol{\delta}^T \boldsymbol{\Sigma} \boldsymbol{\delta}$$

so that

$$f'(z) = 2\hat{\boldsymbol{w}}^T \boldsymbol{\Sigma} \boldsymbol{\delta} + 2z\boldsymbol{\delta}^T \boldsymbol{\Sigma} \boldsymbol{\delta}$$

and $f'(0) = 2\hat{\boldsymbol{w}}^T \boldsymbol{\Sigma} \boldsymbol{\delta}$. It follows that $\hat{\boldsymbol{w}}^T \boldsymbol{\Sigma} \boldsymbol{\delta} = 0$. Because the value of $\boldsymbol{\delta}$ in $\mathcal{V}_0$ is arbitrary, we have shown that if the portfolio with weight vector $\hat{\boldsymbol{w}}$ is on the minimum-risk frontier, then

$$\hat{\boldsymbol{w}}^T \boldsymbol{\Sigma} \boldsymbol{\delta} = 0 \quad \text{for all} \quad \boldsymbol{\delta} \in \mathcal{V}_0.$$

Now suppose that $\hat{\boldsymbol{w}} \in \Re^N$ satisfies $\hat{\boldsymbol{w}}^T \mathbf{1} = 1$ and $\hat{\boldsymbol{w}}^T \boldsymbol{\Sigma} \boldsymbol{\delta} = 0$ for all $\boldsymbol{\delta} \in \mathcal{V}_0$. We must show that this condition implies that the portfolio with the weight vector $\hat{\boldsymbol{w}}$ is on the minimum-risk frontier.

Let $\hat{m} = \hat{\boldsymbol{w}}^T \boldsymbol{\mu}$. We need to show that, for any weight vector $\boldsymbol{w}$ satisfying $\boldsymbol{w}^T \boldsymbol{\mu} = \hat{m}$,

$$\hat{\boldsymbol{w}}^T \boldsymbol{\Sigma} \hat{\boldsymbol{w}} \le \boldsymbol{w}^T \boldsymbol{\Sigma} \boldsymbol{w}.$$

Note that $\boldsymbol{w} - \hat{\boldsymbol{w}} \in \mathcal{V}_0$; it follows that

$$\hat{\boldsymbol{w}}^T \boldsymbol{\Sigma} (\boldsymbol{w} - \hat{\boldsymbol{w}}) = 0. \tag{5.8}$$

Consider $\boldsymbol{w}^T \boldsymbol{\Sigma} \boldsymbol{w}$. To take advantage of (5.8), write

$$\boldsymbol{w}^T \boldsymbol{\Sigma} \boldsymbol{w} = \left(\hat{\boldsymbol{w}} + (\boldsymbol{w} - \hat{\boldsymbol{w}})\right)^T \boldsymbol{\Sigma} \left(\hat{\boldsymbol{w}} + (\boldsymbol{w} - \hat{\boldsymbol{w}})\right)$$

and expand this expression as

$$\hat{\boldsymbol{w}}^T \boldsymbol{\Sigma} \hat{\boldsymbol{w}} + 2\hat{\boldsymbol{w}}^T \boldsymbol{\Sigma} (\boldsymbol{w} - \hat{\boldsymbol{w}}) + (\boldsymbol{w} - \hat{\boldsymbol{w}})^T \boldsymbol{\Sigma} (\boldsymbol{w} - \hat{\boldsymbol{w}}). \tag{5.9}$$

Note that here we have used the fact that, because

$$\hat{\boldsymbol{w}}^T \boldsymbol{\Sigma} (\boldsymbol{w} - \hat{\boldsymbol{w}})$$

is a scalar,

$$\hat{\boldsymbol{w}}^T \boldsymbol{\Sigma} (\boldsymbol{w} - \hat{\boldsymbol{w}}) = (\boldsymbol{w} - \hat{\boldsymbol{w}})^T \boldsymbol{\Sigma} \hat{\boldsymbol{w}}.$$

By (5.8), the cross-product term in (5.9) is zero. Therefore,

$$\boldsymbol{w}^T \boldsymbol{\Sigma} \boldsymbol{w} = \hat{\boldsymbol{w}}^T \boldsymbol{\Sigma} \hat{\boldsymbol{w}} + (\boldsymbol{w} - \hat{\boldsymbol{w}})^T \boldsymbol{\Sigma} (\boldsymbol{w} - \hat{\boldsymbol{w}}).$$

Furthermore, because $\boldsymbol{\Sigma}$ is positive definite,

$$(\boldsymbol{w} - \hat{\boldsymbol{w}})^T \boldsymbol{\Sigma} (\boldsymbol{w} - \hat{\boldsymbol{w}}) \ge 0.$$

It follows that $\boldsymbol{w}^T \boldsymbol{\Sigma} \boldsymbol{w} \ge \hat{\boldsymbol{w}}^T \boldsymbol{\Sigma} \hat{\boldsymbol{w}}$. Because this holds for any $\boldsymbol{w}$ satisfying $\boldsymbol{w}^T \mathbf{1} = 1$ and $\boldsymbol{w}^T \boldsymbol{\mu} = \hat{m}$, it follows that $\hat{\boldsymbol{w}}$ solves the constrained minimization problem (5.6) with $m = \hat{m}$; that is, $\hat{\boldsymbol{w}}$ is on the minimum-risk frontier. □

Proposition 5.1 is an important result in portfolio theory. The importance of the result is not because it is useful for finding specific portfolios on the minimum-risk frontier—there are simple numerical methods available for that purpose—instead, it is important because it gives several useful properties of portfolios on the minimum-risk frontier.

For instance, it follows that portfolios on the minimum-risk frontier are unique in the sense that, if two portfolios on the minimum risk frontier have the same mean return, then they must have the same weight vector. This result is formally stated in the following corollary to Proposition 5.1; the proof is left as an exercise.

Corollary 5.1. *For $j = 1, 2$, let $\hat{R}_{pj}$ denote the return on the portfolio with weight vector $\hat{\boldsymbol{w}}_j$. Suppose that both portfolios are on the minimum-risk frontier and that $E(\hat{R}_{p1}) = E(\hat{R}_{p2})$. Then $\hat{\boldsymbol{w}}_1 = \hat{\boldsymbol{w}}_2$.*

The conclusion of Proposition 5.1 may also be expressed as a property of the covariances of portfolio returns. This result is given in the following corollary; the proof is left as an exercise.

Corollary 5.2. *Let $\hat{R}_p$ denote the return on a portfolio on the minimum-risk frontier and let R_0 denote the return on a zero-investment portfolio that has zero expected return. Then*

$$Cov(R_p, R_0) = 0.$$

Alternatively, let R_1, R_2 denote the returns on two assets satisfying $E(R_1) = E(R_2)$. Then

$$Cov(\hat{R}_p, R_1) = Cov(\hat{R}_p, R_2).$$

Therefore, according to Corollary 5.2, if $\hat{R}_p$ is the return of a portfolio on the minimum-risk frontier and R is the return on any asset, then $\text{Cov}(\hat{R}_p, R)$ is a function of $\text{E}(R)$.

Portfolios Constructed from Portfolios on the Minimum-Risk Frontier

An important feature of the necessary and sufficient conditions given in Proposition 5.1 is that if weight vectors $\hat{\boldsymbol{w}}_1$ and $\hat{\boldsymbol{w}}_2$ are on the minimum-risk frontier then *affine combinations* of $\hat{\boldsymbol{w}}_1, \hat{\boldsymbol{w}}_2$ are on the minimum-risk frontier; that is, weight vectors of the form

$$z\hat{\boldsymbol{w}}_1 + (1-z)\hat{\boldsymbol{w}}_2, \quad -\infty < z < \infty$$

are on the minimum-risk frontier. Note that a weight vector of this form may be viewed as the weight vector of a portfolio constructed from the two portfolios having weight vectors $\hat{\boldsymbol{w}}_1$ and $\hat{\boldsymbol{w}}_2$, respectively.

To establish this result, first note that

$$(z\hat{\boldsymbol{w}}_1 + (1-z)\hat{\boldsymbol{w}}_2)^T\mathbf{1} = z\hat{\boldsymbol{w}}_1^T\mathbf{1} + (1-z)\hat{\boldsymbol{w}}_2^T\mathbf{1} = z(1) + (1-z)(1) = 1$$

using the fact that $\hat{\boldsymbol{w}}_1^T\mathbf{1}$ and $\hat{\boldsymbol{w}}_2^T\mathbf{1}$ are both 1. Furthermore, if $\hat{\boldsymbol{w}}_j^T\boldsymbol{\Sigma}\boldsymbol{\delta} = 0$ for all $\boldsymbol{\delta} \in \mathcal{V}_0$, for $j = 1, 2$ then for all $\boldsymbol{\delta} \in \mathcal{V}_0$

$$\begin{aligned}(z\hat{\boldsymbol{w}}_1 + (1-z)\hat{\boldsymbol{w}}_2)^T\boldsymbol{\Sigma}\boldsymbol{\delta} &= z\hat{\boldsymbol{w}}_1^T\boldsymbol{\Sigma}\boldsymbol{\delta} + (1-z)\hat{\boldsymbol{w}}_2^T\boldsymbol{\Sigma}\boldsymbol{\delta} \\ &= 0.\end{aligned}$$

It now follows from Proposition 5.1 that the portfolio with weight vector $z\hat{\boldsymbol{w}}_1 + (1-z)\hat{\boldsymbol{w}}_2$ is on the minimum-risk frontier.

A formal statement of this result is given in the following lemma. Although the result in Corollary 5.3 applies to two portfolios, clearly it can be extended to a portfolio formed from a finite number of portfolios on the minimum-risk frontier.

Corollary 5.3. *Suppose that* $\hat{\boldsymbol{w}}_1$ *and* $\hat{\boldsymbol{w}}_2$ *are the weight vectors of two portfolios on the minimum-risk frontier. Then, for any* $z \in \Re$, $z\hat{\boldsymbol{w}}_1 + (1-z)\hat{\boldsymbol{w}}_2$ *is the weight vector of a portfolio on the minimum-risk frontier.*

Let $\hat{\boldsymbol{w}}_1$ and $\hat{\boldsymbol{w}}_2$ be the two weight vectors in Corollary 5.3 and let $m_j = \hat{\boldsymbol{w}}_j^T \boldsymbol{\mu}$, $j = 1, 2$; that is, the portfolio with weight vector $\hat{\boldsymbol{w}}_j$ has mean return m_j, $j = 1, 2$. Note that the portfolio with weight vector $z\hat{\boldsymbol{w}}_1 + (1-z)\hat{\boldsymbol{w}}_2$ has mean return $zm_1 + (1-z)m_2$; if $m_1 \neq m_2$, then any real number can be written as $zm_1 + (1-z)m_2$ for some z. Therefore, according to Corollary 5.3, the weight vector of any portfolio on the minimum-risk frontier can be written in terms of the weight vectors $\hat{\boldsymbol{w}}_1$ and $\hat{\boldsymbol{w}}_2$; the details are given in the following result.

Lemma 5.2. *Let* m_1 *and* m_2 *denote distinct real numbers; for* $j = 1, 2$, *let* $\hat{\boldsymbol{w}}_j$ *denote the weight vector of the minimum-risk portfolio with mean return* m_j. *Then, for any given* $m \in \Re$,

$$w_m \hat{\boldsymbol{w}}_1 + (1 - w_m)\hat{\boldsymbol{w}}_2$$

is the weight vector of the minimum-risk portfolio with mean return m, *where*

$$w_m = \frac{m - m_2}{m_1 - m_2}.$$

Lemma 5.2 shows that the entire minimum-risk frontier may be generated from two portfolios; thus, it is sometimes called the *two-fund theorem.* This result shows that, in some respects, portfolio theory for an arbitrary number of assets is essentially the same as portfolio theory for the case of two assets, as discussed in the previous chapter.

For instance, the set of possible variances of portfolios on the minimum-risk frontier is a parabola. It follows that there is a minimum possible variance of portfolios on the minimum-risk frontier and this minimum is achieved by a single portfolio, called the *minimum-variance portfolio.* The properties of the minimum-variance portfolio will be discussed in the following section. Furthermore, if one portfolio on the minimum-variance frontier has a given variance, then, unless that variance is the minimum variance, there is a second portfolio with the same variance, as was illustrated in Figure 4.1.

Calculating the Weight Vector of a Portfolio on the Minimum-Risk Frontier

The results thus far in this section give some important properties of the minimum-risk frontier. We now consider the problem of calculating the weight vector that solves the constrained minimization problem (5.6) for a particular value of m, given values for $\boldsymbol{\Sigma}$ and $\boldsymbol{\mu}$.

A constrained maximization problem of this form is an example of a *quadratic programming problem.* The type of quadratic programming problems we will consider here generally have three components: a quadratic objective function of the form

$$\frac{1}{2}\boldsymbol{x}^T\boldsymbol{D}\boldsymbol{x} - \boldsymbol{d}^T\boldsymbol{x}$$

where $\boldsymbol{x}$ is the decision variable, a vector taking values in $\Re^N$, $\boldsymbol{D}$ is a known $N \times N$ matrix and $\boldsymbol{d}$ is a known $N \times 1$ vector; a set of equality constraints of the form $\boldsymbol{A}^T\boldsymbol{x} = \boldsymbol{b}$ where $\boldsymbol{A}$ is a known $N \times k$ matrix and $\boldsymbol{b}$ is a known $k \times 1$ vector; and a set of inequality constraints on the elements of $\boldsymbol{x}$. Note that the minimization problem (5.6) does not include any inequality constraints; hence, here we consider minimizing the objective function subject to equality constraints. Portfolio problems with inequality constraints will be considered in Section 5.8.

Quadratic programming problems of this type are well-studied and software to obtain numerical solutions is widely available. In R, we can use the function `solve.QP` in the package `quadprog` (Turlach and Weingessel 2013) to solve a general quadratic programming problem as described previously.

The function `solve.QP` has several arguments; `Dmat` corresponds to the matrix $\boldsymbol{D}$, `dvec` corresponds to the vector $\boldsymbol{d}$, `Amat` corresponds to the matrix $\boldsymbol{A}$, and `bvec` corresponds to the vector $\boldsymbol{b}$. There is an additional argument `meq`, which specifies the number of columns of `Amat` that correspond to equality constraints, with the remaining columns corresponding to inequality constraints. Because, in the present context, all constraints are equality constraints, the value given for `meq` is simply the number of columns of `Amat` or, equivalently, the number of rows of `t(Amat)`, the transpose of `Amat`.

The following example describes how to use R to find the weight vectors of portfolios on the minimum-risk frontier.

Example 5.5 Consider the set of four assets described in Example 5.1, with mean return vector $(0.10, 0.20, 0.05, 0.10)^T$ and return covariance matrix

$$\boldsymbol{\Sigma} = \begin{pmatrix} 0.05 & 0.01 & 0.02 & 0 \\ 0.01 & 0.10 & 0.05 & 0.02 \\ 0.02 & 0.05 & 0.20 & 0.10 \\ 0 & 0.02 & 0.10 & 0.20 \end{pmatrix}.$$

These are stored in the R variables `mu` and `Sigma`, respectively.

```
> mu
[1] 0.10 0.20 0.05 0.10
> Sigma
     [,1] [,2] [,3] [,4]
[1,] 0.05 0.01 0.02 0.00
[2,] 0.01 0.10 0.05 0.02
[3,] 0.02 0.05 0.20 0.10
[4,] 0.00 0.02 0.10 0.20
```

Suppose we wish to find the portfolio on the minimum-risk frontier that has mean return 0.2. Then the objective function is

$$\boldsymbol{w}^T \boldsymbol{\Sigma} \boldsymbol{w}$$

and the constraints are $\boldsymbol{w}^T \mathbf{1} = 1$ and $\boldsymbol{w}^T \boldsymbol{\mu} = 0.2$, which may also be written as $\mathbf{1}^T \boldsymbol{w} = 1$ and $\boldsymbol{\mu}^T \boldsymbol{w} = 0.2$.

Therefore, in the notation of `solve.QP`, $\boldsymbol{d}$ is the zero vector of length 4, $\boldsymbol{D}$ is $2\boldsymbol{\Sigma}$, $\boldsymbol{A}$ is a 4×2 matrix with the first column given by a vector of all ones and the second column given by the mean vector $\boldsymbol{\mu}$, and $\boldsymbol{b}$ is the vector $(1, 0.1)$. Thus, the constraint $\boldsymbol{A}^T \boldsymbol{w} = \boldsymbol{b}$ specifies that the portfolio weights sum to 1 and that the mean return on the portfolio is 0.2.

Therefore, the R commands needed to solve this constrained optimization problem are as follows:

```
> library(quadprog)
> A<-cbind(c(1,1,1,1), mu)
> t(A)
    [,1] [,2] [,3] [,4]
     1.0  1.0  1.00  1.0
     0.1  0.2  0.05  0.1
> mrf1<-solve.QP(Dmat=2*Sigma, dvec=mu, Amat=A, bvec=c(1, 0.2),
+ meq=2)
```

Note that `cbind` combines two vectors or matrices by the columns; when the vector `c(1,1,1,1)` is used in this context, it is interpreted as a column vector.

The weight vector that maximizes the objective function is the component `$solution` of the result of the function `solve.QP`; therefore, the solution to the constrained minimization problem is

```
> mrf1$solution
[1]  0.362 0.813 -0.374 0.199
```

Thus, the mean and standard deviation of the return on the portfolio corresponding to the weight vector `mrf1$solution` are given by

```
> sum(mrf1$solution*mu)
[1] 0.2
> (mrf1$solution%*%Sigma%*%mrf1$solution)^.5
      [,1]
[1,] 0.265
```

The calculations are easily repeated for other values of m. For example, for $m = 0.25$,

```
> mrf2<-solve.QP(Dmat=2*Sigma, dvec=mu, Amat=A, bvec=c(1, 0.25),
+ meq=2)
> mrf2$solution
```

```
[1]  0.179 1.198 -0.604 0.227
> sum(mrf2$solution*mu)
[1] 0.25
> (mrf2$solution%*%Sigma%*%mrf2$solution)^.5
      [,1]
[1,] 0.373
```

□

5.4 The Minimum-Variance Portfolio

Choosing from among the portfolios on the minimum risk frontier requires some consideration of the relative importance of the mean return and risk of a portfolio. In this section, we consider the simplest analysis of this type, based on the belief that only risk is important when evaluating a portfolio. Under this assumption, the portfolio with minimum return variance is optimal; we will call this portfolio the *minimum-variance portfolio.*

As in the $N = 2$ case discussed in the previous chapter, the minimum-variance portfolio also provides a useful reference point for evaluating the mean and standard deviation of the returns on portfolios, and it will play a role in several results in this chapter.

Let $\boldsymbol{w}_{mv}$ denote the weight vector of the minimum-variance portfolio and let $R_{mv} = \boldsymbol{w}_{mv}^T \boldsymbol{R}$ denote the corresponding portfolio return. Then

$$\boldsymbol{w}_{mv}^T \boldsymbol{\Sigma} \boldsymbol{w}_{mv} \leq \boldsymbol{w}^T \boldsymbol{\Sigma} \boldsymbol{w} \quad \text{for any} \quad \boldsymbol{w} \in \Re^N \quad \text{such that} \quad \boldsymbol{w}^T \boldsymbol{1} = 1.$$

The following proposition gives a useful characterization of the minimum-variance portfolio, stating that the covariance of the return on the minimum-variance portfolio and any other portfolio is constant, not depending on the portfolio under consideration. The proof proceeds by showing that, if this were not the case, then we could construct a portfolio with variance smaller than that of the minimum-variance portfolio.

Proposition 5.2. *An asset with return $\hat{R}$ is the minimum-variance portfolio if and only if*

$$Cov(\hat{R}, R_p) = Var(\hat{R}) \tag{5.10}$$

for $R_p = \boldsymbol{w}^T \boldsymbol{R}$, for any weight vector $\boldsymbol{w}$.

Proof. First suppose that $\hat{R}$ is the return on the minimum-variance portfolio so that $\hat{R} = R_{mv}$. Let $\boldsymbol{w}$ be the weight vector corresponding to a portfolio with return R_p. For a given real number z, consider the portfolio with weight vector $\boldsymbol{w}_{mv} + z(\boldsymbol{w} - \boldsymbol{w}_{mv})$, which has return $R_{mv} + z(R_p - R_{mv})$. Define

$$f(z) = \text{Var}\left(R_{mv} + z(R_p - R_{mv})\right), \quad -\infty < z < \infty.$$

Because R_{mv} is the return on the minimum-variance portfolio, $f(z)$ is minimized at $z = 0$. Note that $f(z)$ is a quadratic function of z; hence, $f'(0) = 0$.

By expanding the variance used to define $f(z)$,

$$f(z) = \text{Var}(R_{mv}) + 2z\text{Cov}(R_{mv}, R_p - R_{mv}) + z^2\text{Var}(R_p - R_{mv}).$$

It follows that

$$f'(0) = 2\text{Cov}(R_{mv}, R_p - R_{mv})$$

so that

$$\text{Cov}(R_{mv}, R_p - R_{mv}) = 0;$$

using properties of covariance,

$$\begin{aligned}\text{Cov}(R_{mv}, R_p - R_{mv}) &= \text{Cov}(R_{mv}, R_p) - \text{Cov}(R_{mv}, R_{mv}) \\ &= \text{Cov}(R_{mv}, R_p) - \text{Var}(R_{mv})\end{aligned}$$

so that

$$\text{Cov}(R_{mv}, R_p) = \text{Var}(R_{mv}),$$

as stated in the proposition.

Now suppose that $\text{Cov}(\hat{R}, R_p) = \text{Var}(\hat{R})$ holds for any portfolio return R_p. Because

$$\begin{aligned}\text{Var}(R_p) &= \text{Var}\left(\hat{R} + (R_p - \hat{R})\right) \\ &= \text{Var}(\hat{R}) + 2\text{Cov}(\hat{R}, R_p - \hat{R}) + \text{Var}(R_p - \hat{R}) \\ &= \text{Var}(\hat{R}) + 2\left(\text{Cov}(\hat{R}, R_p) - \text{Var}(\hat{R})\right) + \text{Var}(R_p - \hat{R}) \\ &= \text{Var}(\hat{R}) + \text{Var}(R_p - \hat{R}),\end{aligned}$$

it follows that

$$\text{Var}(R_p) \geq \text{Var}(\hat{R}).$$

Because this holds for any portfolio return R_p, $\hat{R}$ must be the return on the minimum-variance portfolio. □

One consequence of the Proposition 5.2 is that the return on the minimum-variance portfolio is uncorrelated with the return on any zero-investment portfolio. Note that, if there were a zero-investment portfolio with weight vector $\boldsymbol{v}$ such that R_{mv} and $R_0 = \boldsymbol{v}^T\boldsymbol{R}$ are correlated, then we could find a constant c such that the portfolio with return $R_{mv} + cR_0$ has a smaller return variance than does R_{mv}.

The following corollary gives a formal statement of this result; the proof is left as an exercise.

Corollary 5.4. *Let R_{mv} denote the return on the minimum-variance portfolio and let R_0 denote the return on a zero-investment portfolio. Then*

$$Cov(R_{mv}, R_0) = 0.$$

The characterization of the minimum-variance portfolio given in Proposition 5.2 can be used to suggest a form for $\boldsymbol{w}_{mv}$, the weight vector of the minimum-variance portfolio.

Let $R_1, R_2, \ldots, R_N$ denote the returns on the N assets under consideration. Then, treating each asset as a portfolio in Proposition 5.2, for any $j = 1, 2, \ldots, N$

$$\text{Cov}(R_{mv}, R_j) = \text{Cov}(R_j, R_{mv}) = \text{Cov}(R_j, \boldsymbol{w}_{mv}^T \boldsymbol{R}) = \boldsymbol{e}_j^T \boldsymbol{\Sigma} \boldsymbol{w}_{mv} = c \tag{5.11}$$

for some constant c, where $\boldsymbol{e}_j$ denotes the jth column of the $N \times N$ identity matrix; that is, it is the vector in $\Re^N$ consisting of all zeros, except for the jth element, which is 1. Of course, by Proposition 5.2, the constant c must be $\text{Var}(R_{mv})$; however, that fact is not needed here.

Therefore, for each $j = 1, 2, \ldots, N$,

$$\boldsymbol{e}_j^T \boldsymbol{\Sigma} \boldsymbol{w}_{mv} = c$$

and combining these, it follows that

$$\boldsymbol{I} \boldsymbol{\Sigma} \boldsymbol{w}_{mv} = c\boldsymbol{1}$$

so that

$$\boldsymbol{\Sigma} \boldsymbol{w}_{mv} = c\boldsymbol{1}.$$

It follows that

$$\boldsymbol{w}_{mv} = c\boldsymbol{\Sigma}^{-1}\boldsymbol{1}. \tag{5.12}$$

Because the weights in $\boldsymbol{w}_{mv}$ must sum to 1, c must be

$$\frac{1}{\boldsymbol{1}^T \boldsymbol{\Sigma}^{-1} \boldsymbol{1}}.$$

The following result uses the Cauchy–Schwarz inequality to show directly that the weight vector of the minimum-variance portfolio is of the form given in (5.12).

Proposition 5.3. *Let* $\boldsymbol{R}$ *denote the return vector for a set of assets and let* $\boldsymbol{\Sigma}$ *denote the covariance matrix of* $\boldsymbol{R}$. *Then the weight vector of the minimum-variance portfolio is given by*

$$\boldsymbol{w}_{mv} = \frac{\boldsymbol{\Sigma}^{-1}\boldsymbol{1}}{\boldsymbol{1}^T \boldsymbol{\Sigma}^{-1} \boldsymbol{1}.}$$

Proof. The variance of the return on a portfolio based on weight vector $\boldsymbol{w}$ is given by $\boldsymbol{w}^T \boldsymbol{\Sigma} \boldsymbol{w}$, which can be written

$$(\boldsymbol{\Sigma}^{\frac{1}{2}} \boldsymbol{w})^T (\boldsymbol{\Sigma}^{\frac{1}{2}} \boldsymbol{w}).$$

Using the Cauchy–Schwarz inequality with $\boldsymbol{x} = \boldsymbol{\Sigma}^{1/2} \boldsymbol{w}$ and $\boldsymbol{y} = \boldsymbol{\Sigma}^{-1/2} \boldsymbol{1}$,

$$\left((\boldsymbol{\Sigma}^{\frac{1}{2}} \boldsymbol{w})^T (\boldsymbol{\Sigma}^{-\frac{1}{2}} \boldsymbol{1})\right)^2 \leq \left((\boldsymbol{\Sigma}^{\frac{1}{2}} \boldsymbol{w})^T (\boldsymbol{\Sigma}^{\frac{1}{2}} \boldsymbol{w})\right) \left((\boldsymbol{\Sigma}^{-\frac{1}{2}} \boldsymbol{1})^T (\boldsymbol{\Sigma}^{-\frac{1}{2}} \boldsymbol{1})\right) \tag{5.13}$$

with equality if $\boldsymbol{\Sigma}^{\frac{1}{2}} \boldsymbol{w} = c\boldsymbol{\Sigma}^{-\frac{1}{2}} \boldsymbol{1}$ for some scalar c.

Note that (5.13) may be written

$$(\boldsymbol{w}^T\mathbf{1})^2 \leq \left(\boldsymbol{w}^T\boldsymbol{\Sigma}\boldsymbol{w}\right)\left(\mathbf{1}^T\boldsymbol{\Sigma}^{-1}\mathbf{1}\right)$$

and because $\boldsymbol{w}^T\mathbf{1} = 1$,

$$\boldsymbol{w}^T\boldsymbol{\Sigma}\boldsymbol{w} \geq \frac{1}{\mathbf{1}^T\boldsymbol{\Sigma}^{-1}\mathbf{1}}$$

with equality if $\boldsymbol{w} = c\boldsymbol{\Sigma}^{-1}\mathbf{1}$; that is, the weight vector of the minimum-variance portfolio must be of the form $c\boldsymbol{\Sigma}^{-1}\mathbf{1}$ for some constant c. Since the weights must sum to 1, we must have

$$c = \frac{1}{\mathbf{1}^T\boldsymbol{\Sigma}^{-1}\mathbf{1}},$$

proving the result. □

Example 5.6 Consider a set of three assets, with mean returns 0.25, 0.125, and 0.3, respectively, and suppose that the returns have covariance matrix

$$\boldsymbol{\Sigma} = \begin{pmatrix} 0.25 & 0.1 & 0.24 \\ 0.1 & 0.16 & 0.096 \\ 0.24 & 0.096 & 0.36 \end{pmatrix}. \tag{5.14}$$

Hence, the asset returns have standard deviations 0.5, 0.4, and 0.6, respectively, and their correlation matrix is

$$\begin{pmatrix} 1 & 0.5 & 0.8 \\ 0.5 & 1 & 0.4 \\ 0.8 & 0.4 & 1 \end{pmatrix} \tag{5.15}$$

The mean vector and covariance matrix may be entered into R using the commands

```
> mu<-c(0.25, 0.125, 0.3)
> Sig<-matrix(c(0.25, 0.1, 0.24, 0.1, 0.16, 0.096, 0.24, 0.096,
+ 0.36),3,3)
> Sig
     [,1]  [,2]  [,3]
[1,] 0.25 0.100 0.240
[2,] 0.10 0.160 0.096
[3,] 0.24 0.096 0.360
```

To calculate the weights of the minimum-variance portfolio, we may use the `solve` function. Let $\boldsymbol{A}$ denote an $m \times m$ invertible matrix and let $\boldsymbol{b}$ denote an $m \times 1$ vector; let `A` and `b` denote the corresponding R variables. Then `solve(A, b)` returns $\boldsymbol{A}^{-1}\boldsymbol{b}$; the function with the second argument omitted, that is, `solve(A)`, returns $\boldsymbol{A}^{-1}$.

Thus, $\boldsymbol{w}_{mv}$ may be calculated by

```
> w0<-solve(Sig, c(1,1,1))
> w_mv<-w0/sum(w0)
> w_mv
[1] 0.243 0.713 0.044
```

□

Example 5.7 Consider a set of N assets with covariance matrix of the form

$$\boldsymbol{\Sigma} = \sigma^2 \begin{pmatrix} 1 & \rho & \dots & \dots & \rho \\ \rho & 1 & \rho & \dots & \rho \\ \vdots & \ddots & \ddots & \ddots & \vdots \\ \rho & \dots & \rho & 1 & \rho \\ \rho & \dots & \dots & \rho & 1 \end{pmatrix} \tag{5.16}$$

where $0 \leq \rho < 1$. Recall that, under this condition, $\boldsymbol{\Sigma}$ is positive-definite; see Example 5.4. Thus, for this covariance matrix, all asset returns have standard deviation σ and the correlation between any two returns is ρ.

To calculate the weight vector of the minimum-variance portfolio we need $\boldsymbol{\Sigma}^{-1}\mathbf{1}$. Recall that in Example 5.4 it was shown that $\mathbf{1}$ is an eigenvector of $\boldsymbol{\Sigma}$, with corresponding eigenvalue $1+(N-1)\rho$. It follows that

$$\mathbf{1} = \boldsymbol{\Sigma}^{-1}\boldsymbol{\Sigma}\mathbf{1} = (1+(N-1)\rho)\boldsymbol{\Sigma}^{-1}\mathbf{1}$$

and, hence, that

$$\boldsymbol{\Sigma}^{-1}\mathbf{1} = \frac{1}{1+(N-1)\rho}\mathbf{1}.$$

Because

$$\mathbf{1}^T\boldsymbol{\Sigma}^{-1}\mathbf{1} = \frac{N}{1+(N-1)\rho},$$

it follows that the weight vector of the minimum-variance portfolio is given by

$$\frac{\boldsymbol{\Sigma}^{-1}\mathbf{1}}{\mathbf{1}^T\boldsymbol{\Sigma}^{-1}\mathbf{1}} = \frac{1}{N}\mathbf{1}$$

so that the equally weighted portfolio is the minimum-variance portfolio.

To find the variance of the minimum-variance portfolio, we use the fact that

$$\left(\frac{1}{N}\mathbf{1}\right)^T \boldsymbol{\Sigma} \left(\frac{1}{N}\mathbf{1}\right) = \frac{\sigma^2}{N^2}(N+N(N-1)\rho) = \frac{\sigma^2}{N} + \left(1-\frac{1}{N}\right)\rho\sigma^2.$$

This is the minimum possible variance for a portfolio based on a return vector with a covariance matrix of the form (5.16). □

5.5 The Efficient Frontier

The minimum-risk frontier consists of those portfolios that have the smallest return standard deviation for a given value of the mean return. However, for some portfolios on the minimum-risk frontier, there is another portfolio with the same risk, but with a larger mean return; see Section 4.4 for a similar situation in the $N = 2$ case.

The efficient frontier consists of those portfolios on the minimum-risk frontier that also have the largest expected return for a given level of risk. For instance, the efficient frontier corresponding to Figure 5.3 is given in Figure 5.4; as shown in the graph, it corresponds to the "top half" of the minimum-risk frontier.

Therefore, when choosing a portfolio based on the mean and standard deviation of the portfolio return, there is never any reason to choose one that is not in the efficient frontier. Like "minimum-risk frontier," the term "efficient frontier" will refer to risk, expected-return pairs, as well as to their corresponding portfolios and weight functions. A portfolio on the efficient frontier will be said to be an *efficient portfolio* and the set of all efficient portfolios is also known as the *efficient set*. The efficient frontier is a fundamental concept in portfolio theory and in this section we consider its properties.

The example illustrated in Figure 4.2 suggests that, if two portfolios on the minimum-risk frontier have the same risk, but different mean returns, then one of the portfolios has a mean return greater than the return on the minimum-variance portfolio and the other has a mean return less than that

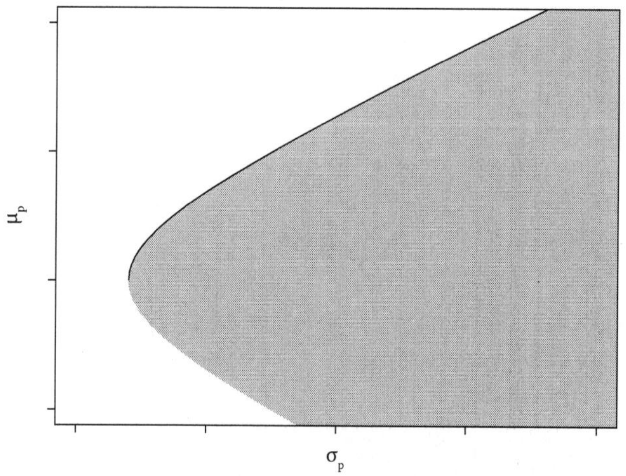

FIGURE 5.4
An example of an efficient frontier.

of the minimum-variance portfolio. The following result shows that this is, in fact, the case.

Lemma 5.3. *Suppose there are two distinct portfolios on the minimum-risk frontier, with returns R_{p1} and R_{p2}, respectively. Let $\mu_j = E(R_{pj})$, $j = 1, 2$ and suppose that $Var(R_{p1}) = Var(R_{p2})$. Then*

$$\mu_{mv} = \frac{1}{2}(\mu_1 + \mu_2)$$

where μ_{mv} denotes the mean return on the minimum-variance portfolio.

It follows that either

$$\mu_1 < \mu_{mv} < \mu_2 \quad or \quad \mu_2 < \mu_{mv} < \mu_1.$$

Proof. Let $\boldsymbol{w}_j$ denote the weight vector corresponding to return R_{pj}, $j = 1, 2$. Because the two portfolios are on the minimum-risk frontier, and they have different mean returns, it follows from Lemma 5.2 that the weight vector of any portfolio on the minimum-risk frontier may be written

$$\boldsymbol{w}_z = z\boldsymbol{w}_1 + (1 - z)\boldsymbol{w}_2$$

for some real number z.

Note that the variance of the return on the portfolio based on weight vector $\boldsymbol{w}_z$ is given by

$$\begin{aligned}
\boldsymbol{w}_z^T \boldsymbol{\Sigma} \boldsymbol{w}_z &= z^2 \boldsymbol{w}_1^T \boldsymbol{\Sigma} \boldsymbol{w}_1 + (1 - z)^2 \boldsymbol{w}_2^T \boldsymbol{\Sigma} \boldsymbol{w}_2 + 2z(1 - z)\boldsymbol{w}_2^T \boldsymbol{\Sigma} \boldsymbol{w}_1 \\
&= \left(z^2 + (1 - z)^2\right) \boldsymbol{w}_1^T \boldsymbol{\Sigma} \boldsymbol{w}_1 + 2z(1 - z)\boldsymbol{w}_2^T \boldsymbol{\Sigma} \boldsymbol{w}_1 \\
&= 2\left(\boldsymbol{w}_1^T \boldsymbol{\Sigma} \boldsymbol{w}_1 - \boldsymbol{w}_2^T \boldsymbol{\Sigma} \boldsymbol{w}_1\right) z^2 + 2\left(\boldsymbol{w}_2^T \boldsymbol{\Sigma} \boldsymbol{w}_1 - \boldsymbol{w}_1^T \boldsymbol{\Sigma} \boldsymbol{w}_1\right) z + \boldsymbol{w}_1^T \boldsymbol{\Sigma} \boldsymbol{w}_1,
\end{aligned}$$

using the fact that $\boldsymbol{w}_1^T \boldsymbol{\Sigma} \boldsymbol{w}_1 = \boldsymbol{w}_2^T \boldsymbol{\Sigma} \boldsymbol{w}_2$.

Note that, because $\boldsymbol{\Sigma}$ is positive definite, the correlation of the returns with weight vectors $\boldsymbol{w}_1$ and $\boldsymbol{w}_2$ is less than one; it follows that

$$\boldsymbol{w}_2^T \boldsymbol{\Sigma} \boldsymbol{w}_1 < \boldsymbol{w}_1^T \boldsymbol{\Sigma} \boldsymbol{w}_1. \tag{5.17}$$

Hence, $\boldsymbol{w}_z^T \boldsymbol{\Sigma} \boldsymbol{w}_z$ is a quadratic function of z with a positive coefficient of z^2. It follows that $\boldsymbol{w}_z^T \boldsymbol{\Sigma} \boldsymbol{w}_z$ can be minimized by solving

$$\frac{d}{dz} \boldsymbol{w}_z^T \boldsymbol{\Sigma} \boldsymbol{w}_z = (4z - 2)\left(\boldsymbol{w}_1^T \boldsymbol{\Sigma} \boldsymbol{w}_1 - \boldsymbol{w}_2^T \boldsymbol{\Sigma} \boldsymbol{w}_1\right) = 0.$$

The result in (5.17) shows that $\boldsymbol{w}_1^T \boldsymbol{\Sigma} \boldsymbol{w}_1 - \boldsymbol{w}_2^T \boldsymbol{\Sigma} \boldsymbol{w}_1 \neq 0$; therefore, $4z - 2 = 0$ or $z = 1/2$.

Thus, the portfolio on the minimum-risk frontier with the smallest variance has weight vector

$$\frac{1}{2}\boldsymbol{w}_1 + \frac{1}{2}\boldsymbol{w}_2;$$

because the minimum variance portfolio is on the minimum-risk frontier, we must have

$$\frac{1}{2}\boldsymbol{w}_1 + \frac{1}{2}\boldsymbol{w}_2 = \boldsymbol{w}_{mv}.$$

Furthermore, we must have

$$\frac{1}{2}\mu_1 + \frac{1}{2}\mu_2 = \mu_{mv}.$$

Clearly, this cannot hold if either μ_1 and μ_2 are both greater than μ_{mv} or if both μ_1 and μ_2 are less than μ_{mv}. The result follows. □

The most important consequence of Lemma 5.3 is the following result describing the efficient frontier; the proof follows immediately from Lemma 5.3.

Proposition 5.4. *The efficient frontier consists of those portfolios on the minimum-risk frontier with a mean return greater than or equal to* μ_{mv}.

Thus, the efficient frontier is the "top half" of the minimum-risk frontier, as suggested previously. It follows that many of the results describing the minimum-risk frontier are easily converted into results for the efficient frontier. The following corollaries give such conversions of Lemma 5.1 and 5.2, respectively.

Corollary 5.5. *Define* $\mathcal{V}_0$ *as in* (5.7). *Consider a vector of portfolio returns* $\boldsymbol{R}$ *with mean vector* $\boldsymbol{\mu}$ *and covariance matrix* $\boldsymbol{\Sigma}$. *The portfolio with weight vector* $\hat{\boldsymbol{w}}$ *is on the efficient frontier if and only if*

$$\hat{\boldsymbol{w}}^T\boldsymbol{\Sigma}\boldsymbol{\delta} = 0 \quad \textit{for all} \quad \boldsymbol{\delta} \in \mathcal{V}_0$$

and

$$\hat{\boldsymbol{w}}^T\boldsymbol{\mu} \geq \mu_{mv}.$$

Corollary 5.6. *Let* $\boldsymbol{w}_0$ *denote the weight vector of a portfolio on the efficient frontier that is not the minimum-variance portfolio and let* $\boldsymbol{w}_{mv}$ *denote the weight vector of the minimum-variance portfolio. Then the portfolio with weight vector*

$$z\boldsymbol{w}_0 + (1-z)\boldsymbol{w}_{mv},$$

where $z \geq 0$, *is on the efficient frontier.*

Proof. By Lemma 5.2, the portfolio with weight vector $z\boldsymbol{w}_0 + (1-z)\boldsymbol{w}_{mv}$ is on the minimum-risk frontier. This portfolio has expected return $\mu_{mv} + z(\mu_0 - \mu_{mv})$, where μ_0 is the expected return on the portfolio with weight vector $\boldsymbol{w}_0$; note that, since that portfolio is on the efficient frontier, $\mu_0 \geq \mu_{mv}$. The result follows. □

5.6 Risk-Aversion Criterion

Choosing a portfolio from the efficient set requires some consideration of the trade-off between higher mean return and greater risk. The minimum variance portfolio deals with this issue by ignoring the mean return completely and simply minimizing risk. In this section, we consider an N-asset version of the risk-aversion criterion analyzed in Section 4.4 that is a function of both the mean return and the variance of the return; we then choose the portfolio that maximizes this function.

Let $\boldsymbol{R}$ denote an N-dimensional vector of asset returns and let $\boldsymbol{w}$ denote the weight vector for a portfolio. Then the portfolio return $\boldsymbol{w}^T\boldsymbol{R}$ has expected return $\boldsymbol{w}^T\boldsymbol{\mu}$ and return variance $\boldsymbol{w}^T\boldsymbol{\Sigma}\boldsymbol{w}$, where $\boldsymbol{\mu}$ and $\boldsymbol{\Sigma}$ denote the mean vector and covariance matrix, respectively, of $\boldsymbol{R}$.

The risk-aversion criterion function is given by

$$\boldsymbol{w}^T\boldsymbol{\mu} - \frac{\lambda}{2}\boldsymbol{w}^T\boldsymbol{\Sigma}\boldsymbol{w} \tag{5.18}$$

where the risk-aversion parameter $\lambda > 0$ is given. This function has the same interpretation as the risk-aversion criterion function considered in Section 4.4. It may be viewed as a penalized mean return, with the penalty based on the variance of the return; the extent to which the variance penalizes the mean return is controlled by the parameter λ. An investor primarily interested in a large mean return with a high tolerance for risk might choose λ to be small. On the other hand, an investor with a strong preference for a low-risk portfolio might choose λ to be large.

Our goal is to choose the weight vector $\boldsymbol{w}$ that maximizes this (5.18), which we denote by $\boldsymbol{w}_\lambda$; the following result gives an explicit expression for $\boldsymbol{w}_\lambda$. For lack of a better term, we will refer to the portfolio with weight vector $\boldsymbol{w}_\lambda$ as the *risk-averse portfolio with parameter* λ.

Proposition 5.5. *Let $\boldsymbol{R}$ denote the return vector for a set of assets, let $\boldsymbol{\mu}$ denote the mean vector of $\boldsymbol{R}$, and let $\boldsymbol{\Sigma}$ denote the covariance matrix of $\boldsymbol{R}$. For a given value of $\lambda > 0$, the weight vector that maximizes*

$$\boldsymbol{w}^T\boldsymbol{\mu} - \frac{\lambda}{2}\boldsymbol{w}^T\boldsymbol{\Sigma}\boldsymbol{w}, \tag{5.19}$$

subject to the restriction $\boldsymbol{w}^T\boldsymbol{1} = 1$, is given by

$$\boldsymbol{w}_\lambda = \boldsymbol{w}_{mv} + \frac{1}{\lambda}\bar{\boldsymbol{v}}$$

where $\boldsymbol{w}_{mv}$ is the weight vector of the minimum variance portfolio,

$$\bar{\boldsymbol{v}} = \boldsymbol{\Sigma}^{-1}(\boldsymbol{\mu} - \mu_{mv}\boldsymbol{1})$$

and μ_{mv} is the mean return on the minimum-variance portfolio.

Proof. To maximize (5.19) subject to the restriction $\boldsymbol{w}^T\mathbf{1} = 1$, we can use the method of Lagrange multipliers. The function (5.19) is modified to

$$\boldsymbol{w}^T\mu - \frac{\lambda}{2}\boldsymbol{w}^T\boldsymbol{\Sigma}\boldsymbol{w} + \theta(\boldsymbol{w}^T\mathbf{1} - 1) \tag{5.20}$$

for $\theta \in \Re$ and then it is maximized over $\boldsymbol{w} \in \Re^N$ for each θ. The solution will depend on θ, which is then chosen so that $\boldsymbol{w}^T\mathbf{1} = 1$.

Because, for a vector $\boldsymbol{q} \in \Re^N$,

$$-\frac{\lambda}{2}(\boldsymbol{w}-\boldsymbol{q})^T\boldsymbol{\Sigma}(\boldsymbol{w}-\boldsymbol{q}) = -\frac{\lambda}{2}\boldsymbol{w}^T\boldsymbol{\Sigma}\boldsymbol{w} + \lambda\boldsymbol{w}^T\boldsymbol{\Sigma}\boldsymbol{q} - \frac{\lambda}{2}\boldsymbol{q}\boldsymbol{\Sigma}\boldsymbol{q}$$

taking

$$\boldsymbol{q} = \frac{1}{\lambda}\boldsymbol{\Sigma}^{-1}\mu + \frac{\theta}{\lambda}\boldsymbol{\Sigma}^{-1}\mathbf{1},$$

it follows that

$$\boldsymbol{w}^T\mu - \frac{\lambda}{2}\boldsymbol{w}^T\boldsymbol{\Sigma}\boldsymbol{w} + \theta(\boldsymbol{w}^T\mathbf{1} - 1) = -\frac{\lambda}{2}(\boldsymbol{w}-\boldsymbol{q})^T\boldsymbol{\Sigma}(\boldsymbol{w}-\boldsymbol{q}) + A$$

where A is a term not depending on $\boldsymbol{w}$. Hence, (5.20) is maximized over $\boldsymbol{w} \in \Re^N$ by

$$\boldsymbol{w}_\lambda(\theta) = \boldsymbol{q} = \frac{1}{\lambda}\boldsymbol{\Sigma}^{-1}\mu + \frac{\theta}{\lambda}\boldsymbol{\Sigma}^{-1}\mathbf{1}. \tag{5.21}$$

To complete the maximization, we choose θ so that $\boldsymbol{w}_\lambda(\theta)^T\mathbf{1} = 1$; because, using (5.21),

$$\boldsymbol{w}_\lambda^T(\theta)\mathbf{1} = \frac{1}{\lambda}\mu^T\boldsymbol{\Sigma}^{-1}\mathbf{1} + \frac{\theta}{\lambda}\mathbf{1}^T\boldsymbol{\Sigma}^{-1}\mathbf{1},$$

it follows that

$$\theta = \frac{\lambda - \mu^T\boldsymbol{\Sigma}^{-1}\mathbf{1}}{\mathbf{1}^T\boldsymbol{\Sigma}^{-1}\mathbf{1}}.$$

Substituting this expression into (5.21) yields

$$\boldsymbol{w}_\lambda = \boldsymbol{w}_{mv} + \frac{1}{\lambda}\boldsymbol{\Sigma}^{-1}\left(\mu - \frac{\mu^T\boldsymbol{\Sigma}^{-1}\mathbf{1}}{\mathbf{1}^T\boldsymbol{\Sigma}^{-1}\mathbf{1}}\mathbf{1}\right)$$

where $\boldsymbol{w}_{mv}$ is the weight vector of the minimum variance portfolio.

The term

$$\frac{\mu^T\boldsymbol{\Sigma}^{-1}\mathbf{1}}{\mathbf{1}^T\boldsymbol{\Sigma}^{-1}\mathbf{1}}$$

appearing in the expression for $\boldsymbol{w}_\lambda$ is the mean return on the minimum variance portfolio,

$$\mu_{mv} = \boldsymbol{w}_{mv}^T\mu = \frac{\mathbf{1}^T\boldsymbol{\Sigma}^{-1}}{\mathbf{1}^T\boldsymbol{\Sigma}^{-1}\mathbf{1}}\mu.$$

Thus, we can write

$$\boldsymbol{w}_\lambda = \boldsymbol{w}_{mv} + \frac{1}{\lambda}\bar{\boldsymbol{v}}$$

where

$$\bar{\boldsymbol{v}} = \boldsymbol{\Sigma}^{-1}(\boldsymbol{\mu} - \mu_{mv}\mathbf{1}) \tag{5.22}$$

as stated in the proposition. □

Thus, the optimal weight vector corresponding to the investor's choice of the risk-aversion parameter λ starts with the weight vector of the minimum variance portfolio and adds $1/\lambda$ times $\bar{\boldsymbol{v}}$. Note that, because the weights in both $\boldsymbol{w}_\lambda$ and $\boldsymbol{w}_{mv}$ must sum to 1, $\bar{\boldsymbol{v}}$ must be the weight vector of a zero-investment portfolio; this may be confirmed directly.

$$\begin{aligned} \bar{\boldsymbol{v}}^T\mathbf{1} &= (\boldsymbol{\mu} - \mu_{mv}\mathbf{1})^T\,\boldsymbol{\Sigma}^{-1}\mathbf{1} \\ &= \boldsymbol{\mu}^T\boldsymbol{\Sigma}^{-1}\mathbf{1} - \mu_{mv}\mathbf{1}^T\boldsymbol{\Sigma}^{-1}\mathbf{1} \\ &= 0 \end{aligned}$$

using the fact that

$$\mu_{mv} = \frac{\boldsymbol{\mu}^T\boldsymbol{\Sigma}^{-1}\mathbf{1}}{\mathbf{1}^T\boldsymbol{\Sigma}^{-1}\mathbf{1}}.$$

Example 5.8 Consider the assets described in Example 5.6, with mean return vector (0.25, 0.125, 0.3) and covariance matrix given by (5.14); these are stored in variables `mu` and `Sig`, respectively. Recall that the weight vector of the minimum-variance portfolio was determined in Example 5.6 and is given by

```
> w_mv
[1] 0.243 0.713 0.044
```

The vector $\bar{\boldsymbol{v}}$ defined in Proposition 5.5 may be calculated by the following commands.

```
> m<-sum(w_mv*mu)
> m
[1] 0.163
> vbar<-solve(Sig, mu - m*c(1,1,1))
> vbar
[1]  0.194 -0.607  0.413
```

The weight vector $\boldsymbol{w}_\lambda$ corresponding to any value of λ may now be calculated using `w_mv` and `v_bar`. For instance, for $\lambda = 1$, $\boldsymbol{w}_\lambda$ is

```
> w_mv + vbar
[1] 0.437 0.106 0.457
```

and for $\lambda = 4$,

```
> w_mv + vbar/4
[1] 0.292 0.561 0.147
```

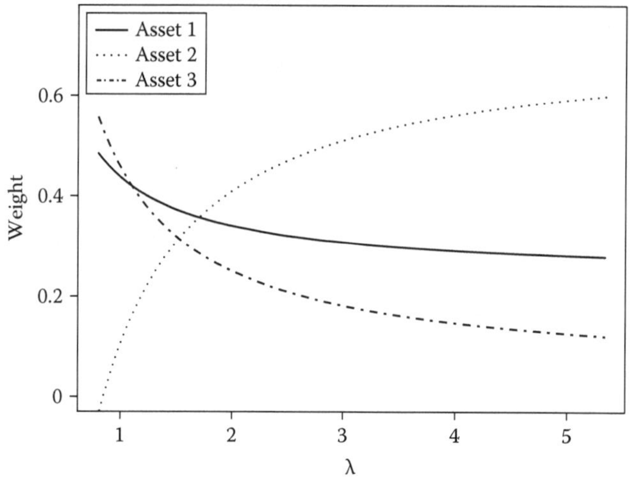

FIGURE 5.5
Weights of the risk-averse portfolio in Example 5.8 as λ varies.

Figure 5.5 contains a plot of the weights for the three assets as λ varies. Note that the weight for asset 1 is relatively constant, while the weight for asset 2 is small, or even negative, for $\lambda < 1$ and increases rapidly as λ increases over the range $(1, 3)$. For $\lambda > 5$, the weights are stable, being approximately equal to the weights of the minimum variance portfolio.

For $\lambda = 1$, the mean return of the portfolio with weight vector $\boldsymbol{w}_\lambda$ is

```
> sum((w_mv + vbar)*mu)
[1] 0.260
```

and the standard deviation of the return is

```
> (t(w_mv + vbar)%*%Sig%*%(w_mv + vbar))^.5
      [,1]
[1,] 0.489
```

For $\lambda = 4$, the mean and standard deviation of the return are 0.187 and 0.386, respectively. Thus, for small λ, there is less of a penalty on the variance of return; it follows that the optimal portfolio has a larger return standard deviation and, hence, a larger mean return. Figure 5.6 contains a plot of the mean and standard deviation of the return on the portfolio with weight vector $\boldsymbol{w}_{mv}$ as λ varies. □

Properties of Risk-Averse Portfolios

Let R_{mv} denote the return on the minimum-variance portfolio. Because the minimum-variance portfolio is on the minimum-risk frontier and the

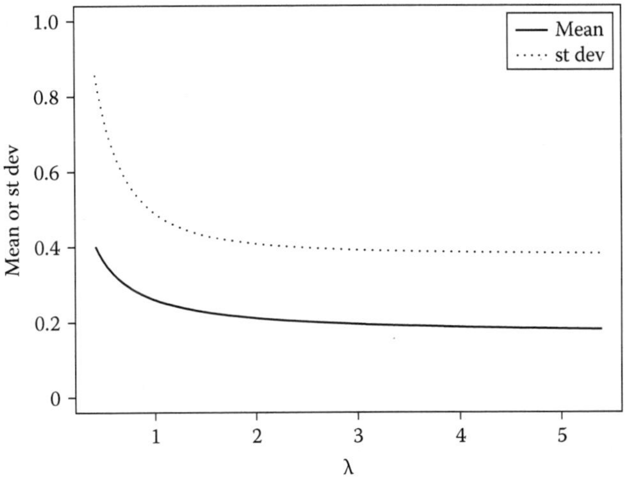

FIGURE 5.6
Mean and standard deviation of the risk-averse portfolio in Example 5.8 as λ varies.

portfolio based on $\bar{\boldsymbol{v}}$ is a zero-investment portfolio, it follows from Corollary 5.4 that

$$\text{Cov}(R_{mv}, \bar{\boldsymbol{v}}^T \boldsymbol{R}) = \boldsymbol{w}_{mv}^T \boldsymbol{\Sigma} \bar{\boldsymbol{v}} = 0.$$

This result may also be established using a direct argument:

$$\begin{aligned}\text{Cov}(R_{mv}, \bar{\boldsymbol{v}}^T \boldsymbol{R}) &= \text{Cov}(\boldsymbol{w}_{mv}^T \boldsymbol{R}, \bar{\boldsymbol{v}}^T \boldsymbol{R}) \\ &= \boldsymbol{w}_{mv}^T \boldsymbol{\Sigma} \bar{\boldsymbol{v}} \\ &= \frac{1}{\mathbf{1}^T \boldsymbol{\Sigma}^{-1} \mathbf{1}} \mathbf{1}^T \boldsymbol{\Sigma}^{-1} \boldsymbol{\Sigma} \boldsymbol{\Sigma}^{-1} (\mu - \mu_{mv} \mathbf{1}) \\ &= \frac{1}{\mathbf{1}^T \boldsymbol{\Sigma}^{-1} \mathbf{1}} (\mathbf{1}^T \boldsymbol{\Sigma}^{-1} \mu - \mu_{mv} \mathbf{1}^T \boldsymbol{\Sigma}^{-1} \mathbf{1}) \\ &= 0.\end{aligned}$$

This fact is useful in computing the mean and variance of the return corresponding to $\boldsymbol{w}_\lambda$.

Corollary 5.7. *Let $\boldsymbol{R}$ denote the return vector for a set of assets, let μ denote the mean vector of $\boldsymbol{R}$, and let $\boldsymbol{\Sigma}$ denote the covariance matrix of $\boldsymbol{R}$. For a given value of $\lambda > 0$, the portfolio with weight vector $\boldsymbol{w}_\lambda$, as given in Proposition 5.5, has mean return*

$$\mu_\lambda = \mu_{mv} + \frac{1}{\lambda} (\mu - \mu_{mv} \mathbf{1})^T \boldsymbol{\Sigma}^{-1} (\mu - \mu_{mv} \mathbf{1})$$

and return variance

$$\sigma_\lambda^2 = \sigma_{mv}^2 + \frac{1}{\lambda}(\mu_\lambda - \mu_{mv})$$

where μ_{mv} *and* σ_{mv} *denote the mean and standard deviation, respectively, of the return on the minimum variance portfolio.*

Proof. Using the expression for $\bar{\boldsymbol{v}}$,

$$\bar{\boldsymbol{v}}^T \mu = (\mu - \mu_{mv}\mathbf{1})^T \boldsymbol{\Sigma}^{-1} \mu;$$

this result, together with the fact that

$$(\mu - \mu_{mv}\mathbf{1})^T \boldsymbol{\Sigma}^{-1}\mathbf{1} = \mu^T \boldsymbol{\Sigma}^{-1}\mathbf{1} - \mu_{mv}\mathbf{1}^T \boldsymbol{\Sigma}^{-1}\mathbf{1} = 0,$$

leads to the expression for μ_λ given in the statement of the corollary.

Note that

$$\begin{aligned}\bar{\boldsymbol{v}}^T \boldsymbol{\Sigma} \bar{\boldsymbol{v}} &= (\mu - \mu_{mv}\mathbf{1})^T \boldsymbol{\Sigma}^{-1}\boldsymbol{\Sigma}\boldsymbol{\Sigma}^{-1}(\mu - \mu_{mv}\mathbf{1}) \\ &= (\mu - \mu_{mv}\mathbf{1})^T \boldsymbol{\Sigma}^{-1}(\mu - \mu_{mv}\mathbf{1});\end{aligned}$$

the expression for σ_λ^2 follows from noting that

$$(\mu - \mu_{mv}\mathbf{1})^T \boldsymbol{\Sigma}^{-1}(\mu - \mu_{mv}\mathbf{1})$$

is simply $\mu_\lambda - \mu_{mv}$, as shown earlier. □

The result in Corollary 5.7 may be used to interpret the risk-aversion parameter λ used to construct the portfolio with weight vector $\boldsymbol{w}_\lambda$. According to the expressions for μ_λ and σ_λ^2 given in Corollary 5.7,

$$\lambda = \frac{\mu_\lambda - \mu_{mv}}{\sigma_\lambda^2 - \sigma_{mv}^2}.$$

Therefore, the value of λ may be chosen to set a desired value for the mean return of the portfolio above that of the minimum variance portfolio as a proportion of the variance of the portfolio above that of the minimum variance portfolio.

It is not surprising that the risk-averse portfolios are on the efficient frontier. However, the converse is also true—every portfolio on the efficient frontier is a risk-averse portfolio with parameter λ, for some $\lambda > 0$. This result is given in the following proposition.

Proposition 5.6. *The portfolio with weight vector* $\boldsymbol{w}_p$ *is on the efficient frontier if and only if either* $\boldsymbol{w}_p = \boldsymbol{w}_\lambda$ *for some* $\lambda > 0$ *or* $\boldsymbol{w}_p = \boldsymbol{w}_{mv}$*. Here* $\boldsymbol{w}_\lambda$ *denotes the weight vector of the risk-averse portfolio with parameter* λ*, as defined in Proposition 5.5.*

Proof. First suppose that $\boldsymbol{w}_p = \boldsymbol{w}_\lambda$ for some $\lambda > 0$. Suppose the portfolio with weight vector $\boldsymbol{w}_p$ is not on the minimum risk frontier; then there is a portfolio with the same mean return but a smaller return variance. However, such a portfolio would have a smaller value of the risk-aversion criterion (for any value of λ), which contradicts the fact that the portfolio with weight vector $\boldsymbol{w}_p$ is the risk-averse portfolio with parameter λ. It follows that the portfolio with weight vector $\boldsymbol{w}_p$ is on the minimum risk frontier. By Corollary 5.7, together with the fact that $\boldsymbol{\Sigma}$ is positive definite, it follows that

$$\mu_\lambda - \mu_{mv} > 0;$$

it follows that the portfolio with weight vector $\boldsymbol{w}_p$ is on the efficient frontier. Note that, because the minimum-variance portfolio is on the efficient frontier, this result also holds if $\boldsymbol{w}_p = \boldsymbol{w}_{mv}$. Therefore, if $\boldsymbol{w}_p = \boldsymbol{w}_\lambda$ for some $\lambda > 0$ or $\boldsymbol{w}_p = \boldsymbol{w}_{mv}$, then the portfolio with weight vector $\boldsymbol{w}_p$ is on the efficient frontier.

Now suppose the portfolio with weight vector $\boldsymbol{w}_p$ is on the efficient frontier, but it is not the minimum-variance portfolio; let $\mu_p = \mathrm{E}(R_p)$ and note that $\mu_p > \mu_{mv}$. According to Corollary 5.7, the risk-averse portfolio with parameter λ has expected return

$$\mu_\lambda = \mu_{mv} + \frac{1}{\lambda}(\mu - \mu_{mv}\mathbf{1})^T \boldsymbol{\Sigma}^{-1}(\mu - \mu_{mv}\mathbf{1}).$$

Because

$$(\mu - \mu_{mv}\mathbf{1})^T \boldsymbol{\Sigma}^{-1}(\mu - \mu_{mv}\mathbf{1}) > 0,$$

there exists a $\lambda_p > 0$ such that the expected return on the portfolio with weight vector $\boldsymbol{w}_{\lambda_p}$ is μ_p. Futhermore, the portfolio with weight vector $\boldsymbol{w}_{\lambda_p}$ is on the efficient frontier. By the uniqueness of portfolios on the efficient frontier, it follows that $\boldsymbol{w}_p = \boldsymbol{w}_\lambda$. Using the fact that the minimum-variance portfolio is on the efficient frontier, it follows that if the portfolio with weight vector $\boldsymbol{w}_p$ is on the efficient frontier, then either $\boldsymbol{w}_p = \boldsymbol{w}_\lambda$ for some $\lambda > 0$ or $\boldsymbol{w}_p = \boldsymbol{w}_{mv}$, proving the result. □

Therefore, the risk-aversion approach gives an alternative way of parameterizing the portfolios on the efficient frontier. Instead of defining efficient portfolios as those with minimum risk for a given mean return $m \geq \mu_{mv}$ or as those with the maximum mean return for a given level of portfolio risk, we can define them as those portfolios maximizing the risk-aversion criterion function for some value of $0 < \lambda \leq \infty$.

Finding $\boldsymbol{w}_\lambda$ Using Quadratic Programming

Although Proposition 5.5 gives an expression for $\boldsymbol{w}_\lambda$, the weight vector of the risk-averse portfolio with risk-aversion parameter λ, it is often more convenient to find $\boldsymbol{w}_\lambda$ numerically. The R function `solve.QP` in the package `quadprog` that was used in Example 5.5 to find the weight vector of a portfolio on the minimum-risk frontier with a given mean return may also be used to compute the weight vector maximizing the risk-aversion criterion function.

Recall that in solve.QP the objective function is of the form

$$\frac{1}{2}\boldsymbol{x}^T\boldsymbol{D}\boldsymbol{x} - \boldsymbol{d}^T\boldsymbol{x},$$

which is minimized with respect to $\boldsymbol{x}$. This is equivalent to maximizing the objective function

$$\boldsymbol{d}^T\boldsymbol{x} - \frac{1}{2}\boldsymbol{x}^T\boldsymbol{D}\boldsymbol{x}.$$

The constraint on the weight vector $\boldsymbol{w}$, $\boldsymbol{w}^T\mathbf{1} = 1$, is easily included using the argument Amat to solve.QP. Equality constraints on $\boldsymbol{x}$ are given by $\boldsymbol{A}^T\boldsymbol{x} = \boldsymbol{b}$ for a given matrix $\boldsymbol{A}$ and a given vector $\boldsymbol{b}$.

The following example illustrates how solve.QP can be used to find the weight vector $\boldsymbol{w}_\lambda$.

Example 5.9 Consider the assets described in Example 5.6 and analyzed in Example 5.8. The assets have mean return vector $(0.25, 0.125, 0.3)$ and return covariance matrix given by (5.14); these are stored in variables mu and Sig, respectively.

The same basic approach used in Example 5.5 can be used here, except that now the objective function is of the form

$$\boldsymbol{w}^T\mu - \frac{\lambda}{2}\boldsymbol{w}^T\boldsymbol{\Sigma}\boldsymbol{w}$$

and in the present context the only constraint is that the portfolio weights sum to 1.

Thus, the arguments of solve.QP that define the objective function are Dmat=lambda*Sig and dvec=mu, where lambda denotes the value of the risk-aversion parameter λ. To specify the constraint that the weights sum to 1, we take Amat to be matrix(rep(1,3), 3, 1); in this command, rep(1,3) is a vector consisting of 1 repeated three times and matrix forms a matrix from that vector. The remaining arguments are bvec=1, which specifies that the weights sum to 1, and meq=1, which indicates that the constraint is an equality constraint.

Consider calculation of the weights of the risk-averse portfolio corresponding to $\lambda = 1$. These may be obtained using the R command

```
> library(quadprog)
> ra<-solve.QP(Dmat=Sig, dvec=mu, Amat=matrix(rep(1,3),3,1),
+ bvec=1, meq=1)
> ra$solution
[1] 0.437 0.106 0.457
```

Note that the result matches that obtained in Example 5.8. □

5.7 The Tangency Portfolio

So far in this chapter, we have considered the problem of choosing a portfolio based on a set of N risky assets. In this section, we consider including a risk-free asset in the portfolio. Let R_f denote the return on the risk-free asset; let $\mu_f = \mathrm{E}(R_f)$ and recall that $\mathrm{Var}(R_f) = 0$. We will assume that $\mu_f < \mu_{mv}$, where μ_{mv} is the mean return on the minimum-variance portfolio.

In selecting the weights for such a portfolio, we may use the same approach utilized in the two-asset case: first choose a portfolio of the N risky assets and then combine that portfolio with the risk-free asset. When selecting the weights of the N risky assets, it is important to use the fact that the resulting portfolio will be combined with the risk-free asset.

In particular, as noted in Section 4.6, the result of Proposition 4.1 continues to hold in this setting. That is, when the portfolio of risky assets is being combined with the risk-free asset, the optimal portfolio is the one that maximizes the Sharpe ratio

$$\frac{\mathrm{E}(R_p) - \mu_f}{(\mathrm{Var}(R_p))^{\frac{1}{2}}},$$

the portfolio known as the *tangency portfolio.*

The following result gives an expression for the weight vector of the tangency portfolio in terms of the mean vector $\boldsymbol{\mu}$ and the covariance matrix $\boldsymbol{\Sigma}$ of the vector $\boldsymbol{R}$ of asset returns.

Proposition 5.7. *Let $\boldsymbol{R}$ denote the return vector for a set of assets, let $\boldsymbol{\mu}$ denote the mean vector of $\boldsymbol{R}$, and let $\boldsymbol{\Sigma}$ denote the covariance matrix of $\boldsymbol{R}$. Then the weight vector for the tangency portfolio is given by*

$$\boldsymbol{w}_T = \frac{\boldsymbol{\Sigma}^{-1}(\boldsymbol{\mu} - \mu_f \mathbf{1})}{\mathbf{1}^T \boldsymbol{\Sigma}^{-1}(\boldsymbol{\mu} - \mu_f \mathbf{1})}$$

where μ_f denotes the expected return on the risk-free asset.

Proof. The Sharpe ratio of the portfolio based on weight vector $\boldsymbol{w}$ is given by

$$\frac{\boldsymbol{w}^T(\boldsymbol{\mu} - \mu_f \mathbf{1})}{(\boldsymbol{w}^T \boldsymbol{\Sigma} \boldsymbol{w})^{\frac{1}{2}}}. \tag{5.23}$$

To find the tangency portfolio, we need to find the weight vector that maximizes (5.23). Define $\boldsymbol{b} = \boldsymbol{\Sigma}^{1/2}\boldsymbol{w}$ and $\boldsymbol{d} = \boldsymbol{\Sigma}^{-1/2}(\boldsymbol{\mu} - \mu_f \mathbf{1})$. Then (5.23) can be written

$$\frac{\boldsymbol{b}^T \boldsymbol{d}}{(\boldsymbol{b}^T \boldsymbol{b})^{\frac{1}{2}}}. \tag{5.24}$$

By the Cauchy–Schwarz inequality,

$$\frac{\boldsymbol{b}^T \boldsymbol{d}}{(\boldsymbol{b}^T \boldsymbol{b})^{\frac{1}{2}}} \leq (\boldsymbol{d}^T \boldsymbol{d})^{\frac{1}{2}}$$

with equality if $\boldsymbol{b} = c\boldsymbol{d}$ for a scalar $c > 0$.

Therefore, (5.24) is maximized over $\boldsymbol{b}$ by $c\boldsymbol{d}$ for any $c > 0$. It follows that (5.23) is maximized when

$$\boldsymbol{\Sigma}^{-1/2}\boldsymbol{w} = c\boldsymbol{\Sigma}^{-\frac{1}{2}}(\boldsymbol{\mu} - \mu_f\mathbf{1})$$

for $c > 0$. That is, $\boldsymbol{w}_T$ is of the form

$$\boldsymbol{w}_T = c\boldsymbol{\Sigma}^{-1}(\boldsymbol{\mu} - \mu_f\mathbf{1})$$

for some $c > 0$. For the weights in $\boldsymbol{w}_T$ to sum to 1, we need

$$c = \frac{1}{\mathbf{1}^T\boldsymbol{\Sigma}^{-1}(\boldsymbol{\mu} - \mu_f\mathbf{1})}.$$

Note that

$$\mathbf{1}^T\boldsymbol{\Sigma}^{-1}(\boldsymbol{\mu} - \mu_f\mathbf{1}) = (\mu_{mv} - \mu_f)\mathbf{1}^T\boldsymbol{\Sigma}^{-1}\mathbf{1} > 0$$

using the fact that $\boldsymbol{\Sigma}$ is positive definite, along with the assumption that $\mu_f < \mu_{mv}$. □

The role of the tangency portfolio here is the same as in the $N = 2$ case: When constructing a portfolio consisting of risky assets plus the risk-free asset, all investors should use the tangency portfolio as their portfolio of risky assets.

Example 5.10 Consider the assets described in Example 5.6, with mean return vector stored in the variable `mu` and covariance matrix of the returns stored in `Sig`:

```
> mu
[1] 0.250 0.125 0.300
> Sig
     [,1]  [,2]  [,3]
[1,] 0.25 0.100 0.240
[2,] 0.10 0.160 0.096
[3,] 0.24 0.096 0.360
```

Suppose that the risk-free asset has return $\mu_f = 0.01$. Then the weight vector of the tangency portfolio is given by

```
> w_T<-solve(Sig, mu-0.01)/sum(solve(Sig, mu-0.01))
> w_T
[1] 0.424 0.148 0.428
```

This tangency portfolio has the Sharpe ratio

```
> sum(w_T*mu)/(w_T%*%Sig%*%w_T)^.5
     [,1]
[1,] 0.532
```

□

An Alternative Characterization of the Tangency Portfolio

The weight vector of the tangency portfolio, $\boldsymbol{w}_{\mathcal{T}}$, is the vector $\boldsymbol{w} \in \Re^N$ that maximizes the Sharpe ratio

$$\frac{\boldsymbol{w}^T(\boldsymbol{\mu} - \mu_f \mathbf{1})}{(\boldsymbol{w}^T \boldsymbol{\Sigma} \boldsymbol{w})^{\frac{1}{2}}}$$

subject to the restriction that $\boldsymbol{w}^T \mathbf{1} = 1$. However, there is another way to describe $\boldsymbol{w}_{\mathcal{T}}$ that is sometimes useful.

Note that, for any scalar $c > 0$,

$$\frac{(c\boldsymbol{w})^T(\boldsymbol{\mu} - \mu_f \mathbf{1})}{((c\boldsymbol{w})^T \boldsymbol{\Sigma} (c\boldsymbol{w}))^{\frac{1}{2}}} = \frac{\boldsymbol{w}^T(\boldsymbol{\mu} - \mu_f \mathbf{1})}{(\boldsymbol{w}^T \boldsymbol{\Sigma} \boldsymbol{w})^{\frac{1}{2}}}. \tag{5.25}$$

Consider two weight vectors $\boldsymbol{w}_1, \boldsymbol{w}_2$ such that

$$\boldsymbol{w}_j^T(\boldsymbol{\mu} - \mu_f \mathbf{1}) > 0, \quad j = 1, 2.$$

Using (5.25),

$$\frac{\boldsymbol{w}_1^T(\boldsymbol{\mu} - \mu_f \mathbf{1})}{(\boldsymbol{w}_1^T \boldsymbol{\Sigma} \boldsymbol{w}_1)^{\frac{1}{2}}} \geq \frac{\boldsymbol{w}_2^T(\boldsymbol{\mu} - \mu_f \mathbf{1})}{(\boldsymbol{w}_2^T \boldsymbol{\Sigma} \boldsymbol{w}_2)^{\frac{1}{2}}}$$

if and only if

$$\frac{(c\boldsymbol{w}_1)^T(\boldsymbol{\mu} - \mu_f \mathbf{1})}{((c\boldsymbol{w}_1)^T \boldsymbol{\Sigma} (c\boldsymbol{w}_1))^{\frac{1}{2}}} \geq \frac{(d\boldsymbol{w}_2)^T(\boldsymbol{\mu} - \mu_f \mathbf{1})}{((d\boldsymbol{w}_2)^T \boldsymbol{\Sigma} (d\boldsymbol{w}_2))^{\frac{1}{2}}} \tag{5.26}$$

for any $c > 0$ and $d > 0$. That is, in maximizing the Sharpe ratio, it is not necessary for the weights to sum to 1; recall that this fact was used in the proof of Proposition 5.7, when we found the vector in $\Re^N$ that maximizes the Sharpe ratio and then rescaled it to sum to 1.

Let

$$\bar{c} = \frac{1}{\boldsymbol{w}_1^T(\boldsymbol{\mu} - \mu_f \mathbf{1})}$$

and

$$\bar{d} = \frac{1}{\boldsymbol{w}_2^T(\boldsymbol{\mu} - \mu_f \mathbf{1})}.$$

Then taking $c = \bar{c}$ and $d = \bar{d}$ in (5.26), it follows that

$$\frac{\boldsymbol{w}_1^T(\boldsymbol{\mu} - \mu_f \mathbf{1})}{(\boldsymbol{w}_1^T \boldsymbol{\Sigma} \boldsymbol{w}_1)^{\frac{1}{2}}} \geq \frac{\boldsymbol{w}_2^T(\boldsymbol{\mu} - \mu_f \mathbf{1})}{(\boldsymbol{w}_2^T \boldsymbol{\Sigma} \boldsymbol{w}_2)^{\frac{1}{2}}}$$

if and only if

$$\frac{(\bar{c}\boldsymbol{w}_1)^T(\boldsymbol{\mu} - \mu_f \mathbf{1})}{((\bar{c}\boldsymbol{w}_1)^T \boldsymbol{\Sigma} (\bar{c}\boldsymbol{w}_1))^{\frac{1}{2}}} \geq \frac{(\bar{d}\boldsymbol{w}_2)^T(\boldsymbol{\mu} - \mu_f \mathbf{1})}{((\bar{d}\boldsymbol{w}_2)^T \boldsymbol{\Sigma} (\bar{d}\boldsymbol{w}_2))^{\frac{1}{2}}}. \tag{5.27}$$

Note that, by definition of $\bar{c}$ and $\bar{d}$,

$$(\bar{c}\boldsymbol{w}_1)^T(\mu - \mu_f \mathbf{1}) = 1$$

and

$$(\bar{d}\boldsymbol{w}_2)^T(\mu - \mu_f \mathbf{1}) = 1.$$

If $\boldsymbol{u}_1$ and $\boldsymbol{u}_2$ satisfy

$$\boldsymbol{u}_j^T(\mu - \mu_f \mathbf{1}) = 1, \quad j = 1, 2$$

then

$$\frac{\boldsymbol{u}_1^T(\mu - \mu_f \mathbf{1})}{(\boldsymbol{u}_1^T \boldsymbol{\Sigma} \boldsymbol{u}_1)^{\frac{1}{2}}} \geq \frac{\boldsymbol{u}_2^T(\mu - \mu_f \mathbf{1})}{(\boldsymbol{u}_2^T \boldsymbol{\Sigma} \boldsymbol{u}_2)^{\frac{1}{2}}}$$

if and only if

$$\boldsymbol{u}_1^T \boldsymbol{\Sigma} \boldsymbol{u}_1 \leq \boldsymbol{u}_2^T \boldsymbol{\Sigma} \boldsymbol{u}_2.$$

That is, we can describe the weight vector of the tangency portfolio as being proportional to the vector $\boldsymbol{u} \in \Re^N$ that minimizes

$$\boldsymbol{u}^T \boldsymbol{\Sigma} \boldsymbol{u} \tag{5.28}$$

subject to the restriction that $\boldsymbol{u}^T(\mu - \mu_f \mathbf{1}) = 1$; the proportionality constant is chosen so that the weights sum to 1.

Example 5.11 Consider the assets described in Example 5.6 and analyzed in Example 5.10; recall that in Example 5.10, the weights of the tangency portfolio were calculated using the expression $\boldsymbol{\Sigma}^{-1}(\mu - \mu_f \mathbf{1})/(\mathbf{1}^T \boldsymbol{\Sigma}^{-1}(\mu - \mu_f \mathbf{1}))$ and were shown to be

```
> w_T
[1] 0.424 0.148 0.428
```

In this example, we calculate the weights of the tangency portfolio using its characterization scalar multiple of the vector that minimizes (5.28) subject to the restriction that $\boldsymbol{u}^T(\mu - \mu_f \mathbf{1}) = 1$.

The return mean vector and covariance matrix are stored in R variables `mu` and `Sig`, respectively; the risk-free rate of return is taken to be 0.01.

To use `solve.QP` to minimize (5.28) subject to the restriction given previously, we construct the constraint matrix `A.tan` by

```
> A.tan<-matrix(mu - 0.01, 3, 1)
```

and use the commands

```
> tan1<-solve.QP(Dmat=2*Sig, dvec=c(0,0,0), Amat=A.tan, bvec=1,
+ meq=1)
> tan1$solution/sum(tan1$solution)
[1] 0.424 0.148 0.428
```

Note that this result matches the one obtained in Example 5.10. □

5.8 Portfolio Constraints

The optimal portfolios described in this chapter are all derived under the assumption that the only restriction on the weight vector $\boldsymbol{w}$ is that the weights sum to 1, $\boldsymbol{w}^T\mathbf{1} = 1$. In practice, analysts often place constraints on the vector of weights available to the investor.

Such constraints often have only minor effects on the basic approaches described in this chapter; for instance, the minimum risk frontier still consists of those portfolios with the smallest risk for a given expected return, but now such portfolios must satisfy the constraints.

On the other hand, constraints may have large effects on specific results and on the numerical solutions to the optimization problems used to calculate the weight vectors of optimal portfolios. For instance, Corollary 5.3, which states that if the portfolios with weight functions $\boldsymbol{w}_1$ and $\boldsymbol{w}_2$ are on the minimum-risk frontier, then the portfolio with weight vector $z\boldsymbol{w}_1 + (1-z)\boldsymbol{w}_2$, for any z, is on the minimum-risk frontier, may not hold since the portfolio with weight vector $z\boldsymbol{w}_1 + (1-z)\boldsymbol{w}_2$ may not satisfy the constraints.

In this section, we consider the calculation of the weight vectors of the risk-averse and tangency portfolios, when those weight vectors are subject to certain types of constraints.

First, consider determination of the weight vector that maximizes the risk-aversion criterion function

$$\boldsymbol{w}^T\mu - \frac{\lambda}{2}\boldsymbol{w}^T\boldsymbol{\Sigma}\boldsymbol{w} \tag{5.29}$$

subject to some commonly used constraints. A constrained maximization problem of this form is another example of a quadratic programming problem. The R function `solve.QP` that was used in Example 5.9 to find the weight vectors of the risk-averse portfolio subject only to the constraint that the weights sum to 1 can be used to solve many constrained minimization problems.

Recall that the function `solve.QP` in the package `quadprog` can be used to maximize an objective function of the form

$$\boldsymbol{d}^T\boldsymbol{x} - \frac{1}{2}\boldsymbol{x}^T\boldsymbol{D}\boldsymbol{x} \tag{5.30}$$

with respect to the vector $\boldsymbol{x}$, which is subject to equality and inequality constraints on $\boldsymbol{x}$. The constraints are of the form

$$\boldsymbol{A}^T\boldsymbol{x}(=,\geq)\boldsymbol{b}$$

for a matrix $\boldsymbol{A}$ and vector $\boldsymbol{b}$, where $(=,\geq)$ denotes either equality or inequality of the form $\geq$ on a component-wise basis.

Inequality constraints may be included in `solve.QP` by specifying appropriately the arguments `Amat`, corresponding to the matrix $\boldsymbol{A}$ described earlier, and `bvec`, corresponding to the vector $\boldsymbol{b}$. The argument `meq` of `solve.QP` indicates the number of equality constraints; these must correspond to the first columns of `Amat`.

The following example illustrates in detail the maximization of the risk-aversion criterion function subject to the constraint that the weight on asset j, w_j, is nonnegative for each j. That is, the portfolio cannot contain a short position on any asset. We will write this constraint as $\boldsymbol{w} \geq \boldsymbol{0}$.

Example 5.12 Consider the assets described in Example 5.6, with mean return vector $(0.25, 0.125, 0.3)$ and covariance matrix given by

$$\boldsymbol{\Sigma} = \begin{pmatrix} 0.25 & 0.1 & 0.24 \\ 0.1 & 0.16 & 0.096 \\ 0.24 & 0.096 & 0.36 \end{pmatrix}.$$

Consider maximization of the risk-aversion criterion function (5.29) as discussed in Example 5.8.

For $\lambda = 0.5$, the weight vector $\boldsymbol{w}_\lambda$ is given by

```
> w_mv + 2*vbar
[1]  0.632 -0.501  0.869,
```

which includes a substantial short position on asset 2. Hence, we might consider maximizing (5.29) when $\lambda = 0.5$, subject to the constraint that all asset weights are nonnegative.

The arguments to `solve.QP` are `Dmat` and `dvec`, which specify the objective function as described in (5.30), and `Amat`, `bvec`, and `meq`, which specify the equality and inequality constraints.

Thus, to maximize the risk-aversion criterion function, `dvec` is taken to be $\boldsymbol{\mu}$, the vector of asset means, and `Dmat` is taken to be $\lambda\boldsymbol{\Sigma}$, where $\boldsymbol{\Sigma}$ is the covariance matrix of the asset returns. The constraints in our problem are

$$\boldsymbol{1}^T\boldsymbol{w} = 1, \quad w_1 \geq 0, \quad w_2 \geq 0, \quad w_3 \geq 0;$$

thus,

$$\boldsymbol{A}^T = \begin{pmatrix} 1 & 1 & 1 \\ 1 & 0 & 0 \\ 0 & 1 & 0 \\ 0 & 0 & 1 \end{pmatrix}$$

and

$$\boldsymbol{b} = (1, 0, 0, 0)^T.$$

The argument `meq` indicates the number of equality constraints; thus, in this example, `meq = 1`, indicating that the constraints are given by

$$\begin{pmatrix} 1 & 1 & 1 \\ 1 & 0 & 0 \\ 0 & 1 & 0 \\ 0 & 0 & 1 \end{pmatrix} \boldsymbol{x} \begin{matrix} = \\ \geq \\ \geq \\ \geq \end{matrix} \begin{pmatrix} 1 \\ 0 \\ 0 \\ 0 \end{pmatrix}.$$

Therefore, the R commands needed to solve this constrained optimization problem are as follows:

```
> library(quadprog)
> A<-cbind(c(1,1,1), diag(3))
> t(A)
     [,1] [,2] [,3]
[1,]    1    1    1
[2,]    1    0    0
[3,]    0    1    0
[4,]    0    0    1
> b<-c(1, 0, 0, 0)
> qpsol<-solve.QP(Dmat=(.5)*Sig, dvec=mu, Amat=A, bvec=b, meq=1)
```

Note that `diag(3)` returns a 3×3 identity matrix and `cbind` combines two vectors or matrices by the columns; when the vector `c(1,1,1)` is used in this context, it is interpreted as a column vector. The weight vector that maximizes the objective function is the component `$solution` of the result of the function `solve.QP`; therefore, for this problem, the weight vector is

```
> qpsol$solution
[1] 0.154 0.000 0.846
```

This weight vector maximizes the risk-aversion criterion function based on $\lambda = 0.5$, subject to the restriction that the weights are nonnegative.

Thus, the expected return of the portfolio that minimizes the risk-aversion criterion function with $\lambda = 0.5$, subject to the no-short-positions constraint, is

```
> sum(qpsol$solution*mu)
[1] 0.292
```

and the return standard deviation is

```
> ((qpsol$solution)%*%Sig%*%qpsol$solution)^.5
      [,1]
[1,] 0.571
```

These may be compared to the mean and standard deviation of the return of the optimizing portfolio that is not subject to the constraint

```
> sum((w_mv + 2*vbar)*mu)
[1] 0.356
> ((w_mv + 2*vbar)%*%Sig%*%(w_mv + 2*vbar))^.5
      [,1]
[1,] 0.727
```

Thus, the no-short-positions constraint leads not only to a lower expected return but also to a lower risk. In terms of the objective function (5.29), for the constrained solution, the value is

```
> 1.17 - (0.5/2)*(1.14^2)
[1] 0.210
```

while for the unconstrained solution, the value is

```
> 1.42 - (0.5/2)*(1.45^2)
[1] 0.224
```

Thus, as expected, the unconstrained optimal value is higher than the constrained optimal value. □

Holding Constraints

Another example of commonly used constraints are *holding constraints* of the form $L_j \leq w_j \leq U_j$, $j = 1, \ldots, N$, where L_j and U_j are lower and upper bounds, respectively, on the proportion of the investment in asset j. These may also be handled using the function `solve.QP`.

Example 5.13 Consider the assets analyzed in Example 5.12. The weight vector that maximizes the risk-aversion criterion function (5.29) for $\lambda = 1$ is given by

```
> w_mv + vbar
[1] 0.437 0.106 0.457
```

Suppose we would like our portfolio to allocate between 25% and 75% of our investment to each asset; that is, we would like to enforce the constraints

$$0.25 \leq w_j \leq 0.75 \quad \text{for} \quad j = 1, 2, 3.$$

To express these constraints as lower bounds on functions of the weights, we write them as

$$w_j \geq 0.25, \quad -w_j \geq -0.75, \quad j = 1, 2, 3.$$

Thus, the matrix `Amat` used in `solve.QP` is taken to be

```
> Ah<-cbind(c(1,1,1), diag(3), -1*diag(3))
> t(Ah)
     [,1] [,2] [,3]
[1,]    1    1    1
[2,]    1    0    0
[3,]    0    1    0
[4,]    0    0    1
[5,]   -1    0    0
[6,]    0   -1    0
[7,]    0    0   -1
```

and the vector bvec is taken to be

```
> bh<-c(1, 0.25, 0.25, 0.25, -0.75, -0.75, -0.75)
```

The solution to the maximization problem is then given by

```
> w1<-solve.QP(Sig, mu, Amat=Ah, bvec=bh, meq=1)$solution
> w1
[1] 0.30 0.25 0.45
```

The corresponding portfolio has a mean return 0.241 and return standard deviation of 0.455. These can be compared to the mean return and return standard deviation of the unconstrained optimal portfolio, given by 0.260 and 0.489, respectively. □

Other types of constraints can be handled in a similar manner. For example, suppose that $N > 5$; we might want the total of the portfolio weights on assets 1 through 5 to be at least 0.50, that is,

$$w_1 + w_2 + w_3 + w_4 + w_5 \geq 0.50.$$

Or if we want the weight on asset 1 to be at least as great as the weight on asset 2, we may use the constraint

$$w_1 - w_2 \geq 0.$$

By properly choosing the argument Amat, solve.QP can solve many constrained optimization problems of this type.

Maximizing the Sharpe Ratio under Nonnegativity Constraints

In Example 5.11, it was shown that the Sharpe ratio may be maximized by minimizing the variance of the portfolio return subject to a restriction on the portfolio mean return; therefore, solve.QP may be used to calculate the weights of the tangency portfolio. That approach is based on the following property of Sharpe ratios:

$$\frac{\boldsymbol{w}_1^T(\boldsymbol{\mu} - \mu_f\mathbf{1})}{(\boldsymbol{w}_1^T\boldsymbol{\Sigma}\boldsymbol{w}_1)^{\frac{1}{2}}} \geq \frac{\boldsymbol{w}_2^T(\boldsymbol{\mu} - \mu_f\mathbf{1})}{(\boldsymbol{w}_2^T\boldsymbol{\Sigma}\boldsymbol{w}_2)^{\frac{1}{2}}}$$

if and only if

$$\frac{(c\boldsymbol{w}_1)^T(\boldsymbol{\mu} - \mu_f\mathbf{1})}{((c\boldsymbol{w}_1)^T\boldsymbol{\Sigma}(c\boldsymbol{w}_2))^{\frac{1}{2}}} \geq \frac{(d\boldsymbol{w}_2)^T(\boldsymbol{\mu} - \mu_f\mathbf{1})}{((d\boldsymbol{w}_2)^T\boldsymbol{\Sigma}(d\boldsymbol{w}_2))^{\frac{1}{2}}} \tag{5.31}$$

for any $c > 0$ and $d > 0$.

Therefore, the same approach can be used to find the portfolio that maximizes the Sharpe ratio under constraints, provided that a weight vector $\boldsymbol{w}$

satisfies the constraints if and only if cw satisfies the constraints for any $c > 0$. Note that this condition is satisfied for nonnegativity constraints of the form $w \geq \mathbf{0}$; however, it is not satisfied for other types of constraints, such as the holding constraints considered in Example 5.13.

The details are described in the following example.

Example 5.14 Consider the set of four assets used in Example 5.1 and let `mu` and `Sigma` denote the R variables containing the mean vector and covariance matrix, respectively, of the returns,

```
> mu
[1] 0.10 0.20 0.05 0.10
> Sigma
     [,1] [,2] [,3] [,4]
[1,] 0.05 0.01 0.02 0.00
[2,] 0.01 0.10 0.05 0.02
[3,] 0.02 0.05 0.20 0.10
[4,] 0.00 0.02 0.10 0.20
```

and take the risk-free rate to be 0.01.

Then the weight vector of the tangency portfolio is given by

```
> solve(Sigma, mu-0.01)/sum(solve(Sigma, mu-0.01))
[1]  0.482  0.559 -0.223  0.182
```

Alternatively, we can calculate this weight vector using `solve.QP`, as described in Example 5.11.

```
> wT<-solve.QP(Dmat=2*Sigma, dvec=rep(0,4), Amat=cbind(mu-0.01),
+ bvec=1, meq=1)$solution
> wT/sum(wT)
[1]  0.482  0.559 -0.223  0.182
```

We can include nonnegativity constraints in the calculation by taking the argument `Amat` to be the matrix

```
> A.shrp<-cbind(mu-0.01, diag(4))
> t(A.shrp)
     [,1] [,2] [,3] [,4]
[1,] 0.09 0.19 0.04 0.09
[2,] 1.00 0.00 0.00 0.00
[3,] 0.00 1.00 0.00 0.00
[4,] 0.00 0.00 1.00 0.00
[5,] 0.00 0.00 0.00 1.00
```

taking `bvec` to be the vector

```
> b.shrp<-c(1, rep(0,4))
> b.shrp
[1] 1 0 0 0 0
```

and taking `meq=1`. Thus, the portfolio with nonnegative weights that maximizes the Sharpe ratio is given by

```
> wT.nn<-solve.QP(Dmat=2*Sigma, dvec=rep(0,4), Amat=A.shrp,
+  bvec=b.shrp, meq=1)$solution
> wT.nn/sum(wT.nn)
[1] 0.425 0.494 0.000 0.081
```

□

Any constraint of the form $\boldsymbol{a}^T\boldsymbol{w} = 0$ or $\boldsymbol{a}^T\boldsymbol{w} \geq 0$, where $\boldsymbol{a}$ is a given vector in $\Re^N$, can be handled by the same method. For example, if $N = 5$, the constraint $w_1 + w_2 + w_3 \geq w_4 + w_5$ satisfies the condition that the constraint holds if and only if $cw_1 + cw_2 + cw_3 \geq cw_4 + cw_5$ for any $c > 0$. Hence, we may use the same general approach used in Example 5.14 to find the weight vector that maximizes the Sharpe ratio subject to this constraint.

5.9 Suggestions for Further Reading

Efficient portfolio theory is one of the cornerstones of quantitative finance and financial engineering, and it is discussed in many texts in these fields; see, for example, Campbell et al. (1997, Section 5.2), Francis and Kim (2013, Chapter 7), and Qian et al. (2007, Chapter 2) for useful introductions to this area. Hult et al. (2012, Chapter 4) provide a detailed treatment of portfolio selection under the risk-aversion criterion. Merton (1972) offers clear proofs of many of the fundamental results of efficient portfolio theory and Markowitz (1987) provides a comprehensive treatment of the subject, although at a more mathematically advanced level. Michaud (1989) presents an interesting discussion of the practical usefulness of the theory; see also Jobson and Korkie (1981).

The method of Lagrange multipliers is a useful technique for solving many constrained optimization problems; see, for example, Stewart (2015, Section 14.8) and Larson and Edwards (2014, Section 13.10) for further discussion.

5.10 Exercises

1. Consider a three-dimensional return vector $\boldsymbol{R}$ with mean vector given by $(0.04, 0.03, 0.05)$ and covariance matrix given by

$$\begin{pmatrix} 0.05 & 0.05 & 0.025 \\ 0.05 & 0.10 & 0.08 \\ 0.025 & 0.08 & 0.075 \end{pmatrix}.$$

Let R_{p1} denote the return on the portfolio with weight vector $(1/3, 1/3, 1/3)$ and let R_{p2} denote the return on the portfolio with weight vector $(0.4, 0.4, 0.2)$.

a. Find the mean and standard deviation of R_{p1}; see Example 5.1.

b. Find the mean and standard deviation of R_{p2}.

c. Find the correlation of R_{p1} and R_{p2}.

d. Based on these results, is one of the portfolios preferable to the other? Why or why not?

2. Consider a four-dimensional return vector $\boldsymbol{R}$ with mean vector given by $(0.02, 0.10, 0.05, 0.06)$ and covariance matrix given by

$$\begin{pmatrix} 0.02 & 0.01 & 0.01 & 0 \\ 0.01 & 0.05 & 0.02 & 0 \\ 0.01 & 0.02 & 0.03 & 0 \\ 0 & 0 & 0 & 0.04 \end{pmatrix}.$$

a. Using the computational method described in Example 5.5, find the portfolio on the minimum-risk frontier with a mean return of 0.05. Find the return standard deviation of the portfolio.

b. Repeat Part (a) for mean returns of 0.06 and 0.07.

c. Based on these results, is it possible to say with certainty that any of these portfolios is not on the efficient frontier? Why or why not? If it is possible, which ones are not on the efficient frontier?

3. (Corollary 5.1) For $j = 1, 2$, let $\hat{R}_{pj}$ denote the return on the portfolio with weight vector $\hat{\boldsymbol{w}}_j$. Suppose that both portfolios are on the minimum-risk frontier and that $\mathrm{E}(\hat{R}_{p1}) = \mathrm{E}(\hat{R}_{p2})$. Show that $\hat{\boldsymbol{w}}_1 = \hat{\boldsymbol{w}}_2$.

4. (Corollary 5.2) Let $\hat{R}_p$ denote the return on a portfolio on the minimum-risk frontier and let R_{p1}, R_{p2} denote the returns on two portfolios having the same expected return. Show that

$$\mathrm{Cov}(\hat{R}_p, R_{p1}) = \mathrm{Cov}(\hat{R}_p, R_{p2}).$$

5. Consider a five-dimensional return vector $\boldsymbol{R}$ with mean vector given by $(0.25, 0.20, 0.30, 0.275, 0.15)$ and covariance matrix given by

$$\begin{pmatrix} 1.0 & 0.40 & 0.60 & 0.5 & 0.30 \\ 0.4 & 0.70 & 0.50 & 0.4 & 0.25 \\ 0.6 & 0.50 & 1.30 & 0.6 & 0.35 \\ 0.5 & 0.40 & 0.60 & 1.0 & 0.30 \\ 0.3 & 0.25 & 0.35 & 0.3 & 0.50 \end{pmatrix}.$$

Find the weight vector of the minimum-variance portfolio; see Example 5.6.

6. Let R_{mv} denote the return on the minimum-variance portfolio and let R_0 denote the return on a zero-investment portfolio. Show that

$$\mathrm{Cov}(R_{mv}, R_0) = 0.$$

Does the converse hold? That is, suppose that a portfolio return R_p satisfies

$$\text{Cov}(R_p, R_0) = 0$$

for the return R_0 on any zero-investment portfolio. Does it follow that R_p is the return on the minimum-variance portfolio? Why or why not?

7. Let R_{mv} denote the return on the minimum-variance portfolio and let R_p denote the return on another portfolio. Find an expression for

$$\frac{\text{Var}(R_p)}{\text{Var}(R_{mv})}$$

in terms of the correlation of R_p and R_{mv}.

8. Consider a market consisting of N assets and let $\boldsymbol{R}$ denote the vector of asset returns; let $\boldsymbol{\mu}$ denote the vector of mean returns and let $\boldsymbol{\Sigma}$ denote the covariance matrix of $\boldsymbol{R}$.

Let $\hat{\boldsymbol{w}}$ denote the weight vector of a portfolio on the efficient frontier and let $\tilde{\boldsymbol{w}}$ denote the weight vector of another portfolio that is not on the efficient frontier. Suppose that the two portfolios have the same mean return and let $\gamma = \text{Var}(\hat{\boldsymbol{w}}^T\boldsymbol{R})/\text{Var}(\tilde{\boldsymbol{w}}^T\boldsymbol{R})$ denote the ratio of the variances of the portfolio returns; note that, since the portfolio with weight vector $\hat{\boldsymbol{w}}$ is on the efficient frontier, $0 < \gamma < 1$.

Find the correlation of returns on the two portfolios as a function of γ.

9. Let λ_1, λ_2 be nonnegative real numbers and let R_j denote the return on the risk-averse portfolio with the risk-aversion parameter λ_j, $j = 1, 2$.

Consider the portfolio with a return of the form

$$R_p = wR_1 + (1 - w)R_2$$

where $0 < w < 1$.

Show that R_p is the return on the risk-averse portfolio with risk-aversion parameter λ_p and find λ_p in terms of λ_1, λ_2, and w.

10. Suppose that the risk-aversion criterion is based on excess returns rather than on standard returns. Therefore, for a given value of $\lambda > 0$, let $\tilde{\boldsymbol{w}}_\lambda$ denote the maximizer of

$$\boldsymbol{w}^T(\boldsymbol{\mu} - \mu_f\boldsymbol{1}) - \frac{\lambda}{2}\boldsymbol{w}^T\boldsymbol{\Sigma}\boldsymbol{w}$$

over $\boldsymbol{w} \in \Re^N$, subject to the restriction that $\boldsymbol{1}^T\boldsymbol{w} = 1$.

Find $\tilde{\boldsymbol{w}}_\lambda$ and show how it relates to $\boldsymbol{w}_\lambda$, the weight vector of the risk-averse portfolio based on parameter λ.

11. Consider a set of three assets with the mean return vector and return covariance matrix as given in Exercise 1.

    Using the approach described in Example 5.8, find $\boldsymbol{w}_{mv}$, the weight vector of the minimum-variance portfolio, and $\bar{\boldsymbol{v}}$, the weight vector of the zero-investment portfolio given in the statement of Proposition 5.5. Use those results to give the weight vectors of the risk-averse portfolio with parameters $\lambda = 1$ and $\lambda = 5$.

12. Consider a set of three assets with the mean return vector and return covariance matrix as given in Exercise 1.

    Find the mean and variance of the return on the risk-averse portfolio based on a risk-aversion parameter λ.

13. Consider the set of five assets with the mean return vector and return covariance matrix as given in Exercise 5.

    Using the R function, `solve.QP`, as in Example 5.9, find the weight vector of the risk-averse portfolio based on the risk-aversion parameter $\lambda = 1$. Find the mean return and return standard deviation of the portfolio.

14. Consider a vector of asset returns with mean vector $\boldsymbol{\mu}$ and covariance matrix $\boldsymbol{\Sigma}$. Find an expression for the Sharpe ratio of the tangency portfolio in terms of $\boldsymbol{\mu}$, $\boldsymbol{\Sigma}$, and μ_f.

15. Consider a market consisting of N assets and let $\boldsymbol{R}$ denote the vector of asset returns. Let $\boldsymbol{\mu}$ denote the vector of mean returns and let $\boldsymbol{\Sigma}$ denote the covariance matrix of $\boldsymbol{R}$. Let μ_f denote the return on the risk-free asset.

    Find conditions on $\boldsymbol{\mu}$ under which the minimum-variance portfolio is the same as the tangency portfolio.

16. Use the general expression for the weight vector of the tangency portfolio given in Section 5.7 to derive the expression for the weight vector of the tangency portfolio for the $N = 2$ case given in Section 4.6.

17. Let $\boldsymbol{w}_{\mathcal{T}}$ denote the weight vector of the tangency portfolio and let $\boldsymbol{w}_{\lambda}$ denote the weight vector of the risk-averse portfolio based on the risk-aversion parameter λ.

    Show that there exists $\lambda_{\mathcal{T}} > 0$ such that $\boldsymbol{w}_{\mathcal{T}} = \boldsymbol{w}_{\lambda_{\mathcal{T}}}$ and give an expression for $\lambda_{\mathcal{T}}$.

18. Consider a set of five assets with the mean return vector and return covariance matrix as given in Exercise 5. Assume that the risk-free rate of return is $\mu_f = 0.01$.

    Find $\boldsymbol{w}_{\mathcal{T}}$, the weight vector of the tangency portfolio; see Example 5.10.

19. Consider a set of six assets with return vector $\boldsymbol{R}$. Suppose that the mean vector of $\boldsymbol{R} - R_f\mathbf{1}$ is given by $(0.04, 0.08, 0.02, 0.10, 0.03, 0.06)$ and that the covariance matrix of $\boldsymbol{R}$ is given by

$$\begin{pmatrix} 0.20 & 0.02 & 0.03 & 0.04 & 0.05 & 0.06 \\ 0.02 & 0.50 & 0.06 & 0.08 & 0.10 & 0.12 \\ 0.03 & 0.06 & 0.20 & 0.12 & 0.15 & 0.18 \\ 0.04 & 0.08 & 0.12 & 0.80 & 0.20 & 0.24 \\ 0.05 & 0.10 & 0.15 & 0.20 & 1.20 & 0.30 \\ 0.06 & 0.12 & 0.18 & 0.24 & 0.30 & 0.80 \end{pmatrix}.$$

    Find $\boldsymbol{w}_T$, the weight vector of the tangency portfolio; see Example 5.10.

20. Consider the set of five assets with the mean return vector and return covariance matrix as given in Exercise 5. Assume that the risk-free rate of return is $\mu_f = 0.01$.

    Find the Sharpe ratio of the tangency portfolio and compare it to the Sharpe ratios of the equally-weighted portfolio and the minimum-variance portfolio.

21. Let $\boldsymbol{w}_\lambda$ denote the weight vector of the risk-averse portfolio with parameter λ, as given by Proposition 5.5. Write $\boldsymbol{w}_\lambda$ in terms of $\boldsymbol{w}_{mv}$, the weight vector of the minimum-variance portfolio, and $\boldsymbol{w}_T$, the weight vector of the tangency portfolio.

22. For the three assets with the mean return vector and return covariance matrix given in Exercise 1, determine the weight vector that maximizes the risk-aversion criterion function with parameter $\lambda = 5$, subject to the constraint that all weights are nonnegative; see Example 5.12.

    Find the mean and variance of the return on the resulting portfolio and compare these to the mean and variance of the return on the risk-averse portfolio based on $\lambda = 5$.

23. Consider the set of five assets with the mean return vector and return covariance matrix as given in Exercise 5.

    Suppose we want to find the portfolio weights that maximize the risk-aversion criterion function with parameter $\lambda = 1$ subject to the constraints that the portfolio weights are all nonnegative and that the sum of the weights given to assets 1 and 2 is equal to the sum of the weights given to assets 4 and 5. That is, in terms of the weight vector $\boldsymbol{w} = (w_1, w_2, w_3, w_4, w_5)^T$, we want to enforce the constraints that $w_j \geq 0$ for $j = 1, 2, \ldots, 5$ and that

$$w_1 + w_2 = w_4 + w_5.$$

    Find the optimal weight vector and calculate the mean return and return standard deviation of the corresponding portfolio.

Compare these to the mean return and return variance of the unconstrained risk-averse portfolio with parameter $\lambda = 1$ found in Exercise 13.

24. Consider the set of six assets with the mean excess return vector and return covariance matrix as specified in Exercise 19. Using the approach described in Example 5.14, find the weight vector of the portfolio that maximizes the Sharpe ratio subject to the restriction that all weights are nonnegative.

25. Consider the set of six assets with the mean excess return vector and return covariance matrix as specified in Exercise 19. Find the weight vector of the portfolio that maximizes the Sharpe ratio subject to the restriction that the weight vector $(w_1, w_2, w_3, w_4, w_5, w_6)^T$ satisfies

$$w_1 + w_2 + w_3 \leq w_4 + w_5 + w_6.$$

6

Estimation

6.1 Introduction

The portfolio theory developed in the previous chapters is based on properties of the distribution of asset returns, specifically their means, standard deviations, and correlations. Of course, in practice, these parameters are all unknown and must be estimated.

The simplest approach to estimating such parameters is to use the corresponding sample versions based on historical data; for instance, we can estimate a mean return by the sample mean of a series of observed returns. Such methods often work fairly well, particularly when analyzing just a few assets. However, in many cases, better estimators are available.

In this chapter, several methods of estimating the parameters needed for portfolio analysis are presented. Other methods, which build on those discussed here, are covered in the following chapters.

6.2 Basic Sample Statistics

The most straightforward approach to estimating parameters of return distributions is to use empirical estimators based on observed returns. Let $R_{j,t}$ denote the return on asset j for period t; here we use monthly returns, although the methods can be applied to other return intervals as well. We assume that T periods of data are available so that $t = 1, 2, \ldots, T$. For $R_{j,t}$, we generally use returns that have been adjusted for dividends and stock splits, as discussed in Section 2.3 and as provided by services such as Yahoo Finance.

When obtaining return data, we must choose the *observation period*, which is sometimes called the *sampling horizon*. The observation period refers to the total time period over which data are collected. The choice of the observation period must balance two competing considerations. A longer period gives us more data points that may yield more accurate estimators. However, this higher accuracy is only available if the parameters being estimated are constant over the period being sampled. For instance, it may be tempting to use 50 years of monthly returns to estimate the mean return on a stock. However, it is unlikely that the return on a stock in 1966 will be relevant to investment decisions made in 2016. From a statistical point-of-view, using a long observation period may lead to a bias in the estimator due to the fact that the parameters are not constant over that period. In practice, observation periods in the range of 3–10 years are typically used. With shorter return intervals, shorter observation periods are sometimes used.

We assume that, over the observation period, the parameters of interest are constant. For instance, consider asset j, for some $j = 1, 2, \ldots, N$; we assume that for each $t = 1, 2, \ldots, T$, $\text{E}(R_{j,t}) = \mu_j$ and $\text{Var}(R_{j,t}) = \sigma_j^2$. Returns on an asset in different time periods are assumed to be uncorrelated so that for $t, s,\ t \neq s$, $\text{Cov}(R_{j,t}, R_{j,s}) = 0$; that is, for a given asset, the sequence of returns consists of uncorrelated random variables all with the same mean and standard deviation. Let ρ_{jk} denote the correlation of $R_{j,t}, R_{k,t}$ for any $j, k = 1, 2, \ldots, N$, $j \neq k$; thus, we assume that returns on different assets in the same time period are correlated. Returns on different assets in different time periods are assumed to be uncorrelated: $\text{Cov}(R_{j,t}, R_{k,s}) = 0$ for $t \neq s$.

In many cases, we are interested in the mean excess return on an asset. Let $R_{f,t}$ denote the return on the risk-free asset at time t. It is important to realize that, even though the return on the risk-free asset has zero variance, the risk-free rate itself changes over time. When estimating the mean excess return on asset j, we assume that the excess returns $R_{j,t} - R_{f,t}$, $t = 1, 2, \ldots, T$ are uncorrelated random variables, each with a given expected value, which we denote by $\mu_j - \mu_f$.

Estimation of Return Means and Standard Deviations

The simplest estimator of μ_j, the expected return on asset j, is given by the sample mean of $R_{j,1}, R_{j,2}, \ldots, R_{j,T}$:

$$\bar{R}_j = \frac{1}{T}\sum_{t=1}^{T} R_{j,t}.$$

The corresponding estimator of the mean excess return, $\mu_j - \mu_f$, is given by

$$\frac{1}{T}\sum_{t=1}^{T}(R_{j,t} - R_{f,t}) = \bar{R}_j - \bar{R}_f$$

where

$$\bar{R}_f = \frac{1}{T}\sum_{t=1}^{T} R_{f,t}.$$

Now consider estimation of the return standard deviations or the return variances. Note that here we have a choice—we have defined σ_j^2 to be the variance of $R_{j,t}$; however, because the return on the risk-free asset has zero variance, σ_j^2 is also the variance of $R_{j,t} - R_{f,t}$. Therefore, to estimate σ_j^2, we may use either the sample variance of the returns

$$R_{j,1}, R_{j,2}, \ldots, R_{j,T}$$

or the sample variance of the excess returns

$$R_{j,1} - R_{f,1}, R_{j,2} - R_{f,2}, \ldots, R_{j,T} - R_{f,T}.$$

Because $R_{f,t}$ changes with t, these estimates will differ.

To some extent, the choice depends on the context. For instance, if we are simply analyzing the properties of the returns on an asset, as we did in Chapter 2, the sample variance of the returns is generally appropriate; on the other hand, if we are using the results to construct an efficient portfolio, as we did in Chapters 4 and 5, then excess returns are often relevant and, in those cases, it would be appropriate to use the sample variance of the excess returns.

Both approaches are used here; of course, given a result for an estimator based on excess returns, it is a simple matter to describe an analogous estimator based on standard returns and vice versa. Note that if values in the series $R_{f,t}$, $t = 1, 2, \ldots, T$ are approximately constant, which is often the case, there will be only minor differences in the two estimates.

The sample variance of the excess returns on asset j is given by

$$S_j^2 = \frac{1}{T-1} \sum_{t=1}^{T} \{R_{j,t} - R_{f,t} - (\bar{R}_j - \bar{R}_f)\}^2;$$

the sample standard deviation S_j is simply the square root of the sample variance, $S_j = \sqrt{S_j^2}$.

Example 6.1 Consider returns on Wal-Mart stock. In Chapter 2, we calculated the monthly returns for the five-year period from January 2010 to December 2014, which we placed in the variable `wmt.m.ret`. Therefore, to calculate the excess returns, we need five years of monthly returns on a risk-free asset.

As noted in Chapter 4, a standard choice for the risk-free return is the return on a three-month Treasury Bill. This can be obtained from the Federal Reserve website,

http://www.federalreserve.gov/releases/h15/data.htm

which contains an Excel spreadsheet with historical values dating back to 1934, found under the "Treasury Bills (secondary market)" heading.

Once the spreadsheet is downloaded, the simplest way to use the data in R is to copy them from Excel by highlighting the relevant values, copying them to the "clipboard" (i.e., by highlighting the cells and clicking "copy"), and using the `scan` function, which reads the data into a vector:

```
> rff<-scan(file="clipboard")
Read 60 items
> head(rff)
[1] 0.06 0.11 0.15 0.16 0.16 0.12
```

Alternatively, the file name may be specified in `scan`; it is often convenient to use `file=file.choose()`, which allows the user to select the file from a menu.

The values in the variable `rff` are annual returns as percentages, which must be converted to proportional monthly returns using the command

```
> rfree<-(1 + rff/100)^(1/12)-1
> head(rfree)
[1] 4.999e-05 9.162e-05 1.249e-04 1.332e-04 1.332e-04 9.995e-05
```

For example, 12 months of compounded returns at a rate of 4.999×10^{-5} yields a yearly rate of 0.06%:

```
> (1 + 4.999e-05)^12
[1] 1.0006
```

The excess returns for Wal-Mart stock can then be obtained as the difference of the Wal-Mart returns and the risk-free rate:

```
> wmt.ex<-wmt.m.ret-rfree
> mean(wmt.ex)
[1] 0.01088
> sd(wmt.ex)
[1] 0.04416
```

Note that the standard deviation of the standard Wal-Mart returns is also approximately 0.04416; retaining more significant digits shows that the standard deviation of the excess returns is 0.0441599 while the standard deviation of the standard returns is 0.0441572, a difference that is meaningless in practice. □

Sample Covariances and Correlations

The same basic approach used to estimate return means and standard deviations may be used to estimate a covariance or correlation of the returns on different assets. Using the excess returns, the sample covariance of the returns on assets j and k is given by

$$S_{jk} = \frac{1}{T-1}\sum_{t=1}^{T}\{R_{j,t} - R_{f,t} - (\bar{R}_j - \bar{R}_f)\}\{R_{k,t} - R_{f,t} - (\bar{R}_k - \bar{R}_f)\}.$$

The correlation of the returns on asset j and k can be estimated by the sample correlation, sometimes called the sample correlation coefficient,

$$\hat{\rho}_{jk} \equiv \frac{S_{jk}}{S_j S_k}.$$

Note that the sample correlation has many of the same properties as the correlation between two random variables; for example, it takes values in the interval $[-1, 1]$ and it is not affected by linear transformations of the data.

Example 6.2 Let `sbux.ex` denote five years of excess monthly returns on Starbucks stock (symbol SBUX), calculated using the same procedure used for `wmt.ex` in Example 6.1. The sample covariance between the excess returns on Wal-Mart and Starbucks stock may be calculated using the `cov` function:

```
> cov(wmt.ex, sbux.ex)
          [,1]
[1,] 0.0002081
```

The corresponding sample correlation may be calculated using the `cor` function:

```
> cor(wmt.ex, sbux.ex)
        [,1]
[1,] 0.07769
```

□

Statistical Properties of the Estimators

The sample mean, standard deviation, and correlation are used in many applications of statistics and their properties are well-known.

Consider the properties of $\bar{R}_j$ and S_j under the assumption that $R_{j,1}, R_{j,2}, \ldots, R_{j,T}$ are independent, identically distributed random variables. Then $\mathrm{E}(\bar{R}_j) = \mu_j$, so that $\bar{R}_j$ is an unbiased estimator of μ_j, and $\mathrm{Var}(\bar{R}_j) = \sigma_j^2/T$. Futhermore, the sampling distribution of $\bar{R}_j$ is approximately normal. More formally, we say that the standardized estimator $\sqrt{T}(\bar{R}_j - \mu_j)/\sigma_j$ converges in distribution to a standard normal distribution as $T \to \infty$, a consquence of the central limit theorem (CLT). The same result holds under the weaker assumption that $R_{j,1}, R_{j,2}, \ldots, R_{j,T}$ are uncorrelated.

Example 6.3 Suppose that the returns on an asset have mean 0.01 and standard deviation 0.05. Then the sample mean return $\bar{R}_j$ has expected value 0.01 and standard deviation $0.05/\sqrt{T}$, where T is the number of observations. For example, for $T = 60$, $\bar{R}_j$ is approximately normally distributed with mean 0.01 and standard deviation

$$\frac{0.05}{\sqrt{60}} = 0.00645.$$

Since the 75th percentile of the standard normal distribution is approximately 2/3, the upper and lower quartiles of the distribution of $\bar{R}_j$ are approximately

$$0.01 \pm \frac{2}{3}(0.00645) = 0.00570 \text{ and } 0.0143.$$

That is, based on a sample of size 60, there is about a 50% chance that the sample mean return will be within 0.0043 of the true mean return. □

The sampling distribution of the sample mean return may be used to construct an approximate confidence interval for the asset's true mean return in the usual way. Let $\bar{R}_j$ and S_j denote the sample mean and sample standard deviation of a series of returns on asset j; then

$$\bar{R}_j \pm 1.96\frac{S_j}{\sqrt{T}}$$

is an approximate 95% confidence interval for μ_j, the mean return on the asset; the same approach may be used to construct an approximate confidence interval for the mean excess return. Recall that the estimated standard deviation of an estimator, $S_j/\sqrt{T}$ in this case, is known as its *standard error.*

When T is small, a confidence interval based on the t-distribution might be used, provided that it is reasonable to assume that the returns are normally distributed. However, when analyzing return data, T is generally large enough that we may use the approximate confidence interval described earlier that does not require normally-distributed returns.

Example 6.4 Let $\mu_W - \mu_f$ denote the mean excess return on a share of Wal-Mart stock. Let $\bar{R}_W$ and $\bar{R}_f$ denote the sample means of the returns on Wal-Mart stock and the risk-free asset, respectively, and let S_W denote the sample standard deviation of the excess returns. Using the results in Example 6.1, $\bar{R}_W - \bar{R}_f = 0.0109$ and $S_W = 0.0442$; therefore, an approximate 95% confidence interval for $\mu_W - \mu_f$ is given by

$$0.0109 \pm 1.96\frac{0.0442}{\sqrt{60}} = 0.0109 \pm 0.0112 = (-0.0003, 0.0221). \qquad \square$$

It may be of interest to compare the returns on two assets. Consider assets i and j; let $R_{i,1}, R_{i,2}, \ldots, R_{i,T}$ denote the returns on asset i and let $R_{j,1}, R_{j,2}, \ldots, R_{j,T}$ denote the returns on asset j. Suppose that we are interested in estimating the difference in the mean returns, $\mu_i - \mu_j$, where $\mu_i = \mathrm{E}(R_{i,t})$ and $\mu_j = \mathrm{E}(R_{j,t})$. To estimate $\mu_i - \mu_j$, we can simply use the difference in the sample means, $\bar{R}_i - \bar{R}_j$.

However, to construct a confidence interval for $\mu_i - \mu_j$, we must take into account the fact that the returns on different assets in the same time period are generally correlated. That is $\mathrm{Cov}(R_{i,t}, R_{j,t}) \neq 0$. The data $R_{i,1}, R_{i,2}, \ldots, R_{i,T}$ and $R_{j,1}, R_{j,2}, \ldots, R_{j,T}$ may be viewed as "matched pairs" data, where the pair is based on the time period. Therefore, we should consider the data to be T pairs of the form $(R_{i,t}, R_{j,t})$, $t = 1, 2, \ldots, T$, where the pairs for different time periods are uncorrelated, but the returns in each pair are correlated.

To construct a confidence interval for $\mu_i - \mu_j$ in such cases, we compute the differences of the pairs' returns, $Y_t = R_{i,t} - R_{j,t}$, $t = 1, 2, \ldots, T$. Then $Y_1, Y_2, \ldots, Y_T$ are uncorrelated random variables with mean $\mu_i - \mu_j$ and standard deviation σ_d; although it is possible to write σ_d in terms of the asset return standard deviations and the correlation of the asset returns because σ_d may be estimated directly, such an expression is not needed.

Therefore, an approximate 95% confidence interval for $\mu_i - \mu_j$ is given by

$$\bar{Y} \pm 1.96 \frac{S_Y}{\sqrt{T}}$$

where $\bar{Y}$ and S_Y are the sample mean and sample standard deviation, respectively, of $Y_1, Y_2, \ldots, Y_T$. Note that $\bar{Y} = \bar{R}_i - \bar{R}_j$ but S_Y cannot be written as a function of the individual return standard deviations. Also note that Y_t is also the difference of the excess returns on the two assets.

Example 6.5 Consider the five years of monthly excess returns on Wal-Mart and Starbucks stock, stored in the variables `wmt.ex` and `sbux.ex`, respectively. Then the differences of the returns may be calculated using

```
> ret.d<-wmt.ex-sbux.ex
```

Note that `ret.d` also contains the differences of the standard returns on the two assets.

The sample mean and standard deviation of the differences are given by

```
> mean(ret.d)
[1] -0.0134
> sd(ret.d)
[1] 0.0722
```

Therefore, an approximate 95% confidence interval for $\mu_W - \mu_S$, the difference in the mean returns of the two stocks, is given by

$$-0.0134 \pm 1.96 \frac{0.0722}{\sqrt{60}} = -0.0134 \pm 0.0183 = (-0.0317, 0.0049). \qquad \square$$

The sample correlation of the returns on two assets, $\hat{\rho}_{ij}$, is also approximately normally distributed, with the mean given by the corresponding true correlation, ρ_{ij}, and standard error $\sqrt{1-\rho_{ij}^2}/\sqrt{T}$. Thus, an approximate 95% confidence interval for ρ_{ij} is given by

$$\hat{\rho}_{ij} \pm 1.96 \frac{\sqrt{1-\hat{\rho}_{ij}^2}}{\sqrt{T}}.$$

Many of the quantities estimated when analyzing return data have similar sampling distributions that can be used to determine standard errors and approximate confidence intervals, and these will be discussed as needed throughout the remainder of the book.

6.3 Estimation of the Mean Vector and Covariance Matrix

In practice, we are often interested in estimating the means, standard deviations, and correlations for the returns, or excess returns, on several assets.

Hence, it is useful to consider estimation of the mean vector and covariance matrix of a vector of asset returns.

Suppose that there are N assets under consideration. Let $\boldsymbol{R}_t$ be the $N \times 1$ vector asset returns at time period t of the form

$$\boldsymbol{R}_t = \begin{pmatrix} R_{1,t} \\ R_{2,t} \\ \vdots \\ R_{N,t} \end{pmatrix}, \quad t = 1, 2, \ldots, T.$$

Let $\boldsymbol{\mu}$ and $\boldsymbol{\Sigma}$ denote the mean vector and covariance matrix, respectively, of $\boldsymbol{R}_t$.

An estimator of $\boldsymbol{\mu}$ is given by the sample mean vector of the returns,

$$\bar{\boldsymbol{R}} = \frac{1}{T} \sum_{t=1}^{T} \boldsymbol{R}_t.$$

The vector of mean excess returns, $\boldsymbol{\mu} - \mu_f \mathbf{1}$, may be estimated by the sample mean vector of the excess returns

$$\bar{\boldsymbol{R}}_E = \bar{\boldsymbol{R}} - \bar{R}_f \mathbf{1} = \begin{pmatrix} \bar{R}_1 - \bar{R}_f \\ \bar{R}_2 - \bar{R}_f \\ \vdots \\ \bar{R}_N - \bar{R}_f \end{pmatrix}.$$

To estimate $\boldsymbol{\Sigma}$, we can use the sample covariance matrix $\boldsymbol{S}$, calculated from either the standard returns or the excess returns; here we use the excess returns. The sample covariance matrix is an $N \times N$ matrix with the (j, k)th element given by S_j^2 if $k = j$ and $\hat{\rho}_{jk} S_j S_k$ if $k \neq j$. The same information, in a form that is easier to interpret, is provided by the asset excess-return sample standard deviations, $S_1, S_2, \ldots, S_N$ together with the corresponding sample correlation matrix, $\hat{\boldsymbol{C}}$, the $N \times N$ matrix with ones on the diagonal, and the (j, k)th element given by $\hat{\rho}_{jk}$ for $j \neq k$.

Data Matrix

When describing the sample mean vector and the sample covariance matrix, it is often convenient to express them in terms of a *data matrix*. The data matrix for the excess returns, which we denote here by $\boldsymbol{X}$, is the $T \times N$ matrix with row t given by the vector of excess returns at time t, $(\boldsymbol{R}_t - R_{f,t}\mathbf{1})^T$:

$$\boldsymbol{X} = \begin{pmatrix} R_{1,1} - R_{f,1} & R_{2,1} - R_{f,1} & \cdots & R_{N,1} - R_{f,1} \\ R_{1,2} - R_{f,2} & R_{2,2} - R_{f,2} & \cdots & R_{N,2} - R_{f,2} \\ \vdots & \vdots & \cdots & \vdots \\ R_{1,T} - R_{f,T} & R_{2,T} - R_{f,T} & \cdots & R_{N,T} - R_{f,T} \end{pmatrix}.$$

Thus, the jth column of $\boldsymbol{X}$ is the time series of excess returns on asset j and the row t of $\boldsymbol{X}$ is the vector of N asset excess returns at time t.

The sample mean vector and the sample covariance matrix have simple expressions in terms of $\boldsymbol{X}$. The (column) vector of sample mean excess returns may be written

$$\bar{\boldsymbol{R}}_E = \bar{\boldsymbol{R}} - \bar{R}_f \mathbf{1} = \frac{1}{T}\boldsymbol{X}^T \mathbf{1}_T. \tag{6.1}$$

The sample covariance matrix has a particularly simple expression in terms of $\boldsymbol{X}$:

$$\boldsymbol{S} = \frac{1}{T-1}(\boldsymbol{X} - \mathbf{1}_T \bar{\boldsymbol{R}}_E^T)^T (\boldsymbol{X} - \mathbf{1}_T \bar{\boldsymbol{R}}_E^T). \tag{6.2}$$

Note that

$$\mathbf{1}_T \bar{\boldsymbol{R}}_E^T = \begin{pmatrix} 1 \\ 1 \\ \vdots \\ 1 \end{pmatrix} \begin{pmatrix} \bar{R}_1 - \bar{R}_f & \bar{R}_2 - \bar{R}_f & \cdots & \bar{R}_N - \bar{R}_f \end{pmatrix}$$

$$= \begin{pmatrix} \bar{R}_1 - \bar{R}_f & \bar{R}_2 - \bar{R}_f & \cdots & \bar{R}_N - \bar{R}_f \\ \bar{R}_1 - \bar{R}_f & \bar{R}_2 - \bar{R}_f & \cdots & \bar{R}_N - \bar{R}_f \\ \vdots & \vdots & \cdots & \vdots \\ \bar{R}_1 - \bar{R}_f & \bar{R}_2 - \bar{R}_f & \cdots & \bar{R}_N - \bar{R}_f \end{pmatrix}$$

so that $\boldsymbol{X} - \mathbf{1}_T \bar{\boldsymbol{R}}_E^T$ is given by

$$\begin{pmatrix} R_{1,1} - R_{f,1} - (\bar{R}_1 - \bar{R}_f) & R_{2,1} - R_{f,1} - (\bar{R}_2 - \bar{R}_f) & \ldots & R_{N,1} - R_{f,1} - (\bar{R}_N - \bar{R}_f) \\ R_{1,2} - R_{f,2} - (\bar{R}_1 - \bar{R}_f) & R_{2,2} - R_{f,2} - (\bar{R}_2 - \bar{R}_f) & \ldots & R_{N,2} - R_{f,2} - (\bar{R}_N - \bar{R}_f) \\ \vdots & \vdots & \vdots & \vdots \\ R_{1,T} - R_{f,T} - (\bar{R}_1 - \bar{R}_f) & R_{2,T} - R_{f,T} - (\bar{R}_2 - \bar{R}_f) & \ldots & R_{N,T} - R_{f,T} - (\bar{R}_N - \bar{R}_f) \end{pmatrix}.$$

Example 6.6 Consider the returns on the stocks of eight large companies, Apple (symbol AAPL), Baxter International (BAX), Coca-Cola (KO), CVS Health Corporation (CVS), Exxon Mobil (XOM), IBM (IBM), Johnson & Johnson (JNJ), and Walt Disney (DIS). These companies were chosen to represent large companies from a variety of industries.

For each stock, five years of monthly excess returns were calculated for the period ending December 31, 2014. In R, each vector of 60 excess returns was stored as a variable with the name of the stock symbol (e.g., `aapl` for Apple).

To calculate the parameters of the distribution of the return vector $\boldsymbol{R}_t$, it is convenient to have all of the data stored in a single matrix, with each column corresponding to a particular stock; this can be done using the `cbind` command:

```
> big8<-cbind(aapl, bax, ko, cvs, xom, ibm, jnj, dis)
```

Then big8 is a 60×8 matrix of excess returns; it corresponds to the data matrix $\boldsymbol{X}$ described previously.

When reading the output from various functions, it is helpful to have the columns of big8 labeled; this may be achieved using the following command:

```
> colnames(big8)<-c("AAPL", "BAX", "KO", "CVS", "XOM", "IBM",
+ "JNJ", "DIS")
> head(big8)
        AAPL     BAX     KO      CVS     XOM      IBM     JNJ
[1,] -0.0886 -0.0186 -0.0483  0.00753 -0.0552 -0.06506 -0.0241
[2,]  0.0653 -0.0116 -0.0283  0.04254  0.0153  0.04353  0.0098
[3,]  0.1483  0.0272  0.0517  0.08313  0.0303  0.00845  0.0348
[4,]  0.1109 -0.1888 -0.0283  0.01211  0.0117  0.00571 -0.0139
[5,] -0.0163 -0.1058 -0.0385 -0.06216 -0.1019 -0.02415 -0.0852
[6,] -0.0209 -0.0309 -0.0168 -0.15344 -0.0562 -0.01431  0.0129
         DIS
[1,] -0.0838
[2,]  0.0571
[3,]  0.1174
[4,]  0.0552
[5,] -0.0930
[6,] -0.0576
```

Descriptive statistics for the returns may now be calculated. Although we could use matrix expressions like the one in (6.1) to obtain such results, a simpler approach is to use the apply command, which applies a function to the margins of a matrix or, more generally, an array. For instance, the command apply(big8, MARGIN=2, FUN=mean) applies the function mean to "margin" of the matrix big8 designated by the MARGIN argument, here given by "2," to denote columns, the second dimension of the matrix big8. As with many other R functions, the argument names can be omitted provided that the order of the arguments is respected.

```
> apply(big8, MARGIN=2, FUN=mean)
   AAPL     BAX      KO     CVS     XOM     IBM     JNJ     DIS
0.02540 0.00740 0.00978 0.02119 0.00825 0.00598 0.01153 0.02075
> apply(big8, 2, sd)
  AAPL    BAX     KO    CVS    XOM    IBM    JNJ    DIS
0.0739 0.0556 0.0412 0.0578 0.0459 0.0458 0.0386 0.0579
```

Therefore, the excess returns on Apple stock, for example, have sample mean 0.0254 and sample standard deviation 0.0739.

Of course, one or both of the results of the aforementioned apply function may be assigned to a variable. For example,

```
> Rbar<-apply(big8, 2, mean)
```

To calculate the sample correlation matrix of the data in the matrix `big8`, we use the `cor` function with the data matrix as the argument.

```
> cor(big8)
      AAPL   BAX    KO   CVS   XOM   IBM   JNJ   DIS
AAPL 1.000 0.193 0.260 0.329 0.303 0.319 0.145 0.346
BAX  0.193 1.000 0.310 0.330 0.327 0.381 0.473 0.196
KO   0.260 0.310 1.000 0.310 0.338 0.197 0.493 0.348
CVS  0.329 0.330 0.310 1.000 0.442 0.244 0.421 0.537
XOM  0.303 0.327 0.338 0.442 1.000 0.520 0.408 0.650
IBM  0.319 0.381 0.197 0.244 0.520 1.000 0.206 0.348
JNJ  0.145 0.473 0.493 0.421 0.408 0.206 1.000 0.323
DIS  0.346 0.196 0.348 0.537 0.650 0.348 0.323 1.000
```

Thus, the excess returns on Apple and Disney stocks have correlation 0.346, for example. Note that the sample correlation matrix, like all correlation matrices, is symmetric.

The sample covariance matrix may be calculated using the `cov` function:

```
> Smat<-cov(big8)
> Smat
         AAPL      BAX       KO      CVS      XOM      IBM
AAPL 0.005460 0.000794 0.000790 0.001405 0.001028 0.001080
BAX  0.000794 0.003088 0.000709 0.001061 0.000835 0.000969
KO   0.000790 0.000709 0.001694 0.000737 0.000638 0.000371
CVS  0.001405 0.001061 0.000737 0.003339 0.001174 0.000646
XOM  0.001028 0.000835 0.000638 0.001174 0.002109 0.001094
IBM  0.001080 0.000969 0.000371 0.000646 0.001094 0.002099
JNJ  0.000413 0.001016 0.000783 0.000939 0.000723 0.000365
DIS  0.001479 0.000631 0.000830 0.001797 0.001729 0.000923
          JNJ      DIS
AAPL 0.000413 0.001479
BAX  0.001016 0.000631
KO   0.000783 0.000830
CVS  0.000939 0.001797
XOM  0.000723 0.001729
IBM  0.000365 0.000923
JNJ  0.001491 0.000722
DIS  0.000722 0.003352
```

Thus, a second way to compute the standard deviations of these eight stocks is to use the square root of the diagonal elements of `Smat`. The diagonal of a square matrix may be extracted using the `diag` command. Hence, the estimated standard deviations are given by

```
> diag(Smat)^.5
  AAPL    BAX     KO    CVS    XOM    IBM    JNJ    DIS
0.0739 0.0556 0.0412 0.0578 0.0459 0.0458 0.0386 0.0579
```

matching the results obtained previously. □

Some Properties of the Sample Covariance Matrix

Some basic properties of the sample covariance matrix $\boldsymbol{S}$ are easily obtained from its expression in terms of the data matrix. For instance, it follows from (6.2) that $\boldsymbol{S}$ is nonnegative definite. To see this, let $\boldsymbol{u}$ be an $N \times 1$ vector. Then $\boldsymbol{u}^T\boldsymbol{S}\boldsymbol{u} = \boldsymbol{d}^T\boldsymbol{d}$ where

$$\boldsymbol{d} = \frac{1}{\sqrt{T-1}}(\boldsymbol{X} - \mathbf{1}_T\bar{\boldsymbol{R}}_E^T)\boldsymbol{u}$$

is a $T \times 1$ vector. Writing $\boldsymbol{d} = (d_1, d_2, \ldots, d_T)^T$, $\boldsymbol{d}^T\boldsymbol{d} = \sum_{t=1}^T d_t^2$; it follows that $\boldsymbol{d}^T\boldsymbol{d} \geq 0$ and, hence, $\boldsymbol{u}^T\boldsymbol{S}\boldsymbol{u} \geq 0$ for any $\boldsymbol{u} \in \Re^N$.

There is another important implication of (6.2) for the properties of $\boldsymbol{S}$. Like $\boldsymbol{X}$, $\boldsymbol{X} - \mathbf{1}_T\bar{\boldsymbol{R}}_E^T$ is a $T \times N$ matrix. Hence, although $\boldsymbol{S}$ is an $N \times N$ matrix, the rank of $\boldsymbol{S}$ is, at most, $\min(N, T)$. Therefore, if $T < N$, that is, if the number of time periods is less than the number of assets, then $\boldsymbol{S}$ cannot be invertible and, hence, it cannot be positive definite. Suppose that we are analyzing five years of monthly returns, $T = 60$. Then for $\boldsymbol{S}$ to be positive definite, we can consider, at most, 60 assets.

In addition to the sample covariance matrix being singular when N is larger than T, if N and T are roughly the same magnitude, as is often the case with financial data, then there are features of the sample covariance matrix that are not, in general, accurate estimators of the corresponding features of $\boldsymbol{\Sigma}$.

The basic issue is somewhat obvious. The covariance matrix $\boldsymbol{\Sigma}$ contains $N(N+1)/2$ parameters. We have NT data points and if $N \doteq T, NT \doteq N^2$. Therefore, we are trying to estimate a large number of parameters with, relatively speaking, a small amount of data. This occurs even though when T is relatively large any one element of $\boldsymbol{S}$, $\boldsymbol{S}_{ij}$, provides an accurate estimator of $\boldsymbol{\Sigma}_{ij}$, the corresponding element of $\boldsymbol{\Sigma}$.

Because the issue arises with the large number of covariances in $\boldsymbol{\Sigma}$, not with the variances, consider the properties of the sample correlation matrix $\widehat{\boldsymbol{C}}$ as an estimator of the $N \times N$ correlation matrix $\boldsymbol{C}$. If T is very large, so that we have many observations, and N/T is near zero, so that the number of assets is small relative to the number of data points, then $\widehat{\boldsymbol{C}} \doteq \boldsymbol{C}$ with high probability; more formally, $\widehat{\boldsymbol{C}}$ converges in probability to $\boldsymbol{C}$ as $T \to \infty$ and N remains fixed. Here, convergence in probability of a sequence of matrices is defined elementwise.

However, if T is large but $q = N/T$ is not near zero, so that N is also large, there are ways in which $\widehat{\boldsymbol{C}}$ is a poor estimator of $\boldsymbol{C}$. For instance, it may be shown that

$$\text{tr}(\widehat{\boldsymbol{C}}^{-1}) \doteq \frac{1}{(1-q)}\text{tr}(\boldsymbol{C}^{-1})$$

where $\text{tr}(\boldsymbol{A})$ denotes the *trace* of a matrix A. Recall that the trace of a matrix is the sum of its diagonal elements; it is also equal to the sum of its eigenvalues.

Although we are not interested directly in the trace of $\boldsymbol{C}^{-1}$, many of the weight vectors we derived in the previous chapter are based on the inverse of the asset return covariance matrix. Hence, this result suggests that, when N/T is not small, such weight vectors are not well-estimated by replacing the return covariance matrix with the corresponding sample covariance matrix. For example, if we have five years of monthly returns on 40 assets, so that $N = 40$ and $T = 60$, then $q = 2/3$ and the trace of $\hat{\boldsymbol{C}}^{-1}$ will be approximately three times the trace of $\boldsymbol{C}^{-1}$, suggesting that $\hat{\boldsymbol{C}}^{-1}$ is, in some respects, a poor estimator of $\boldsymbol{C}^{-1}$.

6.4 Weighted Estimators

One drawback of estimators based on basic sample statistics, such as the ones considered in the previous sections, is that they require a relatively long series of data to have even moderate accuracy. Furthermore, in computing the sample mean and sample standard deviation, the observation at $t = 1$ is treated identically to the observation at $t = T$. For instance, if we are using five years of monthly data, the returns from five years ago receive the same weight in the estimates as do the returns from last month. Thus, the estimates are sensitive to the assumption that the means and standard deviations of the monthly returns are constant over time; if this assumption is questionable, then such estimators may be biased as estimators of the parameters of the return distribution in the current period.

Thus, it is sometimes desirable to use a weighted estimator, in which more recent data receive more weight than do data from the first time period being used. Such an estimator gives some protection against violations of the assumption of constant parameter values over time.

Consider estimation of a mean return μ_j based on a series of returns $R_{j,1}, R_{j,2}, \ldots, R_{j,T}$ on an asset; the same approach can be used to estimate the mean excess return on an asset. An *exponentially weighted moving average (EWMA)* estimator of μ_j is of the form

$$\hat{\mu}_{wj} = \frac{\sum_{t=1}^{T} \gamma^{T-t} R_{j,t}}{\sum_{t=1}^{T} \gamma^{T-t}}$$

for a constant γ, $0 < \gamma < 1$, known as the *decay parameter*. Thus, the estimator is a weighted average of the returns $R_{j,1}, R_{j,2}, \ldots, R_{j,T}$, with the weights proportional to γ^{T-t}; the value of γ used controls how quickly the weights decrease as t decreases.

Note that, under the assumption that $R_{j,1}, R_{j,2}, \ldots, R_{j,T}$ each has mean μ_j,

$$\mathrm{E}(\hat{\mu}_{wj}) = \frac{\sum_{t=1}^{T} \gamma^{T-t} \mathrm{E}(R_{jt})}{\sum_{t=1}^{T} \gamma^{T-t}} = \frac{\sum_{t=1}^{T} \gamma^{T-t} \mu_j}{\sum_{t=1}^{T} \gamma^{T-t}} = \mu_j$$

so that $\hat{\mu}_{wj}$ is an unbiased estimator of μ_j.

If $\mathrm{E}(R_{j,t})$ changes with t, then the weighted estimator will often have smaller bias than does the sample mean return. For instance, suppose that $\mathrm{E}(R_{j,t}) = a + b(T-t)$, $t = 1, 2, \ldots, T$, so that $R_{j,1}$ has an expected value $a + b(T-1)$ and $R_{j,T}$ has an expected value a. Then, it is straightforward to show that $\bar{R}_j$ has expected value

$$a + b\frac{T-1}{2}$$

and, using properties of geometric series, that $\hat{\mu}_{wj}$ has an expected value

$$a + b\left(\frac{\gamma}{1-\gamma} - \frac{\gamma^T}{1-\gamma^T}T\right).$$

Thus, as an estimator of the expected return in period T, $\mathrm{E}(R_{j,T})$, $\bar{R}_j$ has bias

$$b\frac{T-1}{2}$$

and $\hat{\mu}_{wj}$ has bias

$$b\left(\frac{\gamma}{1-\gamma} - \frac{\gamma^T}{1-\gamma^T}T\right).$$

For instance, suppose that $a = 0.02$, $b = 0.01/59$, and $T = 60$ so that the asset mean return is 0.03 in period 1 and it decreases linearly to 0.02 in period 60. Then the sample mean has bias 0.005 as an estimator of $\mathrm{E}(R_{j,T})$ and the EWMA estimator using $\gamma = 0.95$ has a bias of about 0.0027, roughly half that of the sample mean. A smaller value of γ reduces the bias; for instance, using $\gamma = 0.9$, the EWMA estimator has a bias of about 0.0015.

Now consider the variance of the EWMA estimator. Suppose that $R_{j,1}, R_{j,2}, \ldots, R_{j,T}$ each has standard deviation σ_j and are uncorrelated. Then

$$\mathrm{Var}(\hat{\mu}_{wj}) = \frac{\sum_{t=1}^T (\gamma^{T-t})^2 \mathrm{Var}(R_{jt})}{(\sum_{t=1}^T \gamma^{T-t})^2} = \frac{\sum_{t=1}^T (\gamma^{T-t})^2}{(\sum_{t=1}^T \gamma^{T-t})^2}\sigma_j^2$$

so that the variance of $\hat{\mu}_{wj}$ depends on γ.

Note that, when T is reasonably large, in the sense that $\gamma^T \doteq 0$, then

$$\sum_{t=1}^{T} \gamma^{T-t} = \sum_{t=0}^{T-1} \gamma^t \doteq \frac{1}{1-\gamma}; \tag{6.3}$$

it follows that

$$\frac{\sum_{t=1}^T (\gamma^{T-t})^2}{(\sum_{t=1}^T \gamma^{T-t})^2} = \frac{\sum_{t=1}^T (\gamma^2)^{T-t}}{(\sum_{t=1}^T \gamma^{T-t})^2} \doteq \frac{\frac{1}{1-\gamma^2}}{\left(\frac{1}{1-\gamma}\right)^2} = \frac{1-\gamma}{1+\gamma};$$

and, hence,

$$\text{Var}(\hat{\mu}_{wj}) \doteq \frac{1-\gamma}{1+\gamma}\sigma_j^2.$$

Because the variance of the average of M returns is σ_j^2/M, the *effective sample size* corresponding to the decay parameter γ is the value of M satisfying

$$\frac{1}{M} = \frac{1-\gamma}{1+\gamma} \quad \text{or} \quad M = \frac{1+\gamma}{1-\gamma}.$$

For example, an EWMA estimator with $\gamma = 0.75$ has roughly the same variance as the average of $(1+0.75)/(1-0.75) = 7$ observations, while an EWMA estimator based on $\gamma = 0.95$ has an effective sample size of 39. Therefore, there is a price to pay for using a weighted estimator, in terms of the variance of the estimator, for reducing possible bias arising from violation of the constant-mean assumption. Note that, as is often the case in estimation problems, a small value of γ tends to yield a smaller bias but a larger variance; for a large value of γ, the reverse is true. Also note that going back further in time to increase the number of observations has only a minor effect on the variance of a weighted estimator once γ^T is small.

The remaining issue in implementing the weighted estimators is selection of the decay parameter γ. Although it is possible to use a more formal approach, such as choosing γ to maximize some measure of predictive accuracy, γ is often chosen subjectively, looking at properties such as the effective sample size and relying on past experience with weighted estimators in similar estimation problems. In many applications, values of γ in the range from 0.90 to 0.98 are used.

Example 6.7 Consider the monthly returns on Wal-Mart stock, stored in the variable `wmt.m.ret`; see Example 6.1.

To calculate the weighted sample mean of the returns, note that `59:0` is a vector of the form $59, 58, \ldots, 1, 0$; and, hence, using a decay parameter of 0.9, `0.9^(59:0)` is a vector of the form $(0.9)^{59}, (0.9)^{58}, \ldots, (0.9), 1$. Then we may calculate the weighted sample mean using the function `weighted.mean`, which takes a data vector as the first argument and a vector of weights as the second argument. Note that the weights do not need to be normalized to sum to 1; the normalization is done by the function. Hence, the EWMA estimate of the mean return on Wal-Mart stock based on the decay parameter 0.9 is given by

```
> weighted.mean(wmt.m.ret, 0.9^(59:0))
[1] 0.0158
```

The value of the weighted estimator changes with the value of the decay parameter γ and, as γ approaches one, the weighted estimator approaches the sample mean.

```
> weighted.mean(wmt.m.ret, w=0.9^(59:0))
[1] 0.0158
> weighted.mean(wmt.m.ret, w=0.95^(59:0))
[1] 0.01313
> weighted.mean(wmt.m.ret, w=0.97^(59:0))
[1] 0.01231
> weighted.mean(wmt.m.ret, w=0.99^(59:0))
[1] 0.01144
> mean(wmt.m.ret)
[1] 0.01094
```

□

The EWMA approach can also be applied to variances. For a given value of γ, the weighted estimator of the return variance is given by

$$\hat{\sigma}_{wj}^2 = \frac{\sum_{t=1}^{T} \gamma^{T-t}(R_{j,t} - \hat{\mu}_{wj})^2}{\sum_{t=1}^{T} \gamma^{T-t}} \tag{6.4}$$

where $\hat{\mu}_{wj}$ is the EWMA estimate of μ_j based on this same value of γ.

Example 6.8 Consider the monthly returns on Wal-Mart stock stored in the R variable `wmt.m.ret`. Calculation of a weighted estimator of the return variance is simplified by first constructing a weight variable. Suppose we take the decay parameter to be 0.9. Define a variable `wgt` by

```
> wgt<-(0.9^(59:0))/sum(0.9^(59:0))
```

This weight vector can be used to calculate the weighted estimate of the mean return on Wal-Mart stock using the command

```
> muhat_w<-weighted.mean(wmt.m.ret, w=wgt)
> muhat_w
[1] 0.0158
```

and the weighted estimate of the return variance for Wal-Mart stock is given by

```
> weighted.mean((wmt.m.ret-muhat_w)^2, w=wgt)
[1] 0.00261
```

□

Weighted Estimators of the Mean Vector and Covariance Matrix

The EWMA approach may also be applied to estimation of the mean vector and covariance matrix of a set of asset returns. The basic approach is the same as that used to estimate the mean or variance of the return on a single asset; hence, the discussion here focuses on an example.

Example 6.9 Consider Example 6.6, which considered the returns on the stocks of eight large companies. Recall that the data for this example are stored in the variable `big8`.

The calculations needed to estimate the mean vector and covariance matrix of the excess returns may all be performed using the `cov.wt` function, which calculates weighted estimates of the mean vector and covariance matrix based on a data matrix.

Suppose we take the decay parameter to be $\gamma = 0.97$; the corresponding weight vector may be calculated using the command

```
> wgt<-0.97^(59:0)
```

Note that, for use in the function `cov.wt`, the weight vector does not need to be standardized to sum to 1.

To calculate the weighted estimates of the mean return vector and covariance matrix, we use the command

```
> big8.wt<-cov.wt(big8, wt=wgt, method="ML")
```

The default value for the `method` argument is `unbiased`, which applies a multiplicative adjustment to the result to yield an unbiased estimator of the true covariance matrix, analogous to dividing by $T-1$ instead of by T when calculating the unweighted sample covariance matrix. Specifying `method="ML"` returns an estimate that corresponds to (6.4) and does not include such an adjustment.

The result of `cov.wt` is a list containing several components. For example, the estimated mean returns are given in the component `$center` of `big8.wt`:

```
> big8.wt$center
    AAPL       BAX       KO       CVS       XOM       IBM       JNJ
0.024478 0.008672 0.009056 0.025728 0.006367 0.000341 0.013607
     DIS
0.023101
```

These may be compared to the unweighted sample means, as calculated earlier.

```
> Rbar
   AAPL     BAX      KO     CVS     XOM     IBM     JNJ     DIS
0.02540 0.00740 0.00978 0.02119 0.00825 0.00598 0.01153 0.02075
```

In most cases, the weighted and unweighted estimates are similar; however, for IBM, the weighted estimate is much smaller than the weighted estimate, suggesting that the returns on IBM stock may be decreasing over time.

The estimated covariance matrix is available in the component `$cov`. Hence, the estimated weighted sample standard deviations may be obtained by using the `diag` function, which returns the diagonal elements of a matrix.

```
> diag(big8.wt$cov)^.5
  AAPL    BAX     KO    CVS    XOM    IBM    JNJ    DIS
0.0720 0.0462 0.0427 0.0511 0.0434 0.0473 0.0374 0.0505
```

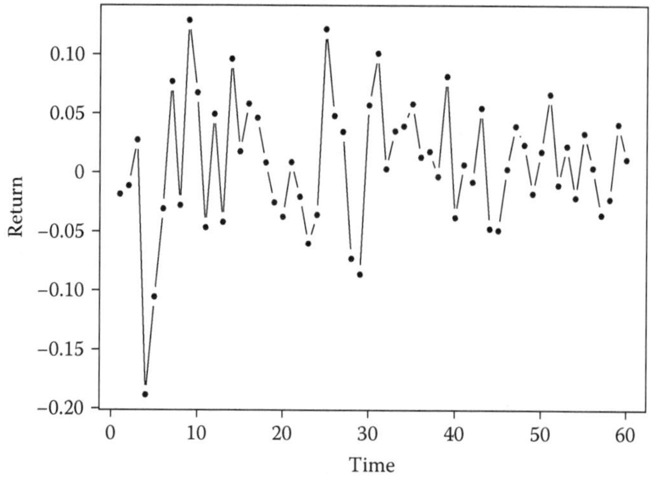

FIGURE 6.1
Plot of excess returns on Baxter stock.

These can be compared to the unweighted estimates.

```
> diag(Sighat)^.5
  AAPL    BAX     KO      CVS     XOM     IBM     JNJ     DIS
0.0739 0.0556 0.0412 0.0578 0.0459 0.0458 0.0386 0.0579
```

Note that, for some stocks, there is a relatively large difference between the weighted and unweighted estimates. For instance, for BAX, the unweighted estimate is about 20% larger than the weighted estimate. This may be explained by a plot of the BAX excess returns versus time, given in Figure 6.1. Note that there is much more variability at the earlier time points than at the later time points. Therefore, the weighted estimator, which gives low weight to the returns at the beginning of the observed time period, is smaller than the unweighted estimator, which treats each return equally.

The weighted and unweighted estimates of the correlations may be obtained using the function cov2cor, which computes a correlation matrix from a covariance matrix.

```
> cov2cor(big8.wt$cov)
      AAPL   BAX    KO     CVS    XOM    IBM    JNJ    DIS
AAPL 1.000 0.199 0.312 0.310 0.247 0.183 0.169 0.254
BAX  0.199 1.000 0.209 0.325 0.298 0.404 0.388 0.153
KO   0.312 0.209 1.000 0.260 0.261 0.171 0.566 0.316
CVS  0.310 0.325 0.260 1.000 0.444 0.137 0.383 0.588
XOM  0.247 0.298 0.261 0.444 1.000 0.420 0.384 0.603
IBM  0.183 0.404 0.171 0.137 0.420 1.000 0.174 0.285
JNJ  0.169 0.388 0.566 0.383 0.384 0.174 1.000 0.301
DIS  0.254 0.153 0.316 0.588 0.603 0.285 0.301 1.000
```

```
> cov2cor(Sighat)
      AAPL   BAX    KO   CVS   XOM   IBM   JNJ   DIS
AAPL 1.000 0.193 0.260 0.329 0.303 0.319 0.145 0.346
BAX  0.193 1.000 0.310 0.330 0.327 0.381 0.473 0.196
KO   0.260 0.310 1.000 0.310 0.338 0.197 0.493 0.348
CVS  0.329 0.330 0.310 1.000 0.442 0.244 0.421 0.537
XOM  0.303 0.327 0.338 0.442 1.000 0.520 0.408 0.650
IBM  0.319 0.381 0.197 0.244 0.520 1.000 0.206 0.348
JNJ  0.145 0.473 0.493 0.421 0.408 0.206 1.000 0.323
DIS  0.346 0.196 0.348 0.537 0.650 0.348 0.323 1.000
```

Although the weighted and unweighted estimates are generally close to each other, there are some exceptions. For these data, the weighted estimates tend to be closer to zero than are the unweighted estimates. □

6.5 Shrinkage Estimators

The estimators of the mean return vector and return covariance matrix based on the sample mean return vector and sample covariance matrix require only weak assumptions regarding the asset returns. However, as we have discussed, when analyzing many assets, such estimators may not work well in practice.

One approach to improving on those simple estimators is to make some further assumptions regarding the return distributions, thus reducing the number of parameters to be estimated. For instance, we might assume that all assets have the same mean return, so that only one mean return needs to be estimated.

Such estimators tend to work well when data follow the assumptions used. However, as in the example in which all assets have the same mean return, the assumptions that lead to the greatest simplification in the analysis are often unrealistic. In such cases, combining the assumption-based estimators with estimators based on simple sample statistics, a procedure known as *shrinkage*, often leads to estimators that have better properties than either the assumption-based estimators or those based on sample statistics.

Let $\boldsymbol{R}_t$ be the $N \times 1$ vector asset returns at time period t, $t = 1, 2, \ldots, T$; let $\boldsymbol{\mu}$ and $\boldsymbol{\Sigma}$ denote the mean vector and covariance matrix of $\boldsymbol{R}_t$. We first consider estimation of the mean vector $\boldsymbol{\mu}$; the techniques are then applied to the more difficult, and in many ways more important, problem of estimating $\boldsymbol{\Sigma}$. The same approach could be used for estimation of the mean excess returns but, to keep the notation simple, the discussion will focus on mean returns; excess returns will be considered in the examples.

In Section 6.3, we considered the estimator of $\boldsymbol{\mu} = (\mu_1, \mu_2, \ldots, \mu_N)^T$ given by the vector of return sample means $\bar{\boldsymbol{R}}$; for this estimator, each μ_j, the mean return on asset j is estimated by the corresponding sample mean $\bar{R}_j$.

Now consider the estimator of $\boldsymbol{\mu}$ based on the assumption that all asset mean returns are equal: $\mu_1 = \mu_2 = \cdots = \mu_N \equiv \mu$ so that the mean vector $\boldsymbol{\mu}$ is of the form

$$\boldsymbol{\mu} = \begin{pmatrix} \mu \\ \mu \\ \vdots \\ \mu \end{pmatrix}.$$

Note that, since the same risk-free rate applies to each asset, the model in which the μ_j are equal is equivalent to one in which the excess mean returns are equal.

Under this model, the common mean μ can be estimated by the sample mean of $\bar{R}_1, \bar{R}_2, \ldots, \bar{R}_N$;

$$\hat{\mu} = \frac{1}{N}\sum_{j=1}^{N} \bar{R}_j.$$

Then each μ_j may be estimated by $\hat{\mu}$, $j = 1, 2, \ldots, N$ or, equivalently, the mean vector $\boldsymbol{\mu}$ can be estimated by the vector

$$\hat{\boldsymbol{\mu}} = \begin{pmatrix} \hat{\mu} \\ \hat{\mu} \\ \vdots \\ \hat{\mu} \end{pmatrix}.$$

The advantage of estimating μ_j by $\hat{\mu}$ rather than by $\bar{R}_j$ is that we may use NT observations to estimate one mean, leading to a smaller standard error for the estimate. Specifically, the variance of $\hat{\mu}$ may be obtained by noting that we may write

$$\hat{\mu} = \frac{1}{T}\sum_{t=1}^{T} \bar{R}_{\bullet t}$$

where $\bar{R}_{\bullet t}$ is the sample mean of all the asset returns in time period t:

$$\bar{R}_{\bullet t} = \frac{1}{N}\sum_{j=1}^{N} R_{j,t}.$$

Note that $\bar{R}_{\bullet t}$ may be written

$$\bar{R}_{\bullet t} = \frac{1}{N}\mathbf{1}^T \boldsymbol{R}_t. \tag{6.5}$$

Under the assumption that the returns in different time periods are uncorrelated, $\bar{R}_{\bullet t}$, $t = 1, 2, \ldots, T$ are uncorrelated; also, under the assumption that $\text{Var}(\boldsymbol{R}_t)$ has covariance matrix $\boldsymbol{\Sigma}$, $\text{Var}(\bar{R}_{\bullet t})$ does not depend on t. It follows that

$$\text{Var}(\hat{\mu}) = \frac{1}{T}\text{Var}(\bar{R}_{\bullet t}).$$

Using the expression (6.5),

$$\mathrm{Var}(\bar{R}_{\cdot t}) = \frac{1}{N^2}\mathbf{1}^T\boldsymbol{\Sigma}\mathbf{1}$$

where $\boldsymbol{\Sigma}$ is the covariance matrix of the vector of asset returns. It follows that

$$\mathrm{Var}(\hat{\mu}) = \frac{1}{N^2}\frac{1}{T}\mathbf{1}^T\boldsymbol{\Sigma}\mathbf{1}. \tag{6.6}$$

Note that $\mathbf{1}^T\boldsymbol{\Sigma}\mathbf{1}$ is simply the sum of all the elements of $\boldsymbol{\Sigma}$.

Example 6.10 Consider the stocks of eight large companies, as described in Example 6.6. The sample mean excess returns of the individual stocks are stored in the variable `Rbar`:

```
> Rbar
   AAPL     BAX      KO     CVS     XOM     IBM     JNJ     DIS
0.02540 0.00740 0.00978 0.02119 0.00825 0.00598 0.01153 0.02075
```

and the overall sample mean excess return is

```
> mean(Rbar)
[1] 0.0138
```

This value provides an estimate of the asset return means under the assumption that all the return means are equal.

Using (6.6), the standard deviation of $\hat{\mu}$ may be estimated by

```
> (sum(Sighat)/(60*8*8))^.5
[1] 0.00439
```

Note that `sum` applied to a matrix returns the sum of all the elements in the matrix.

This standard deviation may be compared to the standard errors of the sample means of the individual asset returns.

```
> (diag(Sighat)/60)^.5
   AAPL     BAX      KO     CVS     XOM     IBM     JNJ     DIS
0.00954 0.00717 0.00531 0.00746 0.00593 0.00591 0.00499 0.00747
```

Hence, the standard error of $\hat{\mu}$ is roughly 50%–90% as large as the standard error of $\bar{R}_j$. □

Of course, the drawback of the estimator $\hat{\mu}$ is that we do not believe that $\mu_1 = \mu_2 = \cdots = \mu_N$ and, if this assumption is not true, then $\hat{\mu}$ is a biased estimator of μ_j. Specifically,

$$\mathrm{E}(\hat{\mu}) = \frac{1}{N}\sum_{i=1}^{N}\mu_i$$

so that the bias of $\hat{\mu}$ as an estimator of μ_j is

$$E(\hat{\mu}) - \mu_j = \frac{1}{N}\sum_{i=1}^{N} \mu_i - \mu_j.$$

If $\mu_1, \mu_2, \ldots, \mu_N$ are nearly equal, then in general this bias will be small; however, it is possible that, for some $j, \hat{\mu}$ will have a large bias.

These results are true for model-based estimators more generally. We expect the variance of a model-based estimator to be smaller than the variance of a simple empirical estimator, in some cases, much smaller. However, if the model assumptions are not valid, the model-based estimators tend to be biased; in some cases, the bias is small but, in others, it may be large.

In the previous example, as well as in (nearly) all examples of this type, we know that the assumption $\mu_1 = \mu_2 = \cdots = \mu_N$ is not literally true. However, we might think that it is approximately true, in the sense that $\mu_1, \mu_2, \ldots, \mu_N$ are "close to" one another. Therefore, we might consider estimating $\boldsymbol{\mu}$ by combining the vector of asset sample mean returns $\bar{\boldsymbol{R}}$ and the vector $\hat{\boldsymbol{\mu}} = (\hat{\mu}, \hat{\mu}, \ldots, \hat{\mu})^T = \hat{\mu}\mathbf{1}$ using a weighted average of the form

$$\psi\hat{\boldsymbol{\mu}} + (1 - \psi)\bar{\boldsymbol{R}}$$

for some weight ψ.

To choose the weight ψ, we consider the distance between $\bar{\boldsymbol{R}}$ and $\hat{\boldsymbol{\mu}}$, relative to the sampling variability in $\bar{\boldsymbol{R}}$; if $\bar{\boldsymbol{R}}$ and $\hat{\boldsymbol{\mu}}$ are "close," then the assumption that all μ_j are equal appears to be reasonable for these data and we give more weight to $\hat{\boldsymbol{\mu}}$.

The squared Euclidean distance between $\boldsymbol{R}$ and $\hat{\boldsymbol{\mu}}$ is given by

$$\sum_{j=1}^{N}(\bar{R}_j - \hat{\mu})^2;$$

and, hence, a standardized measure of this distance is given by

$$\tau^2 = \frac{1}{N}\sum_{j=1}^{N}(\bar{R}_j - \hat{\mu})^2.$$

If the μ_j are approximately equal, then we expect τ^2 to be relatively small; on the other hand, if there is a great deal of variability in the μ_j, then we expect τ^2 to be relatively large.

To choose the value of ψ to use in the estimates of the μ_j, we compare τ^2 to a measure of the sampling variability in a sample mean return. The estimated variance of the sampling distribution of $\bar{R}_j$ is given by S_j^2/T, where

S_j^2 is the sample return variance for asset j. Hence, the average estimated variance of a sample mean return is given by $\overline{S^2}/T$ where

$$\overline{S^2} = \frac{1}{N}\sum_{j=1}^{N} S_j^2.$$

Thus, we take the weight ψ to be

$$\psi = \frac{\overline{S^2}/T}{\overline{S^2}/T + \tau^2} = \frac{1}{1 + T\tau^2/\overline{S^2}}$$

and estimate μ_j by

$$\psi\hat{\mu} + (1-\psi)\bar{R}_j;$$

alternatively, we can think of estimating $\boldsymbol{\mu}$ by

$$\hat{\boldsymbol{\mu}}_S = \psi \begin{pmatrix} \hat{\mu} \\ \hat{\mu} \\ \vdots \\ \hat{\mu} \end{pmatrix} + (1-\psi)\bar{\boldsymbol{R}} = \psi\hat{\boldsymbol{\mu}} + (1-\psi)\bar{\boldsymbol{R}}. \tag{6.7}$$

Hence, if the variation between $\bar{R}_1, \bar{R}_2, \ldots, \bar{R}_N$, as measured by τ^2, is small relative to $\overline{S^2}/T$, we give more weight to $\hat{\boldsymbol{\mu}}$; on the other hand, if τ^2 is large relative to $\overline{S^2}/T$, we give more weight to $\bar{\boldsymbol{R}}$.

An estimator of the form (6.7) is known as a *shrinkage estimator* because it takes the individual asset sample means $\bar{R}_1, \bar{R}_2, \ldots, \bar{R}_N$ and "shrinks" them toward the overall sample mean return $\hat{\mu}$. The hope is that the higher bias from including $\hat{\mu}$ in the estimator is more than offset by the lower variance. A related benefit is that shrinkage estimates tend to be more stable, in the sense that if one of the $\bar{R}_j$ happens to be abnormally large or small, the shrinkage estimate is often more reasonable.

Example 6.11 Consider the returns on eight large companies as described in Example 6.6 and as analyzed in Example 6.10. Recall that the variable `Rbar` contains the sample means of the asset mean excess returns; the sample variances of the asset returns, as well as the estimate of τ^2, may be calculated as follows.

```
> Rbar
   AAPL     BAX      KO     CVS     XOM     IBM     JNJ     DIS
0.02540 0.00740 0.00978 0.02119 0.00825 0.00598 0.01153 0.02075
> S2<-apply(big8, 2, var)
> S2
   AAPL     BAX      KO     CVS     XOM     IBM     JNJ     DIS
0.00546 0.00309 0.00169 0.00334 0.00211 0.00210 0.00149 0.00335
> tau2<-mean((Rbar-mean(Rbar))^2)
> tau2
[1] 4.89e-05
```

These results may then be used to calculate the weight ψ and the vector of estimated asset mean returns.

```
> psi<-(mean(S2)/60)/(tau2 + (mean(S2)/60))
> psi
[1] 0.491
> muhat<-psi*mean(Rbar) + (1-psi)*Rbar
> muhat
  AAPL    BAX     KO    CVS    XOM    IBM    JNJ    DIS
0.0197 0.0105 0.0117 0.0176 0.0110 0.0098 0.0126 0.0173
```

Thus, the estimated mean returns for the eight stocks are weighted averages of the individual sample mean returns and the average mean return for all stocks.

Of course, we do not know which estimator, $\bar{\boldsymbol{R}}$ or the shrinkage estimator $\hat{\boldsymbol{\mu}}_S$ is the more accurate estimator; this issue will be considered in Section 6.7. □

Shrinkage Estimation of a Covariance Matrix

The same basic approach used to construct a shrinkage estimate of the mean vector $\boldsymbol{\mu}$ may be used to estimate the covariance matrix $\boldsymbol{\Sigma}$.

First consider an estimator based on assumptions regarding the form of $\boldsymbol{\Sigma}$. The simplest such assumptions are that the returns all have the same variance, $\text{Var}(R_{j,t}) = \sigma^2$ for all $j = 1, 2, \ldots, N$ and $t = 1, 2, \ldots, T$, and that the returns on different assets are uncorrelated, $\text{Cov}(R_{j,s}, R_{k,t}) = 0$ for all $j \neq k$ and all $t, s = 1, 2, \ldots, T$.

Under these assumptions, $\boldsymbol{\Sigma}$ is of the form

$$\boldsymbol{\Sigma} = \sigma^2 \boldsymbol{I}_N \tag{6.8}$$

where $\boldsymbol{I}_N$ is the $N \times N$ identity matrix and σ^2 is an unknown parameter. This matrix is often referred to as the *target matrix* for the shrinkage estimation.

The parameter σ^2 may be estimated by the average sample variance,

$$\hat{\sigma}^2 = \frac{1}{N} \sum_{j=1}^{N} S_j^2.$$

Under the assumption (6.8), this is an unbiased estimator of σ^2 and, hence, $\hat{\boldsymbol{\Sigma}} = \hat{\sigma}^2 \boldsymbol{I}_N$ is an unbiased estimator of $\boldsymbol{\Sigma}$. Like the estimator of the common mean $\mu, \hat{\sigma}^2$ can be expected to have a relatively small variance compared to those of the individual sample variances $S_1^2, S_2^2, \ldots, S_N^2$. Furthermore, since the covariances and correlations are assumed to be zero, there is no estimation error incurred for these.

Of course, the assumptions on which $\hat{\boldsymbol{\Sigma}}$ is based are very strong, and obviously, they do not hold in practice. Hence, following the method used when estimating the mean return of an asset, we use a shrinkage estimator of the form

$$\psi \hat{\boldsymbol{\Sigma}} + (1 - \psi) \boldsymbol{S}$$

where $\boldsymbol{S}$ is the sample covariance matrix. The same basic reasoning used in estimating a mean return applies here as well: By taking a weighted average of $\hat{\boldsymbol{\Sigma}}$ and $\boldsymbol{S}$, we hope to form an estimator with the best properties of each.

The remaining issue is selection of ψ. A value of ψ close to one leads to an estimator that is close to the model-based estimator $\hat{\boldsymbol{\Sigma}}$, while for a value close to zero the resulting estimator is close to the sample covariance matrix. Although the details are fairly complicated, the basic idea in choosing the value of ψ used in the shrinkage estimate of the covariance matrix is the same as that used in estimating the asset return means. Let a^2 denote an estimate of the squared distance between $\boldsymbol{\Sigma}$ and $\sigma^2\boldsymbol{I}$ and let b^2 denote a measure of the variability in $\boldsymbol{S}$ as an estimator of $\boldsymbol{\Sigma}$. Then the optimal value of ψ is of the form $b^2/(a^2+b^2)$.

To calculate a shrinkage estimate of a covariance matrix in R, using this model, we may use the function `shrinkcovmat.equal`, available in the package `ShrinkCovMat` (Touloumis 2015).

Example 6.12 Consider the returns on eight large companies as described in Example 6.6 and as analyzed in the previous examples of this section. Let $\boldsymbol{\Sigma}$ denote the covariance matrix of the excess returns for the stocks of the eight firms. The argument of `shrinkcovmat.equal` is a data matrix in which the rows represent the different stocks and the columns represent different time periods (i.e., the form of the data matrix is the transpose of the way we have defined the data matrix previously).

```
> library(ShrinkCovMat)
> cov.shrnk<-shrinkcovmat.equal(t(big8))
```

The value of the `shrinkcovmat.equal` function contains a number of components. The one of most interest is the estimated covariance matrix, given by `$Sigmahat`; `$Target` gives the estimated "target" matrix, in this case, the diagonal matrix with diagonal element equal to the average of the return variances, and `$lambdahat` contains the estimated shrinkage parameter, which we have denoted by ψ.

```
> cov.shrnk$Sigmahat
        AAPL     BAX      KO     CVS     XOM
AAPL 0.00503 0.00067 0.00066 0.00118 0.00086
BAX  0.00067 0.00305 0.00059 0.00089 0.00070
KO   0.00066 0.00059 0.00188 0.00062 0.00053
CVS  0.00118 0.00089 0.00062 0.00326 0.00098
XOM  0.00086 0.00070 0.00053 0.00098 0.00223
IBM  0.00091 0.00081 0.00031 0.00054 0.00092
JNJ  0.00035 0.00085 0.00066 0.00079 0.00061
DIS  0.00124 0.00053 0.00070 0.00151 0.00145
```

```
         IBM     JNJ     DIS
AAPL 0.00091 0.00035 0.00124
BAX  0.00081 0.00085 0.00053
KO   0.00031 0.00066 0.00070
CVS  0.00054 0.00079 0.00151
XOM  0.00092 0.00061 0.00145
IBM  0.00222 0.00031 0.00077
JNJ  0.00031 0.00171 0.00061
DIS  0.00077 0.00061 0.00327

> cov.shrnk$Target
        [,1]    [,2]    [,3]    [,4]    [,5]
[1,] 0.00283 0.00000 0.00000 0.00000 0.00000
[2,] 0.00000 0.00283 0.00000 0.00000 0.00000
[3,] 0.00000 0.00000 0.00283 0.00000 0.00000
[4,] 0.00000 0.00000 0.00000 0.00283 0.00000
[5,] 0.00000 0.00000 0.00000 0.00000 0.00283
[6,] 0.00000 0.00000 0.00000 0.00000 0.00000
[7,] 0.00000 0.00000 0.00000 0.00000 0.00000
[8,] 0.00000 0.00000 0.00000 0.00000 0.00000
        [,6]    [,7]    [,8]
[1,] 0.00000 0.00000 0.00000
[2,] 0.00000 0.00000 0.00000
[3,] 0.00000 0.00000 0.00000
[4,] 0.00000 0.00000 0.00000
[5,] 0.00000 0.00000 0.00000
[6,] 0.00283 0.00000 0.00000
[7,] 0.00000 0.00283 0.00000
[8,] 0.00000 0.00000 0.00283
> cov.shrnk$lambdahat
[1] 0.162
```

For interpreting the results, it is often convenient to consider the vector of estimated standard deviations and the estimated correlation matrix. The estimated variances may be obtained by using the **diag** command, which returns the diagonal of a matrix. The correlation matrix may be obtained by the command cov2cor, which takes a covariance matrix as its argument and returns the corresponding correlation matrix.

```
> diag(cov.shrnk$Sigmahat)^.5
  AAPL    BAX     KO    CVS    XOM    IBM    JNJ    DIS
0.0710 0.0552 0.0433 0.0571 0.0472 0.0471 0.0413 0.0572
> cov2cor(cov.shrnk$Sigmahat)
      AAPL   BAX    KO   CVS   XOM   IBM   JNJ   DIS
AAPL 1.000 0.170 0.215 0.291 0.257 0.271 0.118 0.306
BAX  0.170 1.000 0.248 0.282 0.269 0.313 0.373 0.168
```

```
      AAPL   BAX    KO    CVS   XOM   IBM   JNJ   DIS
KO   0.215 0.248 1.000 0.250 0.262 0.152 0.367 0.281
CVS  0.291 0.282 0.250 1.000 0.366 0.202 0.334 0.462
XOM  0.257 0.269 0.262 0.366 1.000 0.413 0.311 0.537
IBM  0.271 0.313 0.152 0.202 0.413 1.000 0.157 0.287
JNJ  0.118 0.373 0.367 0.334 0.311 0.157 1.000 0.256
DIS  0.306 0.168 0.281 0.462 0.537 0.287 0.256 1.000
```

These results may be compared to the sample variances and the sample correlation matrix.

```
> apply(big8, 2, var)^.5
  AAPL    BAX     KO    CVS    XOM    IBM    JNJ    DIS
0.0739 0.0556 0.0412 0.0578 0.0459 0.0458 0.0386 0.0579
> cor(big8)
      AAPL   BAX    KO    CVS   XOM   IBM   JNJ   DIS
AAPL 1.000 0.193 0.260 0.329 0.303 0.319 0.145 0.346
BAX  0.193 1.000 0.310 0.330 0.327 0.381 0.473 0.196
KO   0.260 0.310 1.000 0.310 0.338 0.197 0.493 0.348
CVS  0.329 0.330 0.310 1.000 0.442 0.244 0.421 0.537
XOM  0.303 0.327 0.338 0.442 1.000 0.520 0.408 0.650
IBM  0.319 0.381 0.197 0.244 0.520 1.000 0.206 0.348
JNJ  0.145 0.473 0.493 0.421 0.408 0.206 1.000 0.323
DIS  0.346 0.196 0.348 0.537 0.650 0.348 0.323 1.000
```

Note that the shrinkage estimates of standard deviation are all closer to the average sample variance 0.0532 and the shrinkage correlation estimates are all closer to zero than are the estimates based on the sample covariance matrix. □

6.6 Estimation of Portfolio Weights

In Chapters 4 and 5, expressions for the weight vectors of a number of different portfolios were presented; these expressions are generally functions of the means, standard deviations, and correlations of the returns on the various assets under consideration. For instance, the weight vector for the tangency portfolio is given by

$$\boldsymbol{w}_{\mathcal{T}} = \frac{\boldsymbol{\Sigma}^{-1}(\boldsymbol{\mu} - \mu_f \mathbf{1})}{\mathbf{1}^T \boldsymbol{\Sigma}^{-1}(\boldsymbol{\mu} - \mu_f \mathbf{1})}$$

where $\boldsymbol{\mu} - \mu_f \mathbf{1}$ is the vector of mean excess returns for the assets and $\boldsymbol{\Sigma}$ is the covariance matrix of the return vector of the assets.

Of course, in practice, parameters such as $\boldsymbol{\mu}$ and $\boldsymbol{\Sigma}$ are unknown; and, hence, weight vectors such as $\boldsymbol{w}_{\mathcal{T}}$ are unknown and must be estimated. The simplest approach is to use a *plug-in estimator* in which a weight vector is estimated by simplifying "plugging in" estimators for any unknown parameters; that is, functions of unknown parameters are estimated by replacing any parameter by an appropriate estimator.

For instance, the estimator of $\boldsymbol{w}_T$ using the sample mean of the excess returns and the sample covariance matrix is given by

$$\frac{\boldsymbol{S}^{-1}(\bar{\boldsymbol{R}} - \bar{R}_f \mathbf{1})}{\mathbf{1}^T \boldsymbol{S}^{-1}(\bar{\boldsymbol{R}} - \bar{R}_f \mathbf{1})}.$$

Example 6.13 Consider the stocks of eight large companies as discussed in Example 6.6. Suppose we would like to estimate the weight vector of the minimum-variance portfolio. Recall that the sample covariance matrix of the returns in the data matrix `big8` is stored in the matrix `Smat`. Hence, the estimated weight vector of the minimum-variance portfolio is given by

```
> w_mv<-solve(Smat, rep(1, 8))/sum(solve(Smat, rep(1, 8)))
> w_mv
  AAPL    BAX     KO    CVS    XOM    IBM    JNJ    DIS
 0.030 -0.001  0.268  0.032  0.065  0.269  0.345 -0.008
```

Thus, `w_mv` may be viewed as an estimate of the weight vector of the minimum-variance portfolio constructed from the stocks represented in the data matrix `big8`.

Alternatively, we could use a shrinkage estimator of the covariance matrix to estimate the weights of the minimum-variance portfolio. Recall that the shrinkage estimate based on a target matrix of the form $\sigma^2 \boldsymbol{I}$ is stored in the matrix `cov.shrnk$Sigmahat`; see Example 6.12. The corresponding estimate of the minimum-variance portfolio weight vector is given by

```
>  w_mv.sh<-solve(cov.shrnk$Sigmahat, rep(1, 8))/
+  sum(solve(cov.shrnk$Sigmahat, rep(1, 8)))
> w_mv.sh
 AAPL   BAX    KO   CVS   XOM   IBM   JNJ   DIS
0.041 0.051 0.246 0.048 0.096 0.224 0.274 0.020
```

The two estimates are generally similar but there are some differences; for instance, there are no negative weights in the shrinkage estimate.

For the weights of the tangency portfolio, the estimates based on the sample mean returns and the sample covariance matrix are given by

```
> w_tan<-solve(Smat, Rbar)/sum(solve(Smat, Rbar))
> w_tan
  AAPL    BAX     KO    CVS    XOM    IBM    JNJ    DIS
 0.290 -0.097  0.037  0.258 -0.420  0.020  0.506  0.405
```

and the corresponding estimates based on the shrinkage estimates are given by

```
> w_tan.sh<-solve(cov.shrnk$Sigmahat, muhat)/
+ sum(solve(cov.shrnk$Sigmahat, muhat))
> w_tan.sh
  AAPL    BAX     KO    CVS    XOM    IBM    JNJ    DIS
 0.149  0.006  0.175  0.146 -0.044  0.115  0.300  0.152
```

Here the differences in the weights are greater than we saw for the weights of the minimum-variance portfolio. In general, the weights are less extreme, with the largest differences occurring for the largest positive and largest negative weights. This is not surprising given the nature of shrinkage estimates. □

An alternative implementation of the plug-in approach is to replace any unknown parameters in the objective function used to define the weight vector by appropriate estimators. Then optimizing such an estimated objective function yields an estimate of the corresponding portfolio weight vector.

For instance, to estimate the weight vector of the risk-averse portfolio based on the risk-aversion parameter λ, which maximizes

$$\boldsymbol{w}^T \mu - \frac{\lambda}{2} \boldsymbol{w}^T \boldsymbol{\Sigma} \boldsymbol{w},$$

we can maximize the estimator of the criterion function given by

$$\boldsymbol{w}^T \bar{\boldsymbol{R}} - \frac{\lambda}{2} \boldsymbol{w}^T \boldsymbol{S} \boldsymbol{w}.$$

Example 6.14 Consider estimating the weight vector of a risk-averse portfolio of the stocks represented in the data matrix `big8`. Recall that, given the mean vector and covariance matrix of the returns on the assets under consideration, the weight vector of the risk-averse portfolio may be obtained as the solution to a quadratic programming problem. In R, we can calculate such a solution using the function `solve.QP` in the package `quadprog`; see Example 5.9.

The following commands can be used to estimate the portfolio weights of the risk-averse portfolio with parameter $\lambda = 5$ for the stocks represented in the data matrix `big8`.

```
> library(quadprog)
> mu8<-apply(big8, 2, mean) + mean(rfree)
> A1<-cbind(rep(1, 8))
> ra8.5<-solve.QP(Dmat=5*Smat, dvec=mu8, Amat=A1, bvec=1,
+ meq=1)$solution
> ra8.5
[1]  0.606 -0.213 -0.243  0.532 -1.007 -0.281  0.702  0.905
```

Here mu8 is the vector of sample mean returns; the matrix big8 contains excess returns so that the sample mean of the risk-free rate must be added back in.

Note that the estimated weight vector contains large short positions on four stocks. Hence, we might consider enforcing the requirement that all asset weights be nonnegative. The function solve.QP can also be used to calculate the weight vector that maximizes the risk-aversion criterion subject to such a restriction; see Example 5.12.

```
> A2<-cbind(rep(1, 8), diag(8))
> b2<-c(1, rep(0, 8))
> ra8.5.nn<-solve.QP(Dmat=5*Smat, dvec=mu8, Amat=A2, bvec=b2,
+ meq=1)$solution
> round(ra8.5.nn, 5)
[1] 0.392 0.000 0.000 0.343 0.000 0.000 0.000 0.265
```

The function round is used here so that values very close to 0 are written as 0.

Thus, with the nonnegativity constraint, only three stocks are represented in the risk-averse portfolio with $\lambda = 5$: AAPL, CVS, and DIS. It is interesting to note that the weight of JNJ, which is 0.702 in the unconstrained risk-averse portfolio, is zero in the constrained portfolio. □

It is important to keep in mind that the weight vectors obtained using these procedures are estimates of the underlying "true" weight vector. Hence, it is of interest to determine some measure of the sampling variability of the estimates; methods for studying the sampling distribution of an estimator are discussed in the following section.

6.7 Using Monte Carlo Simulation to Study the Properties of Estimators

In Section 6.2, the statistical properties of some simple estimators were considered, such as the mean and standard deviation of the sampling distribution of the sample mean return on an asset. However, such properties are difficult to derive for more complicated estimators, such as shrinkage estimators or estimators of portfolio weights.

In such cases, an alternative approach to studying the behavior of estimators is to use *Monte Carlo simulation*, in which observations are simulated from a known distribution, estimates of interest are calculated, and those simulated values of the estimates are used to assess the sampling distributions of the estimators. In this section, the Monte Carlo method is used to estimate properties of the sampling distributions of a number of estimators.

The basic approach is illustrated on the following example.

Example 6.15 In Example 6.3, we considered the properties of the sample mean return based on the observation of 60 returns, independently distributed according to a distribution with mean 0.01 and standard deviation 0.05. In this example, we perform a similar analysis, using Monte Carlo simulation in place of the theoretical properties discussed in Section 6.2.

To simulate a sequence of 60 such returns, we may use the command `rnorm`. Specifically,

```
> ret<-rnorm(60, mean=0.01, sd=0.05)
```

draws 60 random variables independently from a normal distribution with mean 0.01 and standard deviation 0.05 and places the result in the variable `ret`, a vector of length 60. The sample mean of this vector,

```
> mean(ret)
[1] 0.0178
```

represents a sample of size one from the distribution of the random variable $\bar{R}_j = \sum_{t=1}^{T} R_{j,t}/60$, where $R_{j,1}, R_{j,2}, \ldots, R_{j,60}$ have the distribution described previously.

Of course, a sample of size one does not provide much information. Hence, we are generally interested in a large sample of such simulated sample means. To simulate many sample means at one time, we may use the following procedure. We begin by simulating a matrix of returns, where each row corresponds to a vector of 60 returns.

```
> ret_mat<-matrix(rnorm(60*10000, mean=0.01, sd=0.05), 10000, 60)
```

Here the function `rnorm` simulates 600,000 independent returns, each normally distributed with mean 0.01 and standard deviation 0.05, and the function `matrix` arranges these in a $10,000 \times 60$ matrix.

Using `apply`, we may now calculate the sample mean of each row:

```
> ret_mean<-apply(ret_mat, 1, mean)
```

The variable `ret_mean` now contains a sample of size $10,000$ from the distribution of $\bar{R}_j$. We may now analyze `ret_mean` as we would any sample of observations. For instance,

```
> mean(ret_mean)
[1] 0.01002
> sd(ret_mean)
[1] 0.00642
```

are estimates of the mean and standard deviation, respectively, of the sampling distribution of $\bar{R}_j$.

Recall that, in this scenario, the mean and standard deviation of this sampling distribution may be calculated exactly and, in Example 6.3, they were shown to be 0.01 and 0.00645, respectively. Thus, the Monte Carlo estimates closely match the true values; generally speaking, even closer agreement could be obtained by increasing the number of Monte Carlo replications. That is, we can achieve closer agreement by increasing the number of rows in the matrix ret_mean.

Other properties of the sampling distribution of $\bar{R}_j$ may be found in the same way. For instance,

```
> summary(ret_mean)
    Min.   1st Qu.    Median      Mean   3rd Qu.      Max.
-0.01260  0.00571  0.01003  0.01002  0.01440  0.03110
```

Thus, estimates of the upper and lower quartiles of the sampling distribution of $\bar{R}_j$ are given by 0.00571 and 0.0144, respectively. Recall that, in Example 6.3, we approximated these by 0.00570 and 0.0143, respectively. The advantage of the Monte Carlo method is that these values were obtained using only numerical methods, without using any analytical approximations. One drawback of the Monte Carlo method is that, if the analysis is repeated, different results will be obtained, although if a large Monte Carlo sample size is used, such as the 10,000 used here, the differences are generally slight. For instance, if the calculations described previously are repeated, the sample mean and standard deviation of the simulated return sample means are 0.01009 and 0.00643, respectively.

A histogram of the simulated mean returns gives some information about the shape of the sampling distribution; such a plot can be produced by the command hist(ret_mean). The result is given in Figure 6.2, and it supports the conclusion of the CLT that the sampling distribution of $\bar{R}_j$ is approximately normal. A normal probability plot, as discussed in Section 2.5, could also be considered. □

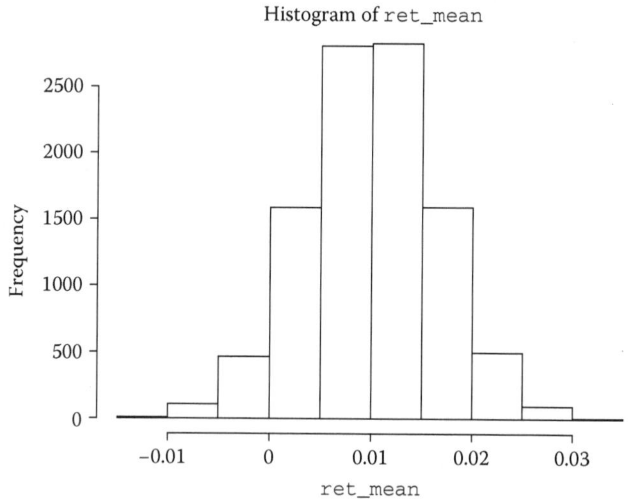

FIGURE 6.2
Histogram of simulated sample mean returns.

Given the form of the estimator—a sample mean—along with the distribution of the returns, which are assumed to be independent and normally distributed, it is not surprising that the distribution of $\bar{R}_j$ is approximately normal; of course, in this case, it is well-known that the distribution of $\bar{R}_j$ is exactly normal. A more extensive study of this type would likely change the assumed distribution of the returns in order to study the effect of distributional assumptions on the properties of the estimator. With Monte Carlo simulation, such changes are generally easy to implement.

Example 6.16 Consider the Monte Carlo simulation described in Example 6.15, which analyzed the properties of the sample mean return based on the observation of 60 returns, independently distributed according to a distribution with mean 0.01 and standard deviation 0.05, but now suppose that the observations are not normally distributed. Empirical studies have shown that the t-distribution is often useful for modeling return data; thus, here we assume that the returns follow a t-distribution with six degrees of freedom, with location and scale parameters chosen to achieve the desired mean and standard deviation.

Changing the distribution used in the Monte Carlo simulation from the normal distribution to the t-distribution focuses on changing the function `rnorm` used in Example 6.15 to the function `rt`, which simulates observations from a t-distribution. However, the properties of the t-distribution are such that some care is needed in order to achieve the required results.

The degrees of freedom in `rt` is specified by the argument `df`; for example, `rt(1, df=6)` returns one observation from a t-distribution with six degrees of freedom. The complications arise when specifying the mean and standard deviation of the distribution. A random variable with a t-distribution with ν degrees of freedom has mean 0 and variance $\nu/(\nu-2)$, provided that $\nu > 2$; thus, a random variable with a t-distribution with six degrees of freedom has mean 0 and standard deviation $\sqrt{1.5}$. It follows that to draw a random sample of size n from a t-distribution with mean 0.01, standard deviation 0.05, and six degrees of freedom, we use the command

```
rt(n, df=6)*(0.05/(1.5^.5))+0.01
```

Therefore, to repeat the Monte Carlo study described in Example 6.15, but with a t-distribution with six degrees of freedom replacing the standard normal distribution, we use the commands

```
> ret_mat_t<-matrix(rt(60*10000, df=6)*(0.05/(1.5^.5))+0.01,
+ 10000, 60)
> ret_mean_t<-apply(ret_mat_t, 1, mean)
> mean(ret_mean_t)
[1] 0.00999
> sd(ret_mean_t)
[1] 0.00641
> summary(ret_mean_t)
```

```
    Min.  1st Qu.   Median     Mean  3rd Qu.     Max.
-0.01340  0.00566  0.01000  0.00999  0.01430  0.03450
```

Note that the results are generally similar to those obtained in Example 6.15, suggesting that the sampling distribution of the sample mean is close to normal even if the returns follow a t-distribution. This is not surprising given that each sample mean is based on 60 observations; the approximation given by the CLT tends to be accurate for that sample size. □

In the examples considered thus far in this section, Monte Carlo simulation was applied to an estimator whose properties may be determined using analytic methods. However, the Monte Carlo method is most useful when such analytic results are not readily available. Many such cases occur when the estimator under consideration is a function of the returns on several assets, so that the correlation structure of the returns plays a role in the sampling distribution.

Simulating a Return Vector

To handle such cases using Monte Carlo simulation, we need a method of simulating a random vector in which the component random variables are correlated. For instance, let $\boldsymbol{R}_1, \boldsymbol{R}_2, \ldots, \boldsymbol{R}_T$ denote N-dimensional random vectors, with $\boldsymbol{R}_t$ representing the vector of asset returns at time t. These random vectors may be viewed as the multivariate analogues of the returns $R_{j,1}, R_{j,2}, \ldots, R_{j,60}$ analyzed in Example 6.15. Note that although it is often reasonable to assume that $\boldsymbol{R}_t$ and $\boldsymbol{R}_s$ are uncorrelated or even independent, for $t \neq s$, the components of $\boldsymbol{R}_t$, which represent the returns on different assets at time t, cannot realistically be modeled as uncorrelated random variables.

Consider simulation of $\boldsymbol{R}_t$ under the assumption that this random vector has mean vector $\boldsymbol{\mu}$ and covariance matrix $\boldsymbol{\Sigma}$ for given values of $\boldsymbol{\mu}$ and $\boldsymbol{\Sigma}$. For the distribution of $\boldsymbol{R}_t$, we first use the multivariate normal distribution; we then will consider a multivariate version of a t-distribution.

A random vector $\boldsymbol{Y} = (Y_1, Y_2, \ldots, Y_N)^T$ has a multivariate normal distribution if any linear function of $\boldsymbol{Y}$, that is, any function of the form $\sum_{j=1}^{N} a_j Y_j$ for constants $a_1, a_2, \ldots, a_N$, has a univariate normal distribution. The parameters of the multivariate normal distribution are its mean vector and its covariance matrix.

To simulate random variates with a multivariate normal distribution in R, we use the function `mvrnorm`, which is available in the package `MASS` (Venables and Ripley 2002); the following example illustrates how `mvrnorm` may be used to conduct a simulation study similar to the one conducted in Example 6.15. The arguments of `mvrnorm` are `n`, the number of samples requested, `mu`, the mean vector of the distribution, and `Sig`, the covariance matrix of the distribution.

Example 6.17 Consider a three-dimensional multivariate normal distribution with mean vector $(0.1, 0.2, 0.3)^T$ and covariance matrix

$$\begin{pmatrix} 0.2 & 0.1 & 0.1 \\ 0.1 & 0.2 & 0.1 \\ 0.1 & 0.1 & 0.2 \end{pmatrix}. \tag{6.9}$$

Thus, each of the three component random variables in the random vector has variance 0.2 and the correlation between any two such random variables is 0.5.

The following command generates one sample from this distribution.

```
> library(MASS)
> mu0<-c(0.1,0.2,0.3)
> Sig0<-matrix(c(0.2,0.1,0.1,0.1,0.2,0.1,0.1,0.1,0.2), 3, 3)
> Sig0
     [,1] [,2] [,3]
[1,]  0.2  0.1  0.1
[2,]  0.1  0.2  0.1
[3,]  0.1  0.1  0.2
> mvrnorm(n=1, mu=mu0, Sig=Sig0)
[1] -0.3335  0.0529  0.0982
```

For $n = 1$, the result of the function is a vector; for $n > 1$, the result is a matrix, with each row corresponding to a simulated random vector. For example, to draw four random vectors from the multivariate normal distribution with mean vector `mu0` and covariance matrix `Sig0`, we use the command

```
> mvrnorm(4, mu=mu0, Sig=Sig0)
        [,1]     [,2]   [,3]
[1,] -0.193  0.1644 -0.273
[2,]  0.297  0.3012 -0.148
[3,]  0.603 -0.0955  1.085
[4,] -0.149  0.2805 -0.331
```

Therefore, if the returns on a set of three assets are modeled as random vectors with mean `mu0` and covariance matrix `Sig0`, and the return vectors in different time periods are independent, then the matrix `ret.sim` represents the returns on the three assets over four time periods, with each row representing a time period. Hence, `ret.sim` is a simulated data matrix of the type described in Section 6.3 and is similar to the observed data matrix `big8`.

Simulated values of functions of these returns may be calculated in the usual way. For instance, simulated return sample means for the three assets are given by

```
> apply(ret.sim, 2, mean)
[1] 0.1393 0.1627 0.0833
```

Now consider replications of this procedure. For the case of a single asset, in which the returns are given in a vector, independent replications of the returns may be stored in a matrix, with each row corresponding to a series of asset returns. When simulating a vector of returns, one replication of the simulation is already a matrix. Although it is possible to handle the replications using a three-dimensional array, a generalization of a matrix (an interesting exercise, for those so inclined), the simplest approach is to use a loop.

For instance, to obtain 10,000 simulated values of the return sample means for the three assets in the example, we can use the following commands:

```
ret_means<-matrix(0, 10000, 3)
for (j in 1:10000){
+         ret_sim<-mvrnorm(4, mu=mu0, Sig=Sig0)
+         ret_means[j, ]<-apply(ret_sim, 2, mean)
+           }
```

The command `ret_means<-matrix(0, 10000, 3)` creates a matrix that will store the results. Each iteration of the loop simulates a vector of sample means of the asset returns and stores them in a row of `ret_means`.

The results in `ret_means` may now be used to estimate properties of the sampling distribution. For instance, the mean of the sampling distribution of $\boldsymbol{R}_t$ is estimated by

```
> apply(ret_means, 2, mean)
[1] 0.0967 0.2007 0.2991
```

which is close to the known mean vector $(0.1, 0.2, 0.3)^T$. □

Many statistical methods tend to work particularly well for normally distributed data. Thus, in conducting a Monte Carlo study, it is often of interest to consider other distributions in addition to the multivariate normal.

The multivariate t-distribution is a multivariate generalization of the t-distribution; its relationship to the univariate t-distribution is similar to the relationship the multivariate normal distribution has to the univariate normal distribution.

To simulate random variates with a multivariate t-distribution in R, we use the function `rmvt`, which is available in the package `mvtnorm` (Genz et al. 2016). In the function `rmvt`, we may specify the number of samples (the argument `n`), the degrees of freedom (the argument `df`), and the "scale" matrix (the argument `sigma`). Let $\boldsymbol{A}$ denote the scale matrix; then the covariance matrix of the distribution is given by $\boldsymbol{\Sigma} = \boldsymbol{A}(\nu/(\nu-2))$, where ν denotes the degrees of freedom of the distribution. The following example illustrates the use of `rmvt`.

Example 6.18 Consider a three-dimensional random vector with the mean vector and covariance matrix given in Example 6.17 and stored in the R variables `mu0` and `Sig0`, respectively. Furthermore, assume that the random vector has a multivariate t-distribution.

To generate one sample from the multivariate t-distribution with six degrees of freedom, mean vector mu0, and covariance matrix Sig0, we use the following command.

```
> library(mvtnorm)
> rmvt(n=1, df=6, sigma=Sig0)/(1.5^.5) + mu0
      [,1]  [,2]   [,3]
[1,] 0.172 0.677 0.0995
```

Note that rmvt draws a random vector from a multivariate distribution with mean vector **0**; adding the vector mu0 to the result modifies the mean vector to mu0.

To simulate several random vectors, we can increase the value of n. However, in order to add the mean vector to the simulated random vectors, we must construct a matrix of mean vectors. For instance, suppose that we wish to simulate four random vectors. Then

```
> rmvt(n=4, df=6, sigma=Sig0)/(1.5^.5)
        [,1]    [,2]   [,3]
[1,]  1.2761 -0.2582 -0.367
[2,] -0.0084  0.0747 -0.676
[3,]  0.7659  0.3709  0.588
[4,]  0.4114  0.1233  0.324
```

returns a data matrix with the correct covariance matrix but each value is simulated from a distribution with mean zero.

To add the mean vector to the result of rmvt, note that matrix(mu0, 4, 3, byrow=T) returns a matrix with each row equal to mu0:

```
> matrix(mu0, 4, 3, byrow=T)
     [,1] [,2] [,3]
[1,]  0.1  0.2  0.3
[2,]  0.1  0.2  0.3
[3,]  0.1  0.2  0.3
[4,]  0.1  0.2  0.3
```

Thus, the command

```
> rmvt(n=4, df=6, sigma=Sig0)/(1.5^.5)+matrix(mu0, 4, 3, byrow=T)
        [,1]    [,2]    [,3]
[1,] -0.0498  0.4837 -0.2124
[2,] -0.4936  0.7804  0.0633
[3,]  0.2842  0.0878  0.3022
[4,] -0.0518 -0.0329 -0.1120
```

returns random variates simulated from the correct distribution.

To verify that the command is working as desired, we may simulate a data matrix with 1000 observations on the random vector,

```
> x<-rmvt(n=1000, df=6, sigma=Sig0)/(1.5^.5)+matrix(mu0, 1000, 3,
+ byrow=T)
```

and verify that the sample mean vector and sample covariance matrix are close to `mu0` and `Sig0`, respectively.

```
> apply(x, 2, mean)
[1] 0.110 0.203 0.293
> cov(x)
       [,1]   [,2]   [,3]
[1,] 0.220 0.1116 0.1063
[2,] 0.112 0.2055 0.0956
[3,] 0.106 0.0956 0.2043
```

□

We now consider two examples illustrating how Monte Carlo simulation may be used to better understand the properties of statistical methods used to analyze financial data.

Using Monte Carlo Simulation to Describe the Sampling Distribution of a Statistic

In Chapters 4 and 5, a number of different approaches to constructing portfolio weights were discussed; these methods are based on the means, standard deviations, and correlations of the returns on the assets under consideration, along with some criteria for choosing the weights. Here we use Monte Carlo simulation to assess the variability in the estimated weights for the tangency portfolio of two assets.

Let μ_j and σ_j denote the mean and standard deviation, respectively, of the return on asset j, $j = 1, 2$ and let ρ denote the correlation of the returns on the two assets. Then, according to the analysis in Section 4.6, the tangency portfolio places weight

$$w_T = \frac{(\mu_1 - \mu_f)\sigma_1^2 - (\mu_2 - \mu_f)\rho_{12}\sigma_1\sigma_2}{(\mu_2 - \mu_f)\sigma_1^2 + (\mu_1 - \mu_f)\sigma_2^2 - [(\mu_1 - \mu_f) + (\mu_2 - \mu_f)]\rho_{12}\sigma_1\sigma_2} \tag{6.10}$$

on asset 1, where μ_f denotes the return on the risk-free asset. Alternatively, we may use the formula for the tangency weights for a vector of assets,

$$\frac{\boldsymbol{\Sigma}^{-1}(\boldsymbol{\mu} - \mu_f\mathbf{1})}{\mathbf{1}^T\boldsymbol{\Sigma}^{-1}(\boldsymbol{\mu} - \mu_f\mathbf{1})},$$

which is easier to evaluate numerically. It is straightforward to show that this formula reduces to (6.10) for $N = 2$.

Example 6.19 Consider two assets such that the excess return on the first asset has mean 0.025 and variance 0.0055 and the excess return on the second asset has mean 0.010 and variance 0.0017; take the covariance of the returns to be 0.0008. These are the observed values (slightly rounded) for Apple stock and Coca-Cola stock based on the `big8` data. Define variables `mu1` and `Sig1` to represent the mean vector and covariance matrix of the excess returns:

```
> mu1<-c(0.025, 0.010)
> Sig1<-matrix(c(0.0055, 0.0008, 0.0008, 0.0017), 2, 2)
```

Thus, the tangency portfolio has weights

```
> solve(Sig1, mu1)/sum(solve(Sig1, mu1))
[1] 0.496 0.504
```

so that this portfolio places roughly half its investment on each asset.

Simulating the data according to a multivariate normal distribution, as described previously in this section, we may draw a sample from the sampling distribution of the weight on Apple stock (note that, since the weights sum to 1, we only need to consider one of the weights).

```
> wgts<-rep(0, 10000)
> sharpe<-rep(0, 10000)
> for (j in 1:10000){
+     ret_sim<-mvrnorm(60, mu=mu1, Sig=Sig1)
+     mean_sim<-apply(ret_sim, 2, mean)
+     sig_sim<-cov(ret_sim)
+     wgt_sim<-solve(sig_sim, mean_sim)/sum(solve(sig_sim,
+     mean_sim))
+     wgts[j]<-wgt_sim[1]
+     sharpe[j]<-sum(mean_sim*wgt_sim)/((wgt_sim%*%sig_sim%*%
+     wgt_sim)^.5)
+ }
```

The vector `wgts` contains a sample of size 10,000 from the sampling distribution of w_T, the weight on Apple stock in the tangency portfolio of Apple and Coca-Cola stocks, each based on 60 observations; the vector `sharpe` contains the estimated Sharpe ratio corresponding to the estimated tangency portfolio. The sampling distribution of w_T can be summarized using the usual functions; for example,

```
> mean(wgts)
[1] 0.64
> median(wgts)
[1] 0.487
```

Hence, although the median weight of 0.487 is close to the true weight of 0.496, the mean weight is considerably larger than the true weight, suggesting a skewed distribution.

The quantiles of the sampling distribution are a useful summary of the distribution; these may be calculated using the `quantile` function.

```
>probvec<-c(0.01, 0.05, 0.10, 0.25, 0.50, 0.75, 0.90, 0.95, 0.99)
> quantile(wgts, prob=probvec)
    1%      5%     10%     25%     50%     75%     90%     95%     99%
-0.439   0.095   0.190   0.324   0.487   0.713   1.057   1.476   3.565
```

These results show that there is considerable variability in the estimated weights. For instance, in more than 10% of the Monte Carlo simulations the weight for Apple stock is greater than one and, hence, the weight for Coca-Cola is negative.

The maximum Sharpe ratios also exhibit a high degree of variability:

```
> quantile(sharpe, prob=probvec)
   1%    5%   10%   25%   50%   75%   90%   95%   99%
0.018 0.189 0.235 0.311 0.399 0.491 0.575 0.629 0.737
```

These values may be compared to the Sharpe ratio of the tangency portfolio based on the true distribution, which is 0.373. □

The analysis in Example 6.19 is based on the assumption that the returns in a given period have a multivariate normal distribution; we may repeat the analysis using the assumption that the returns have a multivariate t-distribution with six degrees of freedom.

Example 6.20 Consider the two assets described in Example 6.19, with mean excess return vector given in `mu1` and return covariance matrix given in `Sig1`.

Simulating the data according to a multivariate t-distribution with six degrees of freedom, as described previously in this section, we may draw a sample from the sampling distribution of the weight on Apple stock using the following commands.

```
> wgts.t<-rep(0, 10000)
> sharpe.t<-rep(0, 10000)
> for (j in 1:10000){
+     ret_sim<-rmvt(60, df=6, sigma=Sig1)/(1.5^.5) + matrix(mu1,
+     60, 2, byrow=T)
+     mean_sim<-apply(ret_sim, 2, mean)
+     sig_sim<-cov(ret_sim)
+     wgt_sim<-solve(sig_sim, mean_sim)/sum(solve(sig_sim,
+     mean_sim))
+     wgts.t[j]<-wgt_sim[1]
+     sharpe.t[j]<-sum(mean_sim*wgt_sim)/((wgt_sim%*%sig_sim%*%
+     wgt_sim)^.5)
+ }
```

The vector `wgts.t` contains a sample of size 10,000 from the sampling distribution of the weight on Apple stock in the tangency portfolio of Apple and

Coca-Cola stocks under the assumption that the returns have a t-distribution. Thus, the sampling distribution of this weight may be summarized by its quantiles:

```
> quantile(wgts.t, prob=probvec)
    1%      5%     10%     25%     50%     75%     90%     95%     99%
-0.638  0.085  0.180  0.320  0.486  0.710  1.054  1.418  3.487
```

Note that the results are very similar to those in Example 6.19.

The quantiles of the sampling distribution of the maximum Sharpe ratio are given by

```
> quantile(sharpe.t, prob=probvec)
    1%      5%     10%     25%     50%     75%     90%     95%     99%
-0.060  0.187  0.236  0.315  0.404  0.500  0.588  0.641  0.746
```

Again, these results are similar to those in Example 6.19. □

The results of the previous two examples suggest that tangency weights based on sample data are not well-determined. There are two general reasons for this. One is the variability in the estimates of the mean vector and the covariance matrix, as we have discussed in this section. The other is that, although the tangency portfolio maximizes the Sharpe ratio, many portfolios have a Sharpe ratio close to the maximum value and, hence, the tangency portfolio itself is not very well-defined. This fact may be illustrated in the following example.

Example 6.21 Consider the framework in Example 6.19. Recall that the tangency portfolio, which maximizes the Sharpe ratio, places weight 0.496 on Apple stock. Figure 6.3 contains a plot of the Sharpe ratio of a portfolio of

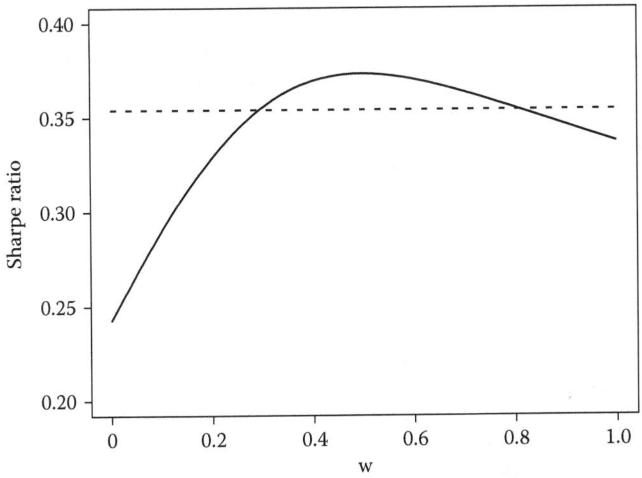

FIGURE 6.3
Plot of the Sharpe ratio versus w in Example 6.21.

Apple and Coca-Cola stocks, using the distribution of returns described in Example 6.19, as a function of w, the weight placed on Apple stock. The dotted line corresponds to a Sharpe ratio of 0.354, which is 95% of the maximum value; given the results in Example 6.19, this is a small difference from the maximum Sharpe ratio relative to the variation in estimating the maximum Sharpe ratio.

The values of w corresponding to a Sharpe ratio of 95% of the maximum value are roughly 0.29 and 0.81. Thus, for a wide range of values of w, the Sharpe ratio is close to the maximum value. □

Using Monte Carlo Simulation to Compare Estimators

In many cases, there are two or more estimators that might be used to estimate a given parameter; and, hence, it is important to use the one that can be expected to yield more accurate estimates. In such cases, Monte Carlo simulation is often useful in evaluating the estimators.

The following example compares the shrinkage estimator of a mean return, as described in Section 6.5, with the estimator based on the sample mean of the return vectors.

Example 6.22 The procedure used to calculate the shrinkage estimate of asset mean returns in Example 6.11, based on the data in the matrix `big8`, may be summarized as follows.

```
> Rbar<-apply(big8, 2, mean)
> S2<-apply(big8, 2, var)
> tau2<-mean((Rbar-mean(Rbar))^2)
> psi<-(mean(S2)/60)/(tau2 + (mean(S2)/60))
> muhat<-psi*mean(Rbar) + (1-psi)*Rbar
```

Suppose we simulate returns using the approach described in Example 6.17, taking `mu` and `Sig` to be the estimates from the `big8` data set; then a simulated value for the vector of shrinkage estimates may be calculated using the aforementioned procedure, with a simulated data matrix replacing `big8`. The following commands repeat such a procedure 10,000 times.

```
> library(MASS)
> shrnk_means<-matrix(0, 10000, 8)
>  sample_means<-matrix(0, 10000, 8)
>  for (j in 1:10000){
+      ret_sim<-mvrnorm(60, mu=Rbar, Sig=Sighat)
+      Rb<-apply(ret_sim, 2, mean)
+      S2<-apply(ret_sim, 2, var)
+      tau2<-mean((Rb-mean(Rb))^2)
+      psi<-(mean(S2)/60)/(tau2 + (mean(S2)/60))
+      muhat<-psi*mean(Rb) + (1-psi)*Rb
+      shrnk_means[j, ]<-muhat
```

```
+       sample_means[j, ]<-Rb
+     }
```

The matrix `shrnk_means` contains a sample of 10,000 from the sampling distribution of the vector of shrinkage estimates of the mean asset returns, with one set of estimates in each row of the matrix. The matrix `sample_means` contains the sample means from each iteration in a similar format.

The results in `shrnk_means` and `sample_means` may be summarized using the usual functions. For instance, the mean vector of the sampling distribution of the shrinkage estimator may be obtained by

```
> apply(shrnk_means, 2, mean)
[1] 0.0211 0.0097 0.0113 0.0183 0.0104 0.0090 0.0122 0.0180
```

and the bias of the shrinkage estimator is given by

```
> apply(shrnk_means, 2, mean)-Rbar
   AAPL     BAX     KO     CVS     XOM     IBM     JNJ     DIS
-0.0043  0.0023  0.0015 -0.0029  0.0022  0.0030  0.0007 -0.0027
```

Thus, the shrinkage estimator appears to be biased; the standard error of the estimated bias is the sample standard deviation of the values in each column of `shrnk_means` divided by the square root of the sample size, which in this context is 10,000, the number of Monte Carlo replications:

```
> apply(shrnk_means, 2, sd)/10000^.5
[1] 7.97e-05 5.67e-05 4.39e-05 6.36e-05 5.01e-05 4.92e-05
    4.36e-05 6.33e-05
```

Therefore, for each asset, the estimated bias is much greater than its standard error; for instance, the ratios of the estimated biases to their standard errors, that is, the t-statistics for testing the hypothesis of no estimator bias, are given by

```
 AAPL   BAX    KO   CVS   XOM   IBM   JNJ   DIS
-53.6  40.9  33.6 -45.4  42.9  60.5  16.1 -43.3
```

Hence, based on these results, we conclude that the shrinkage estimators are biased; this is to be expected because the shrinkage estimate of an asset mean return is a weighted average of the sample mean return for that asset and the sample mean return for all assets.

A similar analysis can be done on the sample mean returns. The estimated biases are given by

```
> apply(sample_means, 2, mean)-Rbar
    AAPL         BAX          KO         CVS         XOM         IBM
-2.86e-05  3.98e-05 -7.07e-05 -4.33e-05 -4.45e-05 -1.64e-05
    JNJ         DIS
-5.85e-05 -4.29e-05
```

and the ratios of the estimated biases to their standard errors are

```
> (apply(sample_means, 2, mean)-Rbar)/(apply(sample_means, 2,
+ sd)/10000^.5)
  AAPL    BAX     KO    CVS    XOM    IBM    JNJ    DIS
-0.300  0.558 -1.354 -0.576 -0.748 -0.279 -1.161 -0.573
```

As expected, none of the estimated biases is statistically significant at the 5% level.

To compare the accuracies of these estimators, we can look at the *mean squared error* (MSE) of the estimators. For an estimator $\hat{\theta}$ of a real-valued parameter θ, the MSE is given by

$$\text{MSE}(\hat{\theta}) = \text{E}\left((\hat{\theta} - \theta)^2\right);$$

the MSE is therefore a type of expected squared-distance between the estimator and the parameter we are trying to estimate. Thus, a smaller value of the MSE indicates a more accurate estimator, on average.

It may be shown that

$$\text{MSE}(\hat{\theta}) = \text{bias}(\hat{\theta})^2 + \text{Var}(\hat{\theta}),$$

where $\text{bias}(\hat{\theta}) = \text{E}(\hat{\theta}) - \theta$ is the bias of the estimator. Hence, the MSE combines the bias of an estimator with the variability of its sampling distribution, as measured by the variance. Often the square root of the MSE is reported (the "root mean squared error" or RMSE), for the same reason that the standard deviation is often preferred to the variance as a measure of variability.

For the shrinkage estimators, the MSE may be estimated using the results in `shrnk_means`.

```
> shrnk.mse<-(apply(shrnk_means, 2, mean)-Rbar)^2 +
+ apply(shrnk_means, 2, var)
```

Monte Carlo estimates of the RMSE of the estimators are therefore

```
> shrnk.mse^.5
   AAPL     BAX      KO     CVS     XOM     IBM     JNJ     DIS
0.00904 0.00613 0.00463 0.00699 0.00546 0.00575 0.00442 0.00689
```

The analogous results for the sample mean returns are

```
> mean.mse<-(apply(sample_means, 2, mean)-Rbar)^2 +
+ apply(sample_means, 2, var)
> mean.mse^.5
   AAPL     BAX      KO     CVS     XOM     IBM     JNJ     DIS
0.00954 0.00714 0.00522 0.00752 0.00595 0.00587 0.00504 0.00749
```

The ratios of the RMSE values for the shrinkage estimators to the RMSE values for the sample mean returns are given by

```
> sort((shrnk.mse/mean.mse)^.5)
  BAX   JNJ    KO   XOM   DIS   CVS  AAPL   IBM
0.858 0.876 0.887 0.917 0.921 0.929 0.948 0.980
```

Note that the `sort` function puts the vector in increasing order.

Thus, for all of the stocks, the shrinkage estimator has a smaller estimated RMSE than does the sample mean estimator. That is, the biases of the shrinkage estimators are more than offset by their smaller standard deviations, leading to more accurate estimates, on average.

It is important to keep in mind that the analysis in this example, like all analyses based on Monte Carlo simulation, are simply estimates of the true properties of the estimators under consideration. In particular, if the Monte Carlo analysis is repeated, the results will change; however, with a large Monte Carlo sample size, such as the 10,000 used here, the changes tend to be small. For instance, if the analysis in this example is repeated, the new estimates of the ratios of the RMSE values are given by

```
  BAX   JNJ    KO   DIS   XOM   CVS  AAPL   IBM
0.865 0.880 0.888 0.915 0.919 0.920 0.938 0.979
```

Note that, although the ratios have all changed from the original estimates, the changes are minor and the general conclusions regarding the estimators do not change. □

6.8 Suggestions for Further Reading

Parameter estimation is a central topic in statistics. Rice (2007, Chapter 8) presents a good, general discussion of parameter estimation. Johnson and Wichern (2007, Chapter 3) consider properties of the sample mean vector and sample covariance matrix based on observation of a random vector; see also Jobson and Korkie (1980). Fabozzi et al. (2006, Chapter 8) discuss estimation from the perspective of finance. The properties of a sample covariance matrix when the number of variables is close to the number of time periods observed are studied in an area of statistics known as *random matrix theory*; see, for example, Fabozzi et al. (2006, Chapter 8) for a useful summary. The result in Section 6.2 on the trace of the inverse of the sample correlation matrix is attributed to Bouchaud and Potters (2011), who present further discussion of the implications of such a result.

EWMA estimators are discussed by Miller (2012, Chapter 10) and by Alexander (2008, Section II.3.8). Shrinkage estimation, also known as *empirical Bayes* estimation, is an important statistical method; see Efron and Morris (1970) for a useful introduction and Carlin and Louis (2000) for a book-length treatment. The shrinkage estimator of the mean vector presented here is attributed to DeMiguel et al. (2013) who show that it performs well in portfolio analysis; see Fabozzi et al. (2006, Chapter 9) for an alternative approach. Jorion (1986) and Ledoit and Wolf (2004) discuss shrinkage estimation of expected returns and the return covariance matrix, respectively, in the context of portfolio analysis.

Monte Carlo simulation is an extremely useful technique for understanding the properties of statistical methods in complex settings. Evans and Rosenthal (2004, Chapter 4) offer a good introduction to the use of Monte Carlo methods in understanding sampling distributions. Ross (2013) provides a detailed introduction to simulation in general; see Hammersley and Handscomb (1964) for a more advanced introduction to Monte Carlo methods. The Monte Carlo methods described here are closely related to the statistical method known as the *parametric bootstrap*; see, for example, Efron and Tibshirani (1993, Section 6.5) and Davison and Hinkley (1997, Section 2.2). The suggestion to model asset returns as random variables having a t-distribution with six degrees of freedom is attributed to Praetz (1972); see also Blattberg and Gonedes (1974) and Gray and French (1990). The sensitivity of optimal portfolio weights to estimation error is discussed by Best and Grauer (1991) and DeMiguel et al. (2009).

6.9 Exercises

1. Consider the returns on two stocks, Papa John's International, Inc. (symbol PZZA), and Bed Bath & Beyond, Inc. (BBBY). For each stock, calculate five years of monthly returns for the period ending December 31, 2015.

   a. Calculate the sample mean of the returns on each stock.
   b. Calculate the sample standard deviation of the returns on each stock.
   c. Calculate the sample correlation of the returns on the two stocks.

2. Repeat Exercise 1 using five years of daily returns for the period ending December 31, 2015.

   Compare the sample means and sample standard deviations for the daily returns to the corresponding values for the monthly returns. Do the relationships between monthly and daily values discussed in Section 2.5 appear to hold at least approximately? For the comparisons, round the results to three significant figures.

   Compare the sample correlation of the daily returns to the sample correlation of the monthly returns.

3. Using data on three-month Treasury Bills obtained from the Federal Reserve website, calculate five years of monthly risk-free rates

for the period ending December 31, 2015. Calculate five years of monthly excess returns for this period for Papa John's and Bed Bath & Beyond stock.

a. Calculate the sample mean of the excess returns on each stock.
b. Calculate the sample standard deviation of the excess returns on each stock.
c. Calculate the sample correlation of the returns on the two stocks using the excess returns.
d. Compare the results obtained in Parts (b) and (c) to those obtained in Exercise 1. For the comparison, round the results to three significant figures.

4. Calculate approximate 95% confidence intervals for the mean monthly excess return on Papa John's stock and the mean monthly excess return on Bed Bath & Beyond stock. Using the procedure discussed in Example 6.5, calculate an approximate 95% confidence interval for the difference in mean monthly excess returns on these two stocks.

5. Using a decay parameter of 0.93, calculate weighted estimates of the mean and standard deviation of the monthly excess return on Papa John's stock; see Examples 6.7 and 6.8. Compare the results to the (unweighted) sample mean and sample standard deviation.

6. Construct a data matrix consisting of five years of monthly excess returns on five stocks, Papa John's International, Inc. (symbol PZZA), Bed Bath & Beyond, Inc. (BBBY), Netflix, Inc. (NFLX), Time Warner, Inc. (TWX), and Verizon Communications, Inc. (VZ); use returns for the time period ending December 31, 2015. Add column names corresponding to the stock symbols to the data matrix.

   a. Calculate the sample mean of the excess returns of each stock.
   b. Calculate the sample mean of the risk-free rate $\bar{R}_f$ and use that to calculate the sample mean of the standard returns of each stock.
   c. Calculate the sample standard deviation of the excess returns of each stock.
   d. Calculate the sample covariance matrix of the excess returns of the five stocks.
   e. Calculate the sample correlation matrix of the excess returns of the five stocks.

7. Consider the data matrix constructed in Exercise 6, consisting of five years of monthly excess returns on five stocks. Let $\mu_1, \mu_2, \ldots, \mu_5$ denote the respective mean returns on the stocks. Using the procedure described in Example 6.11, construct shrinkage estimates of

$\mu_1 - \mu_f, \mu_2 - \mu_f, \ldots, \mu_5 - \mu_f$ where μ_f is the mean return on the risk-free asset.

8. Let $\boldsymbol{S}$ denote an $N \times N$ sample covariance matrix. Show that $\psi\sigma^2\boldsymbol{I} + (1 - \psi)\boldsymbol{S}$ is positive-definite for any $\sigma^2 > 0$, and any ψ.

9. Consider the data matrix constructed in Exercise 6, consisting of five years of monthly excess returns on five stocks. Let $\boldsymbol{\Sigma}$ denote the covariance matrix of the returns on the five stocks. Using the procedure described in Example 6.12, construct a shrinkage estimate of $\boldsymbol{\Sigma}$ using a matrix of the form $\sigma^2\boldsymbol{I}$ as the target matrix.

   Compare the shrinkage estimates of the asset return standard deviations to the sample standard deviations. Compare the corresponding shrinkage estimate of the correlation matrix of the assets to the sample correlation matrix.

10. Consider the data matrix constructed in Exercise 6, consisting of five years of monthly excess returns on five stocks. Using the sample covariance matrix based on the excess returns, estimate the weights of the minimum-variance portfolio for the five stocks corresponding to symbols PZZA, BBBY, NFLX, TWX, and VZ; see Example 6.13.

11. Consider the data matrix constructed in Exercise 6, consisting of five years of monthly excess returns on five stocks. For these stocks, estimate the weights of the risk-averse portfolio based on a risk-aversion parameter $\lambda = 8$; see Example 6.14.

12. Repeat the calculation in Exercise 11, adding the restriction that all portfolio weights must be nonnegative.

13. Consider the data matrix constructed in Exercise 6, consisting of five years of monthly excess returns on five stocks. Using the shrinkage estimate of the return covariance matrix calculated in Exercise 9, estimate the weights of the minimum-variance portfolio for the five stocks corresponding to symbols PZZA, BBBY, NFLX, TWX, and VZ; see Example 6.13.

14. Consider the data matrix constructed in Exercise 6, consisting of five years of monthly excess returns on five stocks. Using the sample means and the sample covariance matrix based on the excess returns, estimate the weights of the tangency portfolio for the five stocks corresponding to symbols PZZA, BBBY, NFLX, TWX, and VZ.

15. Consider the data matrix constructed in Exercise 6, consisting of five years of monthly excess returns on five stocks. Using the shrinkage estimate of the mean excess return vector calculated in Exercise 7 and the shrinkage estimate of the return covariance matrix calculated in Exercise 9, estimate the weights of the tangency portfolio for the five stocks corresponding to symbols PZZA, BBBY, NFLX, TWX, and VZ.

16. Consider an asset with return R; let $\mu = \mathrm{E}(R)$ and $\sigma^2 = \mathrm{Var}(R)$; let R_f denote the return on the risk-free asset and let $\mu_f = \mathrm{E}(R_f)$.

    Let R_p denote a portfolio return of the form $wR + (1-w)R_f$. Suppose that w is chosen to achieve a return standard deviation of σ^* for some given $\sigma^* > 0$; then w, the weight given to the risky asset in the portfolio, must be taken to be σ^*/σ.

    Now suppose that σ is unknown and must be estimated based on a sample of 60 observed returns on the risky asset; let S denote the sample standard deviation of these returns. Then the estimated weight given to the risky asset in the portfolio is σ^*/S and the actual standard deviation of the portfolio return based on this estimate is

$$\bar{\sigma} = \frac{\sigma^*}{S}\sigma.$$

    The purpose of this exercise is to use Monte Carlo simulation to study the properties of the quantity $\bar{\sigma}$ in this setting. Take $\sigma^* = 0.02$ and suppose that $\sigma = 0.04$; because the analysis uses only the standard deviations, we may take $\mu = 0$ without loss of generality.

    a. Using the function `rnorm`, construct a $10,000 \times 60$ matrix of observations drawn independently from a normal distribution with mean 0 and standard deviation 0.04; see Example 6.15. Using the `apply` function, compute the sample standard deviation of each row.
    b. Using the vector of sample standard deviations, calculate a vector of values of $\bar{\sigma}$, using $\sigma^* = 0.02$ and $\sigma = 0.04$. This vector represents a sample from the distribution of $\bar{\sigma}$.
    c. Using the `quantile` function, estimate the quantiles of the distribution of $\bar{\sigma}$ corresponding to probabilities 0.01, 0.05, 0.10, 0.20, 0.50, 0.80, 0.90, 0.95, 0.99. Summarize the results.
    d. Repeat Parts (a)–(c) using a t-distribution with six degrees of freedom in place of the normal distribution. To do this, the function `rnorm` is replaced by the function `rt`; see Example 6.16.

17. The purpose of this exercise is to use Monte Carlo simulation to compare the properties of the weighted estimator of a mean return and the estimator based on the sample mean for the case in which the true mean returns and the true returns standard deviations increase over time.

    Consider 60 periods of returns on a given asset. Suppose that, for $t = 1, 2, \ldots, 60$, the mean return in period t is $0.01 + (0.01/59)(t-1)$ and the return standard deviation in period t is $0.02 + (0.02/59)(t-1)$. Hence, the mean return is a linear function of t, taking the value 0.01 in period 1 and the value 0.02 in period 60; the return standard deviation is 0.02 in period 1 and 0.04 in period 60.

A matrix of return values corresponding to this scenario may be simulated using the command

```
> ret_mat<-matrix(rnorm(60*1000, mean=0.01 + (0.01/59)*
+ (0:59), sd=0.02 + (0.02/59)*(0:59)), 1000, 60,
+ byrow=T)
```

Note that the `rnorm` function recycles the values specified in the argument `mean`. Because of the way in which they are recycled, we must populate the matrix of returns by row, instead of by column, which is the default; this is achieved by the argument `byrow=T`.

a. Construct a simulated return matrix using the command given previously.
b. Consider a decay parameter of $\gamma = 0.93$. By constructing a vector of the form γ^{60-t} for $t = 1, 2, \ldots, 60$, using a loop, calculate the $10,000$ EWMA estimates of the return standard deviation.
c. Using the return matrix from Part (a) together with the `apply` function, calculate a vector of $10,000$ sample mean returns.
d. Calculate the sample mean and standard deviation of the EWMA estimates calculated in Part (b) and the sample means calculated in Part (c).
e. Consider the estimates obtained by the two methods as estimates of the return standard deviation in period 60, 0.04. Using the results from the Monte Carlo simulation, estimate the bias and RMSE of each estimator. Based on these results, which estimator is preferable?
f. Estimate the RMSE for EWMA estimators based on different values of the decay parameter, $\gamma = 0.90$ and $\gamma = 0.95$. Which EWMA estimator has the smallest estimated RMSE?

18. The goal of this exercise is to use Monte Carlo simulation to study the behavior of estimates of the weights for tangency and minimum-variance portfolios. Consider a three-dimensional vector of asset returns with excess mean vector of the form $c(1, 1, 1)^T$ for some c and covariance matrix of the form

$$\sigma^2 \begin{pmatrix} 1 & \rho & \rho \\ \rho & 1 & \rho \\ \rho & \rho & \rho \end{pmatrix}$$

for some $\sigma^2 > 0$ and some $0 < \rho < 1$.

Specifically, consider the case in which $c = 0.02$, $\sigma = 0.05$, and $\rho = 0.2$.

a. Find the weights of the tangency and minimum-variance portfolios for the mean vector and covariance matrix described earlier.

b. Use the approach described in Example 6.17 to repeatedly simulate 60 observations from this return distribution; for each set of simulated excess returns, estimate the mean vector and covariance matrix of the returns. Use these estimates to calculate the weights of the tangency and minimum-variance portfolios. Repeat the procedure 10,000 times so that the result is a $10,000 \times 3$ matrix of tangency portfolio weights and a $10,000 \times 3$ matrix of minimum-variance portfolio weights.

c. Using `apply`, calculate the means and standard deviations of the three columns of the matrix of tangency weights; repeat the calculation for the matrix of minimum-variance weights. Does the estimator of tangency weights appear to be unbiased? Does the estimator of the minimum-variance weights appear to be unbiased? Does either of the estimators appear to be more accurate than the other?

d. Plot the first two columns of the matrix of tangency weights (note that because the weights sum to 1, one of the three columns is redundant). Use the limits $-1/2$ and $7/6$ for both the x and y axes (use the `xlim` and `ylim` arguments to the `plot` function). How would you expect the plot to appear if the weights are estimated extremely accurately? Based on the plot, what can you conclude about the sampling distribution of the estimated tangency portfolio weights?

e. Repeat the previous question using the estimated minimum-variance weights. Are there any important differences between the plot of estimated tangency weights and the plot of estimated minimum-variance weights?

7

Capital Asset Pricing Model

7.1 Introduction

In Chapters 4 and 5, we considered portfolio theory in which information about the means, variances, and correlations of asset returns is used to construct portfolios that are optimal according to certain criteria. In this chapter, we turn this around—we analyze an optimal portfolio and see what this optimality implies about the distribution of the asset returns. This analysis leads to important properties of the relationship between the returns on a given asset and the returns on the optimal portfolio.

According to the theory described in Sections 4.6 and 5.7, an investor choosing a portfolio of risky assets to combine with the risk-free asset should always choose the tangency portfolio. This is true for any level of risk preferred by the investor as follows: to achieve low levels of risk, more of the investment is placed in the risk-free asset, while investors able to tolerate higher levels of risk place more of their investment in the tangency portfolio, even borrowing to do so, if desired. Because, according to this theory, all investors use the same combination of risky assets, that is, the tangency portfolio, the market as a whole gives us useful information about the tangency portfolio.

The *market portfolio* is a portfolio of assets in which the weight placed on asset j is equal to the investment in asset j, as a proportion of the total investment in the market. The market portfolio may be viewed as a type of "consensus portfolio" for all investors.

According to portfolio theory, all investors should use the tangency portfolio so that this consensus portfolio should be identical to the tangency portfolio. Therefore, we do not need to calculate the weights of the tangency portfolio, we can observe them by calculating the investment in each asset in the market. Furthermore, the equivalence of the market and tangency portfolios has important implications for the relationship between the returns on an asset and the returns on the market portfolio, which is summarized in the *capital asset pricing model* (*CAPM*). This model is a starting point for a number of models describing the behavior of asset returns.

Of course, such an analysis must be based on a number of assumptions. Specifically, we assume the following:

- Asset prices are in equilibrium, with supply equaling demand for each asset.

- Investors make their investment decisions based on the expected returns and the standard deviation of the returns on their investments. This information is freely available to all investors.
- All investors hold a combination of the portfolio that maximizes the Sharpe ratio, that is, the tangency portfolio, and the risk-free asset. The proportion of the investment in the risk-free asset varies by investor.

7.2 Security Market Line

Consider a portfolio consisting of the tangency portfolio together with the risk-free asset. The following result shows that the Sharpe ratio of such a portfolio is equal to the Sharpe ratio of the tangency portfolio.

Lemma 7.1. *Let $\mu_{\mathcal{T}}$ and $\sigma_{\mathcal{T}}$ denote the mean and standard deviation, respectively, of the return on the tangency portfolio. Let R_p denote the return on a portfolio consisting of the tangency portfolio together with the risk-free asset, with positive weight placed on the tangency portfolio, and let μ_p and σ_p denote the mean and standard deviation, respectively, of R_p. Then*

$$\frac{\mu_p - \mu_f}{\sigma_p} = \frac{\mu_{\mathcal{T}} - \mu_f}{\sigma_{\mathcal{T}}} \tag{7.1}$$

where μ_f is the return on the risk-free asset.

Proof. For the portfolio with return R_p, let w denote the proportion of the portfolio invested in the tangency portfolio. Then

$$\mu_p = w\mu_{\mathcal{T}} + (1-w)\mu_f$$

and

$$\sigma_p = w\sigma_{\mathcal{T}};$$

recall that, by assumption, $w > 0$. The result follows. □

The proof also follows from noting that any such portfolio has (σ_p, μ_p) falling on the line connecting $(0, \mu_f)$ and $(\sigma_{\mathcal{T}}, \mu_{\mathcal{T}})$.

Note that the condition that the portfolio in Lemma 7.1 places positive weight on the tangency portfolio is the condition that the investor does not take a short position in the tangency portfolio in order to buy the risk-free asset.

According to the argument given in the introduction to this chapter, the return on the tangency portfolio may be viewed as the return on the market portfolio. Therefore, we may write the result (7.1) as

$$\frac{\mu_p - \mu_f}{\sigma_p} = \frac{\mu_m - \mu_f}{\sigma_m} \tag{7.2}$$

where μ_m, σ_m are the mean and standard deviation, respectively, of the return on the market portfolio. This equation may also be written as

$$\mu_p = \mu_f + \frac{\mu_m - \mu_f}{\sigma_m}\sigma_p; \tag{7.3}$$

this form emphasizes the relationship between the expected return on the portfolio and the portfolio risk. The relationship given in (7.3) is known as the *capital market line.*

In this section, we show that a similar relationship holds for the return on any asset, such as a single stock or a portfolio. Based on the assumptions discussed in Section 7.1, the market portfolio is equivalent to the tangency portfolio, and hence, it maximizes the Sharpe ratio among all portfolios. Therefore, modifying the weight given to asset i in the market portfolio must decrease the Sharpe ratio. This fact may be used to derive a relationship between the expected return on asset i and the expected return on the market portfolio.

Proposition 7.1. *Let R_i denote the return on a given asset, let $\mu_i = E(R_i)$ and let $\sigma_i^2 = Var(R_i)$. Let R_m denote the return on the market portfolio, let $\mu_m = E(R_m)$, let $\sigma_m^2 = Var(R_m)$, and let ρ_i denote the correlation of R_i and R_m. Then*

$$\mu_i - \mu_f = \rho_i \frac{\sigma_i}{\sigma_m}(\mu_m - \mu_f) \tag{7.4}$$

where μ_f is the return on the risk-free asset.

Proof. Consider a new portfolio, formed by combining asset i with the market portfolio. Let w_i denote the weight given to asset i so that the new portfolio has return $R_p = w_i R_i + (1 - w_i)R_m$. It follows that

$$\mu_p(w_i) \equiv \mathrm{E}(R_p) = w_i\mu_i + (1 - w_i)\mu_m$$

and

$$\sigma_p^2(w_i) \equiv \mathrm{Var}(R_p) = w_i^2\sigma_i^2 + (1 - w_i)^2\sigma_m^2 + 2w_i(1 - w_i)\rho_i\sigma_i\sigma_m.$$

Viewing this as a two-asset portfolio, we know that the tangency portfolio occurs at $w_i = 0$, because R_m is the return on the tangency portfolio. That is,

$$\frac{\mu_p(w_i) - \mu_f}{\sigma_p(w_i)},$$

the Sharpe ratio of the portfolio with return R_p is maximized at $w_i = 0$. Based on the analysis in Section 4.6, we know that $\mu_p(w_i)$ and $\sigma_p(w_i)$ must satisfy the tangency condition at $w_i = 0$:

$$\frac{d\mu_p(w_i)/dw_i|_{w_i=0}}{d\sigma_p(w_i)/dw_i|_{w_i=0}} = \frac{\mu_p(0) - \mu_f}{\sigma_p(0)} = \frac{\mu_m - \mu_f}{\sigma_m}. \tag{7.5}$$

Here

$$\frac{d\mu_p(w_i)}{dw_i} = \mu_i - \mu_m$$

and

$$\frac{d\sigma_p^2(w_i)}{dw_i} = 2w_i\sigma_i^2 - 2(1-w_i)\sigma_m^2 + 2(1-2w_i)\rho_i\sigma_i\sigma_m$$

so that

$$\left.\frac{d\sigma_p^2(w_i)}{dw_i}\right|_{w_i=0} = 2\rho_i\sigma_i\sigma_m - 2\sigma_m^2.$$

Note that

$$\frac{d\sigma_p^2(w_i)}{dw_i} = 2\sigma_p(w_i)\frac{d\sigma_p(w_i)}{dw_i};$$

therefore,

$$\left.\frac{d\sigma_p(w_i)}{dw_i}\right|_{w_i=0} = \frac{\left.\frac{d\sigma_p^2(w_i)}{dw_i}\right|_{w_i=0}}{2\sigma_p(0)} = \frac{\rho_i\sigma_i\sigma_m - \sigma_m^2}{\sigma_p(0)}.$$

Because $\mu_p(0) = \mu_m$ and $\sigma_p(0) = \sigma_m$,

$$\left.\frac{d\sigma_p(w_i)}{dw_i}\right|_{w_i=0} = \rho_i\sigma_i\sigma_m$$

and the tangency condition (7.5) may be written

$$\frac{\mu_i - \mu_m}{\rho_i\sigma_i - \sigma_m} = \frac{\mu_m - \mu_f}{\sigma_m}.$$

Rearranging this expression,

$$\mu_i - \mu_m = \left(\rho_i\frac{\sigma_i}{\sigma_m} - 1\right)(\mu_m - \mu_f)$$

or

$$\mu_i - \mu_f = \rho_i\frac{\sigma_i}{\sigma_m}(\mu_m - \mu_f) \tag{7.6}$$

as stated in (7.4) in the proposition. □

The result given in Proposition 7.1 is known as the capital asset pricing model, often abbreviated as CAPM. It may also be written as

$$\frac{\mu_i - \mu_f}{\sigma_i} = \rho_i\frac{\mu_m - \mu_f}{\sigma_m} \tag{7.7}$$

so that the Sharpe ratio of a given asset is equal to the Sharpe ratio of the market portfolio times the correlation of the asset's return with the return on the market portfolio.

That is, according to Proposition 7.1, the only way for an asset to have a large Sharpe ratio is for its returns to be highly correlated with the market returns. On the other hand, an asset with returns that are approximately uncorrelated with the market returns necessarily has a small Sharpe ratio.

Note that, because $\rho_i \leq 1$, the relationship in (7.7) is consistent with the assumption that the market portfolio has the largest possible Sharpe ratio.

Example 7.1 Suppose that the monthly return on the market portfolio has an expected value $\mu_m = 0.025$ and a standard deviation $\sigma_m = 0.04$, and suppose that the risk-free rate of return is $\mu_f = 0.005$. Consider an asset with a return with expected value and standard deviation of μ_i and σ_i, respectively, and let ρ_i denote the correlation of the asset's return with the market return. Then

$$\frac{\mu_i - 0.005}{\sigma_i} = \rho_i \frac{0.025 - 0.005}{0.04} = 0.5\rho_i$$

so that the Sharpe ratio of the asset is $\rho_i/2$. □

Define

$$\beta_i = \rho_i \frac{\sigma_i}{\sigma_m} = \frac{\text{Cov}(R_i, R_m)}{\text{Var}(R_m)}. \tag{7.8}$$

Then the relationship given in Proposition 7.1 may also be written

$$\mu_i - \mu_f = \beta_i(\mu_m - \mu_f);$$

this equation is known as the *security market line* (SML). It shows that the expected excess return on an asset is proportional to its value of β_i. Thus, the parameter β_i describes an important property of an asset and analysts often refer to the "beta" of an asset; the interpretation of beta will be discussed in detail in the following section.

Relationship to Linear Regression Analysis

The parameter $\beta_i = \rho_i(\sigma_i/\sigma_m)$ is closely related to the slope parameter in a linear regression analysis. Consider the problem of finding constants a, b to minimize

$$\text{E}[(R_i - R_f - a - b(R_m - R_f))^2] \tag{7.9}$$

where R_f denotes the return on the risk-free asset; recall that, as discussed in Section 4.5, we will use R_f when referring to returns, as in the aforementioned expression, and will use μ_f when referring to properties of the distribution of returns, as in the SML, for example.

This criterion can be described as the mean squared error (MSE) of $a + b(R_m - R_f)$ as an approximation to the excess return $R_i - R_f$; hence, the goal is to find a and b so that $a + b(R_m - R_f)$ best approximates $R_i - R_f$ in a certain sense. Note that this may be viewed as a "population-level" least-squares problem.

Recall that, for any random variable X, $\mathrm{E}(X^2) = \mathrm{E}(X)^2 + \mathrm{Var}(X)$ and, for any constant c, $\mathrm{Var}(c+X) = \mathrm{Var}(X)$. It follows that

$$\mathrm{E}[(R_i - R_f - a - b(R_m - R_f)^2] = \mathrm{E}\left[(R_i - R_f - a - b(R_m - R_f))^2\right] + \mathrm{Var}(R_i - bR_m) \quad (7.10)$$

Let $\hat{a}, \hat{b}$ denote the values of a, b, respectively, that minimize (7.9), or equivalently, the expression in (7.10). Then, given $\hat{b}$, $\hat{a}$ minimizes $\mathrm{E}[R_i - R_f - a - \hat{b}(R_m - R_f)]^2$ with respect to a. It follows that

$$\hat{a} = \mathrm{E}(R_i) - \mu_f - \hat{b}\mathrm{E}(R_m - R_f) = \mu_i - \mu_f - \hat{b}(\mu_m - \mu_f)$$

so that

$$\mathrm{E}\left[\left(R_i - R_f - \hat{a} - \hat{b}(R_m - R_f)\right)^2\right] = 0.$$

Therefore, $\hat{b}$ is the value of b that minimizes $\mathrm{Var}(R_i - bR_m)$. Note that

$$\mathrm{Var}(R_i - bR_m) = \sigma_i^2 + b^2\sigma_m^2 - 2b\mathrm{Cov}(R_i, R_m)$$

is a quadratic function of b, with a positive coefficient for b^2. It follows that $\hat{b}$ solves

$$\left.\frac{d\mathrm{Var}(R_i - bR_m)}{db}\right|_{b=\hat{b}} = 0$$

so that

$$\hat{b} = \frac{\mathrm{Cov}(R_i, R_m)}{\mathrm{Var}(R_m)} = \beta_i,$$

as defined in (7.8).

Because $\hat{b} = \beta_i$,

$$\hat{a} = \mu_i - \mu_f - \beta_i(\mu_m - \mu_f).$$

Therefore, according to Proposition 7.1, $\hat{a} = 0$. It follows that

$$\mathrm{E}[(R_i - R_f - \beta_i(R_m - R_f))^2] \leq \mathrm{E}[(R_i - R_f - a - b(R_m - R_f))^2] \quad (7.11)$$

for any a and b. That is, the linear function of $R_m - R_f$ that best approximates $R_i - \mu_f$ in the sense of MSE is $\beta_i(R_m - R_f)$.

7.3 Implications of the CAPM

The relationship, given the SML, gives the most obvious conclusion of the CAPM: The excess return on an asset is equal to the excess return on the market portfolio times the asset's value of beta; alternatively, using (7.7), the Sharpe ratio of a given asset is equal to the Sharpe ratio of the market portfolio times the correlation of the asset's return with the market return. However, there are a number of different implications of this result and, in this section, we consider several of these.

The CAPM, as given in Proposition 7.1, describes a relationship between the expected return on a portfolio and the expected return on a market portfolio in terms of the standard deviations of the returns and their correlation. However, that result also implies a relationship for the returns themselves.

Corollary 7.1. *Let R_i denote the return on an asset, let R_m denote the return on the market portfolio, let R_f denote the return on the risk-free asset, and let $\beta_i = Cov(R_i, R_m)/Var(R_m)$. Then we may write*

$$R_i - R_f = \beta_i(R_m - R_f) + Z_i$$

for a random variable Z_i that has mean 0 and that is uncorrelated with R_m.

Proof. Note that Z_i may be written

$$Z_i = R_i - R_f - \beta_i(R_m - R_f), \tag{7.12}$$

where β_i is as given in the statement of the corollary.

Then, according to Proposition 7.1, Z_i has expected value 0:

$$\mathrm{E}(Z_i) = \mu_i - \mu_f - \beta_i(\mu_m - \mu_f) = 0.$$

Furthermore, using properties of covariance,

$$\begin{aligned}\mathrm{Cov}(Z_i, R_m) &= \mathrm{Cov}(R_i - \mu_f - \beta_i(R_m - \mu_f), R_m)\\ &= \mathrm{Cov}(R_i, R_m) - \beta_i \mathrm{Var}(R_m) = 0\end{aligned}$$

so that Z_i is uncorrelated with the market return. □

It is important to note that it is always true that

$$R_i - \mu_f = \beta_i(R_m - \mu_f) + Z_i$$

where Z_i and R_m are uncorrelated; the fact that $\beta_i = \mathrm{Cov}(R_i, R_m)/\mathrm{Var}(R_m)$ implies that $\mathrm{Cov}(Z_i, R_m) = 0$. The role of the CAPM is to show that $\mathrm{E}(Z_i) = 0$.

Example 7.2 As in Example 7.1, suppose that the return on the market portfolio has an expected value $\mu_m = 0.025$ and a standard deviation $\sigma_m = 0.04$ and suppose that the risk-free rate of return is $\mu_f = 0.005$. Consider an asset with a return with a mean and standard deviation of $\mu_i = 0.02$ and $\sigma_i = 0.05$, respectively, and suppose that $\beta_i = 0.75$.

Let R_i and R_m denote the returns on the asset and the market portfolio, respectively. Then according to (7.12)

$$R_i - 0.005 = 0.75(R_m - 0.005) + Z_i.$$

That is, the excess return on asset i can be expressed as 0.75 times the excess return on the market plus a random quantity that is uncorrelated with the market return and that has expected value zero. □

The random variable Z_i defined by (7.12) is uncorrelated with the market return; however, it may be correlated with other economic variables. In Chapter 10, we will consider models that extend the relationship in (7.12) by including additional variables.

Note that, according to the expression for $R_i - R_f$ given in Corollary 7.1, along with the fact that Z_i and R_m are uncorrelated, we may write the variance of R_i in terms of two components,

$$\text{Var}(R_i) = \text{Var}(\beta_i R_m) + \text{Var}(Z_i) = \beta_i^2 \text{Var}(R_m) + \text{Var}(Z_i).$$

The term $\beta_i^2 \text{Var}(R_m)$ represents the component of $\text{Var}(R_i)$ that is "due to the market" or that "may be explained by the market"; that is, because the returns on an individual asset are, in general, correlated with the market return, and the market return fluctuates, some of the variation of an asset's returns can be explained by this variation in the market return. The second component, $\text{Var}(Z_i)$, may be interpreted as the nonmarket component of $\text{Var}(R_i)$.

Thus,

$$\beta_i^2 \text{Var}(R_m)/\text{Var}(R_i)$$

denotes the proportion of the variance of R_i that is "explained by the market"; it is important to keep in mind that when we say "due to the market" or "explained by the market," we are referring specifically to an asset return's *linear* relationship with the market return. Because

$$\beta_i^2 \text{Var}(R_m)/\text{Var}(R_i) = \rho_i^2,$$

this measure of the proportion of the variance of R_i that is due to the market is simply one of the standard interpretations of the correlation coefficient ρ_i.

Example 7.3 As in the previous example in this section, suppose that the return on the market portfolio has $\mu_m = 0.025$ and $\sigma_m = 0.04$ and that a given asset has $\mu_i = 0.02$, $\sigma_i = 0.05$, and $\beta_i = 0.75$. Then the component of the variance of asset i that is explained by the market is

$$\beta_i^2 \sigma_m^2 = (0.75)^2 (0.04)^2 = (0.03)^2 = 0.0009$$

and the proportion of the variance of asset i that is due to the market is

$$0.0009/(0.05)^2 = 0.36.$$ □

Therefore, the random variable Z_i in the relationship

$$R_i - R_f = \beta_i (R_m - R_f) + Z_i$$

increases the risk of asset i beyond the level attributable to the asset's relationship with the market. As noted earlier, decomposing the variance in this way does not require the CAPM. The role of the CAPM is to show that Z_i has zero mean; that is, the additional variance as a result of Z_i does not lead to an increase in the expected return of the asset. This idea is explored further as follows.

Relationship between Risk and Reward

Consider two assets with returns R_i and R_j, respectively, such that $\mu_i \equiv \mathrm{E}(R_i) > \mu_f$ and $\mu_j \equiv \mathrm{E}(R_j) > \mu_f$ and let β_i, β_j denote the respective values of beta for those assets. Because of the relationship between the expected return on an asset and the expected return on a market portfolio, as given by the CAPM, and the relationship between the variance of an asset's return and the variance of the market return, as discussed earlier, there is a relationship between the risk of an asset and the corresponding "reward," as measured by the asset's expected return.

Suppose that $\mathrm{Var}(R_j) > \mathrm{Var}(R_i)$ so that asset j is "riskier" than asset i. If the additional risk of asset j is attributable entirely to the difference in the assets' market components of variance, then

$$\mathrm{Var}(R_j) - \mathrm{Var}(R_i) = (\beta_j^2 - \beta_i^2)\mathrm{Var}(R_m);$$

hence, it follows that $\beta_j > \beta_i$. Note that, because μ_i, μ_j, and μ_m are all greater than μ_f, β_i, and β_j must be positive. Therefore,

$$\mu_j - \mu_f = \beta_j(\mu_m - \mu_f) > \beta_i(\mu_m - \mu_f) = \mu_i - \mu_f;$$

so that $\mu_j > \mu_i$. That is, an investor who assumes additional market risk by investing in asset j is rewarded with a higher expected return.

On the other hand, suppose that the additional risk of asset j is attributable entirely to the difference in the assets' nonmarket components of variance. If the market components of the variances of assets i and j are equal, then $\beta_i^2\sigma_m^2 = \beta_j^2\sigma_m^2$ so that $\beta_i = \beta_j$. It follows that $\mu_i = \mu_j$; that is, there is no "reward" for the additional nonmarket risk.

Now suppose that the difference between $\mathrm{Var}(R_j)$ and $\mathrm{Var}(R_i)$ is because of differences in both the market and the nonmarket components of the variances. Then the same argument holds, except that $\mu_j - \mu_i$ depends only on the difference between the market components of variance.

Specifically,

$$\begin{aligned}\mu_j - \mu_i &= (\beta_j - \beta_i)(\mu_m - \mu_f) \\ &= (\beta_j\sigma_m - \beta_i\sigma_m)\frac{\mu_m - \mu_f}{\sigma_m}.\end{aligned}$$

Note that $\beta_i\sigma_m$ and $\beta_j\sigma_m$ are the square roots of the market components of variance for assets i and j, respectively. We will refer to $\beta_i\sigma_m$ as the *market component of risk* for the asset; this market component may also be written as $\rho_i\sigma_i$.

Thus, the difference $(\mu_j - \mu_i)$ is proportional to the difference in the market components of risk for the two assets. This consequence of the CAPM is often summarized by saying that there is a reward for assuming risk but only for the market component of risk; there is no benefit in investing in an asset that has a large nonmarket component of risk.

Example 7.4 Suppose that the return on the market portfolio has $\mu_m = 0.025$ and $\sigma_m = 0.04$; let $\mu_f = 0.005$. Consider an asset with a return that has mean μ_i, standard deviation σ_i, and correlation with the market return of ρ_i. Then

$$\beta_i = \rho_i \frac{\sigma_i}{\sigma_m} = 25\rho_i\sigma_i$$

and, hence,

$$\mu_i = \mu_f + \beta_i(\mu_m - \mu_f) = 0.005 + 25\rho_i\sigma_i(0.025 - 0.005) = 0.005 + \frac{1}{2}\rho_i\sigma_i.$$

Assume that $\rho_i > 0$. Let γ_i^2 denote the component of the variance of the return on asset i that is due to the market, so that $\gamma_i^2 = \rho_i^2\sigma_i^2$. Then

$$\mu_i = 0.005 + \frac{1}{2}\gamma_i.$$

That is, the expected return on the asset is a linear function of its market component of risk, γ_i. □

Clearly, this type of relationship holds in general.

Corollary 7.2. *Let R_i denote the return on an asset and let μ_i and σ_i denote the mean and standard deviation, respectively, of R_i. Assume that $\mu_i > \mu_f$, where μ_f denotes the expected return on the risk-free asset. Let R_m denote the return on the market portfolio, let μ_m and σ_m denote the mean and standard deviation, respectively, of R_m, and let ρ_i denote the correlation of R_i and R_m. Then*

$$\mu_i - \mu_f = \left(\frac{\mu_m - \mu_f}{\sigma_m}\right)(\rho_i\sigma_i). \tag{7.13}$$

Note that $\rho_i\sigma_i$ is the market component of risk for the asset so that, according to the corollary, the excess return on an asset is proportional to its market component of risk.

7.4 Applying the CAPM to a Portfolio

Suppose that there are N assets in the market, with returns $R_1, R_2, \ldots, R_N$, and suppose that the SML holds for all assets:

$$\mu_i - \mu_f = \beta_i(\mu_m - \mu_f) \tag{7.14}$$

where $\mu_i = \text{E}(R_i)$, $\beta_i = \text{Cov}(R_i, R_m)/\text{Var}(R_m)$, $\mu_m = \text{E}(R_m)$, R_m is the return on the market portfolio, and μ_f is the risk-free rate of return.

Consider a portfolio based on weights $w_1, w_2, \ldots, w_N$ and let

$$R_p = \sum_{i=1}^{N} w_i R_i$$

denote its return. Then β_p, the value of beta for the portfolio, may be written

$$\begin{aligned}\beta_p &= \frac{\mathrm{Cov}(R_p, R_m)}{\mathrm{Var}(R_m)} = \frac{\mathrm{Cov}(\sum_{i=1}^N w_i R_i, R_m)}{\mathrm{Var}(R_m)} \\ &= \frac{\sum_{i=1}^N \mathrm{Cov}(w_i R_i, R_m)}{\mathrm{Var}(R_m)} = \frac{\sum_{i=1}^N w_i \mathrm{Cov}(R_i, R_m)}{\mathrm{Var}(R_m)} \\ &= \frac{\sum_{i=1}^N w_i \beta_i \mathrm{Var}(R_m)}{\mathrm{Var}(R_m)} = \sum_{i=1}^N w_i \beta_i .\end{aligned}$$

Because

$$\mu_p = \mathrm{E}(R_p) = \sum_{i=1}^N w_i \mu_i,$$

it follows from (7.14) that

$$\begin{aligned}\mu_p - \mu_f &= \sum_{i=1}^N w_i(\mu_i - \mu_f) = \sum_{i=1}^N w_i \beta_i (\mu_m - \mu_f) \\ &= \beta_p(\mu_m - \mu_f).\end{aligned}$$

That is, the SML holds for the portfolio as well.

Therefore, when we say that the CAPM holds for a given set of assets it follows that it holds for all portfolios constructed from those assets as well.

Example 7.5 Consider four assets, with returns R_1, R_2, R_3, and R_4, respectively, and let R_m denote the return on the market portfolio. Suppose that standard deviations of the returns on the four assets are $0.02, 0.05, 0.01$, and 0.04, respectively, and that the standard deviation of R_m is 0.01. Let ρ_i denote the correlation of R_i and R_m, for $i = 1, 2, 3$, and 4 and suppose that $\rho_1 = 0.6$, $\rho_2 = 0.1$, $\rho_3 = 0.8$, and $\rho_4 = 0.2$. Then the values of beta for the four assets are given by

$$\beta_1 = (0.6)\frac{0.02}{0.01} = 1.2, \quad \beta_2 = (0.1)\frac{0.05}{0.01} = 0.5, \quad \beta_3 = (0.8)\frac{0.01}{0.01} = 0.8,$$

and

$$\beta_4 = (0.2)\frac{0.04}{0.01} = 0.8.$$

Let R_p denote the return on an equally-weighted portfolio of the four assets; then β_p, the value of beta for the portfolio is given by

$$\beta_p = 0.25\beta_1 + 0.25\beta_2 + 0.25\beta_3 + 0.25\beta_4 = 0.825.$$

Alternatively, the value of β_p may be obtained from the properties of R_p. Note that

$$\mathrm{Cov}(R_i, R_m) = \beta_i \sigma_m = 0.01\beta_i, \quad i = 1, 2, 3, \text{and } 4$$

where σ_m is the standard deviation of R_m. Hence,

$$\text{Cov}(R_1, R_m) = 0.012, \quad \text{Cov}(R_2, R_m) = 0.005, \quad \text{Cov}(R_3, R_m) = 0.008,$$

and

$$\text{Cov}(R_4, R_m) = 0.008.$$

Using properties of covariance, it follows that

$$\begin{aligned}\text{Cov}(R_p, R_m) &= (0.25)(0.012) + (0.25)(0.005) + (0.25)(0.008) + (0.25)(0.008)\\ &= 0.00825\end{aligned}$$

and, hence, that

$$\beta_p = \frac{\text{Cov}(R_p, R_m)}{\sigma_m} = \frac{0.00825}{0.01} = 0.825,$$

matching the result obtained previously. □

7.5 Mispriced Assets

For a given asset with return R_i, let

$$\alpha_i = \mu_i - \mu_f - \beta_i(\mu_m - \mu_f) \tag{7.15}$$

where $\beta_i = \text{Cov}(R_i, R_m)/\sigma_m^2$, R_m is the return on the market portfolio, $\mu_m = \text{E}(R_m)$, $\sigma_m^2 = \text{Var}(R_m)$, $\mu_i = \text{E}(R_i)$, and μ_f is the return on the risk-free asset. According to the CAPM,

$$\alpha_i = 0.$$

However, suppose that $\alpha_i > 0$; that is, suppose that for a given asset the conclusion of Proposition 7.1 does not hold. As in Section 7.2, let R_p denote the return on a portfolio consisting of the market portfolio and asset i, with return of the form $R_p = w_i R_i + (1 - w_i)R_m$, for some weight w_i. Let

$$\mu_p(w_i) = \text{E}(R_p) = w_i\mu_i + (1 - w_i)\mu_m$$

and

$$\begin{aligned}\sigma_p^2(w_i) = \text{Var}(R_p) &= w_i^2\sigma_i^2 + (1 - w_i)^2\sigma_m^2 + 2w_i(1 - w_i)\text{Cov}(R_i, R_m)\\ &= w_i^2\sigma_i^2 + (1 - w_i)^2\sigma_m^2 + 2w_i(1 - w_i)\beta_i\sigma_m^2.\end{aligned}$$

Define

$$f(w_i) = \frac{\mu_p(w_i) - \mu_f}{\sigma_p(w_i)}$$

to be the Sharpe ratio of this portfolio, as a function of w_i. Then, using the results in Section 7.2,

$$f'(w_i) = \frac{d\mu_p(w_i)/dw_i}{\sigma_p(w_i)} - \frac{\mu_p(w_i) - \mu_f}{\sigma_p(w_i)} \frac{d\sigma_p(w_i)/dw_i}{\sigma_p(w_i)}$$

and

$$f'(0) = \frac{\mu_i - \mu_m}{\sigma_m} - \frac{\mu_m - \mu_f}{\sigma_m} \frac{d\sigma_p(w_i)/dw_i|_{w_i=0}}{\sigma_p(0)}.$$

We have seen that

$$d\sigma_p(w_i)/dw_i|_{w_i=0} = \frac{\rho_i \sigma_i \sigma_m - \sigma_m^2}{\sigma_m}$$

so that, using the fact that $\rho_i = \beta_i \sigma_m / \sigma_i$, we may write

$$\begin{aligned} f'(0) &= \frac{\mu_i - \mu_m}{\sigma_m} - \frac{\mu_m - \mu_f}{\sigma_m}(\beta_i - 1) \\ &= \frac{\mu_i - \mu_m - \beta_i(\mu_m - \mu_f)}{\sigma_m} \\ &= \frac{\alpha_i}{\sigma_m}. \end{aligned}$$

Therefore, if $\alpha_i > 0$, then $f'(0) > 0$ so that adding a small investment in asset i to the market portfolio increases the Sharpe ratio. Stated another way, the market portfolio does not contain enough of asset i to maximize the Sharpe ratio.

Let Q_i denote the number of shares of asset i in the market and let P_i denote the price of one share of asset i. Then the weight given to asset i in the market portfolio is

$$\frac{Q_i P_i}{C}$$

where C denotes the total investment in the market, known as the *market capitalization.*

When $\alpha_i > 0$, the weight given to asset i in the market portfolio is too small; that is, the ratio $Q_i P_i / C$ is too small. Therefore, P_i, the price of asset i, should be higher on average. It follows that, according to the CAPM, an asset with $\alpha_i > 0$ is mispriced and its price is too low. Conversely, the price of an asset with $\alpha_i < 0$ is too high; according to the CAPM, its price should be lower on average.

Example 7.6 Suppose that R_m, the return on the market portfolio, has mean 0.025 and standard deviation 0.04 and that the risk-free rate is $\mu_f = 0.005$. Consider an asset with return R_i that has mean 0.02 and standard deviation 0.08, and suppose that the correlation of R_i and R_m is $\rho_i = 0.30$.

Then

$$\beta_i = \rho_i \sigma_i / \sigma_m = (0.30)(0.08)/(0.04) = 0.60$$

and, hence, according to the CAPM,

$$\mu_i = \mu_f + \beta_i(\mu_m - \mu_f) = 0.005 + 0.60(0.025 - 0.005) = 0.017.$$

However, $\mu_i = 0.02$, so that

$$\alpha_i = \mu_i - \mu_f - \beta_i(\mu_m - \mu_f) = 0.02 - 0.017 = 0.003.$$

Therefore, the price of asset i is too low. □

The CAPM given in Proposition 7.1 follows from the assumptions presented in Section 7.1 as follows: Asset prices are in equilibrium, investments decisions are based on the means and standard deviations of the returns, and all investors hold a combination of the tangency portfolio and the risk-free asset. Therefore, if the conclusion of Proposition 7.1 does not hold, then one or more of the assumptions must be incorrect.

For instance, it may be that market prices are not in equilibrium. This suggests that if $\alpha_i > 0$, then the price of asset i needs to increase in order to reach equilibrium, at which point α_i will be 0. This leads to an expected return for asset i that is larger than the expected return given by the CAPM. The case of $\alpha_i < 0$ is similar except that we expect the return on asset i to be lower than what is implied by the CAPM.

Alternatively, it may be that prices are in equilibrium but that investors hold inefficient portfolios so that the market portfolio is inefficient in the sense that it does not maximize the Sharpe ratio. Thus, if $\alpha_i > 0$, the demand for asset i is lower than it would be if the market portfolio were efficient, leading to a price for asset i that is too low.

The Role of the Efficiency of the Market Portfolio

The analysis in this section shows that, if the CAPM does not hold for asset i, that is, if α_i as defined previously is not 0, then the market portfolio can be improved by including more or less of asset i. On the other hand, if the CAPM does hold for asset i, then changing the weight of asset i in the market portfolio cannot increase its Sharpe ratio. This suggests the following converse to Proposition 7.1: If the SML holds for all assets in the market, then the market portfolio must have the maximum Sharpe ratio.

Proposition 7.2. *Consider a set of assets with returns $R_1, R_2, \ldots, R_N$ and let R_m denote the return on the market portfolio, which is not necessarily equivalent to the tangency portfolio. Suppose that the SML (7.4) holds for each asset:*

$$\mu_i - \mu_f = \beta_i(\mu_m - \mu_f) \tag{7.16}$$

where $\mu_i = E(R_i)$, $\beta_i = Cov(R_i, R_m)/Var(R_m)$, and $\mu_m = E(R_m)$.

Let R_p denote the return on a portfolio based on weights $w_1, w_2, \ldots, w_N$ so that

$$R_p = \sum_{i=1}^{N} w_i R_i$$

and let μ_p and σ_p denote the mean and standard deviation, respectively, of R_p. Then

$$\frac{\mu_p - \mu_f}{\sigma_p} \leq \frac{\mu_m - \mu_f}{\sigma_m} \tag{7.17}$$

with equality if and only if R_p and R_m have correlation one. That is, the market portfolio has the maximum possible Sharpe ratio.

Proof. Using the form of the SML given by (7.7), together with the properties of portfolios discussed in Section 7.4, it follows that

$$\frac{\mu_p - \mu_f}{\sigma_p} = \rho_p \frac{\mu_m - \mu_f}{\sigma_m} \tag{7.18}$$

where ρ_p denotes the correlation of R_p and R_m.

The result now follows from the fact that $\rho_p \leq 1$. □

The CAPM shows that if a given portfolio is efficient in the sense that it maximizes the Sharpe ratio, then the SML holds for all assets with respect to that efficient portfolio. Proposition 7.2 shows that if the SML holds for all assets in the market with respect to a given market portfolio, then that market portfolio must maximize the Sharpe ratio. Therefore, there is a sense in which the CAPM, as stated in Proposition 7.1, is actually a statement about the efficiency of the market portfolio.

The result in Proposition 7.2 may be stated in an alternative form, which is given in the following corollary; the proof of Proposition 7.2 may be easily adapted to prove this result.

Corollary 7.3. *Let R_p^* denote a given portfolio and for any arbitrary asset with return R define*

$$\alpha(R) = E(R) - \mu_f - \beta(R)(\mu_p^* - \mu_f)$$

where $\mu_p^* = E(R_p^*)$ and $\beta(R) = Cov(R, R_p^*)/Var(R_p^*)$. Consider the set of assets for which $\alpha(R) = 0$. Then the asset with return R_p^* is in this set and it has the maximum Sharpe ratio among all portfolios formed from assets in this set.

7.6 The CAPM without a Risk-Free Asset

The CAPM discussed in this chapter is based on the assumption that all the investors choose a combination of the risk-free asset and a portfolio of risky assets. According to efficient portfolio theory, this portfolio of risky assets is the tangency portfolio for all investors. Thus, the market portfolio is equivalent to the tangency portfolio so that the market portfolio has the properties of the tangency portfolio. It is important to note that the optimality of the tangency portfolio in this context is based on the fact that investors combine their portfolio of risky assets with the risk-free asset.

In this section, we consider a version of the CAPM that holds without relying on the existence of a risk-free asset. There are two important implications of this change for the CAPM. The more obvious one is that we cannot include the risk-free rate, μ_f, in the model. The second, less obvious, change is that the tangency portfolio is no longer the unambiguous optimal portfolio and, hence, we may no longer assume that the market portfolio is equivalent to the tangency portfolio.

Instead, we assume that each investor holds a portfolio of risky assets that lies on the efficient frontier, but these portfolios may vary by investor. According to Propositions 5.2, portfolios constructed from assets lying on the efficient frontier are also on the efficient frontier provided that the mean return of the portfolio is greater than the mean return on the minimum variance portfolio. Hence, we may still assume that the market portfolio lies on the efficient frontier. However, it is not necessarily equal to the tangency portfolio.

Let R_m denote the return on the market portfolio. We assume that if there is another portfolio, with return R_p, such that $\mathrm{E}(R_p) = \mathrm{E}(R_m)$, then $\mathrm{Var}(R_p) \geq \mathrm{Var}(R_m)$; alternatively, if $\mathrm{Var}(R_p) = \mathrm{Var}(R_m)$, then $\mathrm{E}(R_p) \leq \mathrm{E}(R_m)$. Note that these assumptions state simply that the market portfolio lies on the efficient frontier.

The proof of the CAPM given in Proposition 7.1 is based on the fact that the market portfolio has the maximum Sharpe ratio among all assets and, hence, modifying the weight given to any asset cannot increase the Sharpe ratio. For the version of the CAPM considered in this section, we use a similar argument based on the efficiency of the market portfolio.

Let R_i denote the return on asset i. Suppose that we can construct a portfolio consisting of asset i together with the market portfolio that has the same expected return as the market portfolio; then the variance of that portfolio must be at least as large as that of the market portfolio. We may try to use this fact to establish a relationship similar to that in the SML.

However, it is clear that such an approach cannot work—unless asset i has the same expected return as the market portfolio, a portfolio including both asset i and the market portfolio cannot have the same expected return as the market portfolio. Hence, we need to include a third asset in the portfolio. Because we would like the eventual result to focus on the relationship between the return on asset i and the return on the market portfolio, we might consider an asset with a return that is uncorrelated with the market return.

Let R_z denote the return on an asset satisfying $\mathrm{Cov}(R_z, R_m) = 0$ and $\mathrm{E}(R_z) \neq \mathrm{E}(R_m)$. At the end of this section it will be shown that such a portfolio exists. Note that $\mathrm{Cov}(R_z, R_m) = 0$ implies that the value of beta corresponding to R_z is zero; therefore, the asset with return R_z is known as the *zero-beta portfolio.*

Proposition 7.3. *Let R_m denote the return on the market portfolio and let R_z denote the return on the corresponding zero-beta portfolio; let $\mu_m = E(R_m)$ and $\mu_z = E(R_z)$. Consider an asset with return R_i; let $\mu_i = E(R_i)$ and let $\beta_i = Cov(R_i, R_m)/Var(R_m)$. Then*

$$\mu_i - \mu_z = \beta_i(\mu_m - \mu_z). \tag{7.19}$$

Proof. For a real number θ, consider the zero-investment portfolio with return

$$R_i + (\theta - 1)R_m - \theta R_z = \theta(R_i - R_z) + (1 - \theta)(R_i - R_m); \tag{7.20}$$

hence, this portfolio places weight 1 on asset i, weight $\theta - 1$ on the market portfolio, and weight $-\theta$ on the zero-beta portfolio. Note that the expected value of (7.20) is

$$\theta(\mu_i - \mu_z) + (1 - \theta)(\mu_i - \mu_m).$$

Let

$$\theta_0 = \frac{\mu_m - \mu_i}{\mu_m - \mu_z}$$

and let

$$R_0 = \theta_0(R_i - R_z) + (1 - \theta_0)(R_i - R_m).$$

Then

$$\begin{aligned} \mathrm{E}(R_0) &= \frac{\mu_m - \mu_i}{\mu_m - \mu_z}(\mu_i - \mu_z) + \left(1 - \frac{\mu_m - \mu_i}{\mu_m - \mu_z}\right)(\mu_i - \mu_m) \\ &= \frac{(\mu_m - \mu_i)(\mu_i - \mu_z) + (\mu_i - \mu_z)(\mu_i - \mu_m)}{\mu_m - \mu_z} = 0. \end{aligned}$$

Thus, R_0 is the return on a zero-investment portfolio that has zero expected return. Because the market portfolio is on the efficient frontier, it now follows from Corollary 5.2 that $\mathrm{Cov}(R_0, R_m) = 0$. Note that

$$\begin{aligned} \mathrm{Cov}(R_0, R_m) &= \mathrm{Cov}(R_i + (\theta_0 - 1)R_m - \theta_0 R_z, R_m) \\ &= \mathrm{Cov}(R_i, R_m) + (\theta_0 - 1)\mathrm{Var}(R_m) \\ &= (\beta_i - (1 - \theta_0))\,\mathrm{Var}(R_m). \end{aligned} \tag{7.21}$$

Therefore,

$$\beta_i = 1 - \theta_0$$

and, using the expression for θ_0,

$$\beta_i = 1 - \frac{\mu_m - \mu_i}{\mu_m - \mu_z} = \frac{\mu_i - \mu_m}{\mu_m - \mu_z},$$

proving the result. □

That is, a form of the SML holds with μ_z replacing μ_f. The pricing model given by (7.19) is known as the *zero-beta CAPM.*

Note that Proposition 7.3 requires only that the market portfolio is on the efficient frontier, which is weaker than the condition that the market portfolio is the tangency portfolio required in Proposition 7.1. Hence, one might consider the possibility of weakening the conditions of Proposition 7.1 to require only that the market portfolio is on the efficient frontier, using the method of proof used in Proposition 7.3 with the risk-free asset playing the role of the zero-beta portfolio. However, in Proposition 7.3, it is important to keep in mind that the efficiency of the market portfolio is with respect to all assets under consideration; if a risk-free asset is available, then the market portfolio must be efficient with respect to portfolios that include the risk-free asset. Thus, such efficiency requires that the market portfolio is equivalent to the tangency portfolio; that is, it is not possible to use the approach of Proposition 7.3 to weaken the conditions used to establish the SML in Proposition 7.1.

Existence of the Zero-Beta Portfolio

Proposition 7.3 is based on the existence of the zero-beta portfolio; thus, we now show that such a zero-beta portfolio always exists, provided that the market portfolio is not the same as the minimum-variance portfolio. It may be shown that the market portfolio is equivalent to the minimum-variance portfolio if and only if all investors hold the minimum-variance portfolio.

Lemma 7.2. *Let R_m denote the return on the market portfolio, with variance σ_m^2, and let R_{mv} denote the return on the minimum-variance portfolio, with variance σ_{mv}^2. If $\sigma_m^2 > \sigma_{mv}^2$, then there exists a portfolio with return R_z such that $Cov(R_z, R_m) = 0$ and $E(R_z) \neq E(R_m)$.*

The return R_z may be written

$$R_z = \frac{1}{\sigma_{mv}^2 - \sigma_m^2}(\sigma_{mv}^2 R_m - \sigma_m^2 R_{mv}).$$

Proof. Consider the portfolio with return $\phi R_m + (1-\phi)R_{mv}$ for some real number ϕ. Recall that, according to Proposition 5.2, the covariance of R_{mv} with the return on any other portfolio is equal to $\text{Var}(R_{mv})$. Therefore,

$$\text{Cov}(R_{mv}, R_m) = \text{Var}(R_{mv}) = \sigma_{mv}^2.$$

It follows that, for any real number ϕ,

$$\begin{aligned}\text{Cov}(\phi R_m + (1-\phi)R_{mv}, R_m) &= \phi\text{Var}(R_m) + (1-\phi)\text{Var}(R_{mv}) \\ &= \phi\sigma_m^2 + (1-\phi)\sigma_{mv}^2.\end{aligned}$$

Let

$$\phi_z = \frac{\sigma_{mv}^2}{\sigma_{mv}^2 - \sigma_m^2}$$

and define

$$R_z = \phi_z R_m + (1-\phi_z)R_{mv}.$$

Then

$$\text{Cov}(R_z, R_m) = \phi_z\sigma_m^2 + (1-\phi_z)\sigma_{mv}^2 = 0.$$

Note that because $\sigma_{mv}^2 < \sigma_m^2$, $\phi_z < 0$ and the efficiency of the market portfolio implies that $\text{E}(R_{mv}) < \text{E}(R_m)$. It follows that

$$\text{E}(R_z) = \text{E}(R_m) + (1-\phi_z)[\text{E}(R_{mv}) - \text{E}(R_m)] < \text{E}(R_m);$$

that is, $\text{E}(R_z) \neq \text{E}(R_m)$, as required. □

7.7 Using the CAPM to Describe the Expected Returns on a Set of Assets

In Section 7.3, we considered several different interpretations of the CAPM. These interpretations are based on an analysis of the properties of a single asset return and how that return relates to the return on the market portfolio. In this section, another interpretation is given, based on analyzing the expected returns of a set of assets.

Consider a set of K assets, with returns $R_1, R_2, \ldots, R_K$, and let

$$\boldsymbol{R}_K = \begin{pmatrix} R_1 \\ R_2 \\ \vdots \\ R_K \end{pmatrix}$$

denote the corresponding return vector; $\boldsymbol{R}_K$ might include all the returns on all stocks in a given market, the returns on a subset of those stocks, or the returns on a set of portfolios.

Let $\mu_k = \mathrm{E}(R_k)$, $k = 1, 2, \ldots, K$, and let

$$\boldsymbol{\mu}_K = \begin{pmatrix} \mu_1 \\ \mu_2 \\ \vdots \\ \mu_K \end{pmatrix}$$

be the corresponding vector of expected returns. According to the classical form of the CAPM, as given by Proposition 7.1,

$$\mu_k = \mu_f + \beta_k(\mu_m - \mu_f), \quad k = 1, 2, \ldots, K \tag{7.22}$$

where μ_m is the expected return of the market portfolio, μ_f is the risk-free rate, and

$$\beta_k = \frac{\mathrm{Cov}(R_k, R_m)}{\mathrm{Var}(R_m)}, \quad k = 1, 2, \ldots, K.$$

Let

$$\boldsymbol{\beta}_K = \begin{pmatrix} \beta_1 \\ \beta_2 \\ \vdots \\ \beta_K \end{pmatrix}.$$

Then the set of K equations given by 7.22 may be written

$$\boldsymbol{\mu}_K = \mu_f \mathbf{1}_K + (\mu_m - \mu_f)\boldsymbol{\beta}_K. \tag{7.23}$$

That is, the vector of expected asset returns may be written as a linear function of the vector of betas and the vector of all ones.

According to the relationship in (7.23), the differences in the values of μ_k for the different assets may be described in terms of the differences in $\beta_1, \beta_2, \ldots, \beta_K$. For instance, suppose we plot the points $(\beta_1, \mu_1), (\beta_2, \mu_2), \ldots, (\beta_K, \mu_K)$. In such a plot, the points will fall on a line with slope $\mu_m - \mu_f$ and intercept μ_f; see Figure 7.1 for a hypothetical example.

Example 7.7 Consider the four assets described in Example 7.5. Recall that, for these assets, the values of beta are given by

$$\beta_1 = 1.2, \quad \beta_2 = 0.5, \quad \beta_3 = 0.8, \quad \text{and} \quad \beta_4 = 0.8.$$

Thus, the beta vector for the assets is

$$\begin{pmatrix} 1.2 \\ 0.5 \\ 0.8 \\ 0.8 \end{pmatrix}.$$

Suppose that $\mu_f = 0.002$ and $\mu_m = 0.01$. Then the vector of expected returns on the four assets may be written

$$0.002 \begin{pmatrix} 1 \\ 1 \\ 1 \\ 1 \end{pmatrix} + (0.01 - 0.002) \begin{pmatrix} 1.2 \\ 0.5 \\ 0.8 \\ 0.8 \end{pmatrix}$$

so that $\beta_1, \beta_2, \beta_3,$ *and* β_4 fall on the line with slope $0.01 - 0.002 = 0.008$ and intercept 0.002, as described previously; alternatively, this relationship may be described by stating that the vector of expected excess returns is proportional to the vector of asset betas. □

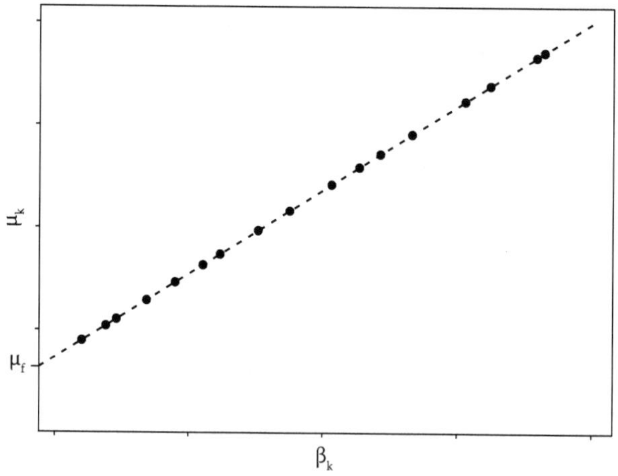

FIGURE 7.1
Hypothetical Plot of μ_k versus β_k.

A relationship similar to (7.23) is implied by the zero-beta form of the CAPM:

$$\mu_K = \mu_z \mathbf{1}_K + (\mu_m - \mu_z)\boldsymbol{\beta}_K. \tag{7.24}$$

In (7.23), it is generally assumed that μ_f is known, while in (7.24), μ_z is generally considered to be an unknown parameter.

7.8 Suggestions for Further Reading

The CAPM is one of the central results of modern portfolio theory. The classical CAPM, given here as Proposition 7.1, is attributed to Sharpe (1964), Lintner (1952), and Mossin (1966). Roll (1977) and Ross (1977) show that the CAPM follows directly from the assumption that the market portfolio is efficient. The zero-beta form of the CAPM is attributed to Black (1972).

Francis and Kim (2013, Chapter 12) present a detailed discussion of the CAPM and its derivation. Francis and Kim (2013, Chapter 13) discuss a number of extensions of the CAPM, including the zero-beta CAPM. Campbell et al. (1997, Chapter 5) describe the CAPM from a more statistical perspective, including empirical tests of the CAPM. Fabozzi et al. (2006, Chapter 7) discuss asset pricing models in general, including the CAPM as a special case.

7.9 Exercises

1. Consider an asset with expected return 0.04 and suppose that the return on the market portfolio is 0.06. Assuming that the SML holds for the asset and that the risk-free return is 0.004, find the value of beta for the asset.

2. Use the relationship given by the CAPM, as stated in (7.7), along with the assumption that the market portfolio is equivalent to the tangency portfolio, to establish the result in Lemma 7.1.

3. Consider an asset with return R and let R_m denote the return on the market portfolio. Let $\mu = \text{E}(R)$, $\mu_m = \text{E}(R_m)$, $\sigma^2 = \text{Var}(R)$, and $\sigma_m^2 = \text{Var}(R_m)$, and let μ_f denote the return on the risk-free asset. Suppose that $\sigma = \sigma_m/2$ and that

$$\frac{\mu - \mu_f}{\sigma} = \frac{1}{2}\frac{\mu_m - \mu_f}{\sigma_m}.$$

   Assuming that the SML holds for this asset, find its value of β.

4. Consider an asset with return R and let R_m denote the return on the market portfolio. Suppose that R and R_m are related by

$$R = 0.002 + 0.9R_m + \epsilon$$

where ϵ is a random variable satisfying $E(\epsilon) = 0$ and

$$\text{Cov}(R_m, \epsilon) = 0.$$

Assuming that the SML holds for this asset, find the value of μ_f, the return on the risk-free asset.

5. Consider two assets, with returns R_1 and R_2, respectively, and let R_m denote the return on the market portfolio. For $j = 1, 2$, let $\mu_j = E(R_j)$, let $\sigma_j^2 = \text{Var}(R_j)$, and let ρ_j denote the correlation of R_j and R_m. Let μ_f denote the expected return on the risk-free asset and assume that $\mu_j > \mu_f$ for $j = 1, 2$. Assume that the SML holds for both assets.

   For each of the sets of parameter values given as follows, state that asset 1 has the greater mean return, that asset 2 has the greater mean return, or that it is not possible to determine the greater mean return based on the information given.

   a. Suppose that $\rho_1 = 0.6$, $\sigma_1 = 0.2$, $\rho_2 = 0.5$, and $\sigma_2 = 0.3$.
   b. For $j = 1, 2$, let SR_j denote the Sharpe ratio of asset j. Suppose that
   $$SR_1 = 0.9 \quad \text{and} \quad SR_2 = 1.2.$$
   c. For $j = 1, 2$, let $\gamma_j^2 = \rho_j^2 \sigma_j^2$; note that γ_j^2 is the market component of variance for the return on asset j. Suppose that $\gamma_1^2 = 0.5$ and $\gamma_2^2 = 0.8$.

6. Consider two assets with returns R_1 and R_2, respectively, and let R_m denote the return on the market portfolio. Suppose that the SML holds for both assets, with $\beta = 0.80$ for asset 1 and $\beta = 0.90$ for asset 2. Does it follow that the correlation of R_2 and R_m is greater than the correlation of R_1 and R_m? Why or why not?

7. Let R_{mv} denote the return on the minimum-variance portfolio and let R_m denote the return on the market portfolio. Suppose that the correlation of R_{mv} and R_m is 0.4. Find the value of beta in the SML for the minimum-variance portfolio.

8. Consider an asset with return R_i. Suppose that the variance of R_i is 0.04 and that the market component of the variance of R_i is 0.03. Let μ_f denote the risk-free return and assume that $\mu_i \equiv E(R_i) > \mu_f$.

   a. Find the correlation of R_i and R_m, the return on the market portfolio.
   b. If the Sharpe ratio of the market portfolio is 0.12, find $\mu_i - \mu_f$.

9. Consider a set of N assets and let R_λ^* denote the return on the risk-averse portfolio based on risk-aversion parameter λ. Let R_i denote the return on a given asset in this set. Find $E(R_i) - E(R_\lambda^*)$, in terms of $\text{Cov}(R_i, R_\lambda^*)$, $\text{Var}(R_\lambda^*)$, and λ.

10. Consider an asset with return R, where R has standard deviation 0.03 and suppose that the return on the market has standard deviation 0.02. If the value of beta for the asset is 0.6, find the market component of Var(R) and find the proportion of Var(R) that is due to the market.

11. Consider an asset for which the SML holds with $\beta = 1.1$. If the return on the market portfolio has standard deviation 0.04, what is the smallest value that the return standard deviation for the asset can take?

12. Consider two assets with mean returns μ_1 and μ_2, respectively, and assume that $\mu_j > \mu_f$, $j = 1, 2$, where μ_f denotes the risk-free rate.

    If the market component of variance of asset 1 is larger than the market component of variance of asset 2, does it follow that the Sharpe ratio of asset 1 is greater than the Sharpe ratio of asset 2? Why or why not?

13. Consider a market of N assets, where $N \geq 5$. Let R_1, R_2, R_3, and R_4 denote the returns on four of the assets and let R_m denote the return on the market portfolio. Suppose that the covariance matrix of $(R_1, R_2, R_3, R_4, R_m)^T$ is given by

$$\begin{pmatrix} 0.012 & 0.004 & 0.005 & 0.0066 & 0.0055 \\ 0.004 & 0.008 & 0.0036 & 0.005 & 0.004 \\ 0.005 & 0.0036 & 0.012 & 0.0054 & 0.0045 \\ 0.0066 & 0.005 & 0.0054 & 0.01 & 0.006 \\ 0.0055 & 0.004 & 0.0045 & 0.006 & 0.005 \end{pmatrix}. \qquad (7.25)$$

    For instance, the $\text{Cov}(R_1, R_m) = 0.0055$ and $\text{Var}(R_m) = 0.0050$.

    Suppose that the mean excess return on the market portfolio is 0.04 and that the SML holds for all assets.

    a. Calculate beta for each of the four assets.
    b. Find the minimum-variance portfolio subject to the restriction that beta for the portfolio is 1. Calculate the return variance and the expected excess return for the portfolio.
    c. Find the minimum-variance portfolio subject to the restriction that beta for the portfolio lies in the interval $[0.95, 1.05]$. Calculate the return variance and the expected excess return for the portfolio.
    d. Find the minimum-variance portfolio subject to the restrictions that beta for the portfolio lies in the interval $[0.95, 1.05]$ and that the portfolio weights are nonnegative. Calculate the return variance and the expected excess return for the portfolio.

14. Suppose the market portfolio has mean return $\mu_m = 0.075$ and return standard deviation $\sigma_m = 0.14$; suppose that μ_f, the risk-free return, is 0.005.

Suppose that asset i has mean return μ_i and return standard deviation σ_i and let ρ_i denote the correlation of the return on asset i with the return on the market portfolio. Based on the CAPM, determine if the asset is overpriced, underpriced, or priced correctly if the asset parameters take the following parameter values:

a. $\mu_i = 0.035$, $\sigma_i = 0.3$, and $\rho_i = 0.2$
b. $\mu_i = 0.045$, $\sigma_i = 0.2$, and $\rho_i = 0.6$
c. $\mu_i = 0.075$, $\sigma_i = 0.15$, and $\rho_i = 0.8$

15. Consider a market portfolio with return R_m. Suppose that there exists an asset that is not included in the market portfolio but for which the SML with respect to the market portfolio holds. Thus, if R_0 denotes the return on this asset,

$$R_0 - R_f = \beta_0(R_m - R_f) + Z$$

where $\mathrm{E}(Z) = 0$, $\mathrm{Cov}(Z, R_m) = 0$, and R_f denotes the return on the risk-free asset.

Find the tangency portfolio consisting of the market portfolio (viewed as a single asset) and the asset with return R_0. Interpret the result.

16. Consider a market without a risk-free asset and let R_m denote the return on the market portfolio, which is assumed to be efficient. Let R_z denote the return on the zero-beta portfolio defined in Proposition 7.3. That is,

$$R_z = \phi_z R_m + (1 - \phi_z) R_{mv}$$

where R_{mv} is the return on the minimum-variance portfolio and

$$\phi_z = \frac{\sigma_{mv}^2}{\sigma_{mv}^2 - \sigma_m^2}.$$

Here $\sigma_{mv}^2 = \mathrm{Var}(R_{mv})$ and $\sigma_m^2 = \mathrm{Var}(R_m)$.

a. Is the zero-beta portfolio with return R_z on the minimum-risk frontier? Why or why not?
b. Is the zero-beta portfolio with return R_z on the efficient frontier? Why or why not?

17. Consider a market without a risk-free asset and let R_m denote the return on the market portfolio, which is assumed to be efficient. Suppose that there are two zero-beta portfolios with respect to R_m and let R_{z1} and R_{z2} denote their respective returns. Show that

$$\mathrm{E}(R_{z1}) = \mathrm{E}(R_{z2}).$$

8

The Market Model

8.1 Introduction

The capital asset pricing model (CAPM) describes a relationship between the expected return on an asset and the expected return on a "market portfolio," which is assumed to be equivalent to the tangency portfolio. The value of the parameter β for an asset gives important information regarding both the expected return and the risk of an asset.

However, the CAPM describes a theoretical relationship that is based on a number of assumptions that are difficult, or impossible, to verify. In this chapter, we consider the *market model*, a statistical model for the relationship between observed asset returns and the observed returns on a type of market portfolio. This model is a form of linear regression model that can be estimated using standard techniques. It is consistent with the CAPM in many respects and, hence, the estimates from the market model give useful information regarding the statistical properties of asset returns.

8.2 Market Indices

A key component of the CAPM is the market portfolio, consisting of all marketable assets. However, in many respects, the market portfolio is a hypothetical concept rather than an observable feature of the market. For instance, because some investors might prefer to invest in real estate or art instead of stocks, these assets are part of the market and should be included in the market portfolio. It might even be argued that because an investor might sell stocks to pay for a child's education, certain types of human capital must also be included.

Therefore, it is clear that we cannot hope to accurately measure the return on a true market portfolio. However, we can use the return on a stock market index as a proxy for the return on a market portfolio. Hence, in this section, we consider the properties of such indices.

Cap-Weighted Indices

Suppose that a market index is to be based on N assets, with respective prices $P_{1,t}, P_{2,t}, \ldots, P_{N,t}$ at time t; here $P_{j,t}$ is the price of one share of asset j at time t. Let Q_j denote the number of shares of asset j available in the market, $j = 1, 2, \ldots, N$. Then the amount invested in asset j at time t is $Q_j P_{j,t}$; this is known as the *market capitalization* of asset j at time t. The total capitalization of the entire market, or the *market cap*, at time t is given by

$$\sum_{j=1}^{N} Q_j P_{j,t}.$$

A *cap-weighted index* based on these assets is of the form

$$I_t = \frac{\sum_{j=1}^{N} Q_j P_{j,t}}{D_t}, \quad t = 1, 2, \ldots$$

where D_t is a divisor used to rescale the total capitalization. The divisor is modified periodically so that the index provides a continuous measure of the value of these assets and it is not affected by certain corporate actions, such as mergers, or changes to the set of assets used to form the index.

For simplicity, assume that the divisor is constant over time. Then the return on the index at time $t + 1$ is given by

$$\frac{I_{t+1}}{I_t} - 1 = \frac{\sum_{j=1}^{N} Q_j P_{j,t+1}}{\sum_{j=1}^{N} Q_j P_{j,t}} - 1.$$

Note that

$$\frac{\sum_{j=1}^{N} Q_j P_{j,t+1}}{\sum_{j=1}^{N} Q_j P_{j,t}} - 1 = \frac{\sum_{j=1}^{N} Q_j (P_{j,t+1} - P_{j,t})}{\sum_{j=1}^{N} Q_j P_{j,t}} = \frac{\sum_{j=1}^{N} Q_j P_{j,t} R_{j,t+1}}{\sum_{j=1}^{N} Q_j P_{j,t}}$$

where

$$R_{j,t+1} = \frac{P_{j,t+1}}{P_{j,t}} - 1$$

denotes the return on asset j at time $t + 1$.

Let

$$w_{j,t}^c = \frac{Q_j P_{j,t}}{\sum_{j=1}^{N} Q_j P_{j,t}}$$

denote the proportion of the market cap corresponding to asset j at time t. Note that

$$\sum_{j=1}^{N} w_{j,t}^c = 1.$$

The weights $w_{j,t}^c$, $j = 1, 2, \ldots, N$ are known as *capitalization weights* or simply *cap weights*, at time t.

The return on the index at time $t + 1$ may be written

$$\frac{I_{t+1}}{I_t} - 1 = \sum_{j=1}^{N} w_{j,t}^c R_{j,t+1},$$

which is identical to the return on the portfolio with weights $w_{j,t}^c$, $j = 1, 2, \ldots, N$. If the divisor changes from time t to time $t+1$, then that change also plays a role in the return on the index.

In this discussion, the weights $w_{j,t}^c$, $j = 1, 2, \ldots, N$, are based on the total number of shares of the asset, Q_j, along with the prices of the assets. However, not all shares of an asset are always available to investors; for instance, for stocks, there may be blocks of shares held by directors of the company. Shares that are available to investors are said to be part of the *float*. Therefore, shares held by directors are generally not part of the float.

Some indices exclude shares that are not part of the float from the index calculation. Let $\tilde{Q}_j$ denote the number of available shares of asset $j, j = 1, \ldots, N$. Then

$$\tilde{I}_t = \frac{\sum_{j=1}^N \tilde{Q}_j P_{t,j}}{D_t}$$

represents a *float-adjusted* index. It has the same general properties as a cap-weighted index. The ratio $\tilde{Q}_j/Q_j$ is known as the *investable weight factor* of the asset.

A commonly used cap-weighted index is the Standard & Poors (S&P) 500 index, which is based on 500 large-capitalization stocks, representing about 80% of the total market capitalization. Other cap-weighted indices include the Russell 3000 index, which is based on the 3000 stocks with the largest capitalizations, representing about 98% of the total market capitalization; the Russell 1000 index, which includes the 1000 largest stocks, in terms of capitalization, of those used in the Russell 3000 index and that represents about 92% of the total market capitalization; and the Wilshire 5000 index, which is based on the stocks of all publicly traded companies trading on a U.S. stock exchange. The S&P 500 index, the Russell 3000 index, and the Russell 1000 index are all float-adjusted; the Wilshire 5000 index is not float-adjusted, although there is a float-adjusted version available.

Example 8.1 Data on stock market indices are available from Yahoo Finance. For instance, the S&P 500 index is available using the symbol ^GSPC; in general, the symbols used for stock market indices start with the character ^. Hence, returns on the S&P 500 index may be calculated using the same method used to calculate the returns on a stock. Suppose that the monthly excess returns on the S&P 500 index have been calculated for the time period January 2010 to December 2014 and are stored in the R variable `sp500`.

```
> mean(sp500)
[1] 0.0109
> sd(sp500)
[1] 0.0376
```

The monthly excess returns on the Russell 1000 (Yahoo Finance symbol ^RUI), Russell 3000 (^RUA), and Wilshire 5000 (^W5000) indices for the same period are stored in the variables `r1000`, `r3000`, and `w5000`, respectively, and the matrix `indices` contains all four of the indices considered; the first few rows of this matrix are given as follows:

```
> head(indices)
       SP500   R1000   R3000   W5000
[1,] -0.0370 -0.0370 -0.0370 -0.0344
[2,]  0.0284  0.0305  0.0316  0.0323
[3,]  0.0587  0.0597  0.0613  0.0615
[4,]  0.0146  0.0174  0.0205  0.0207
[5,] -0.0821 -0.0815 -0.0811 -0.0812
[6,] -0.0540 -0.0573 -0.0591 -0.0562
```

It is clear from these few values that the returns on these four indices are generally, but not always, similar. Therefore, it is not surprising that the means and standard deviations of the returns on the four indices are generally close.

```
> apply(indices, 2, mean)
 SP500  R1000  R3000  W5000
0.0109 0.0111 0.0112 0.0112
> apply(indices, 2, sd)
 SP500  R1000  R3000  W5000
0.0376 0.0383 0.0391 0.0390
```

Furthermore, the returns are highly correlated.

```
> cor(indices)
      SP500 R1000 R3000 W5000
SP500 1.000 0.999 0.997 0.996
R1000 0.999 1.000 0.999 0.999
R3000 0.997 0.999 1.000 1.000
W5000 0.996 0.999 1.000 1.000
```

Thus, the *smallest* correlation among the four indices is 0.996, between the S&P 500 index and the Wilshire 5000 index; this is not surprising given that, of the four indices, the Wilshire 5000 represents the most stocks while the S&P 500 represents the fewest.

It is worth noting that, even though the return means and standard deviations of the different indices are in close agreement and the indices are highly correlated, often there is considerable variation in the returns on the different indices in a given time period. For instance, in period 4, the returns on the four indices are 0.0146, 0.0174, 0.0205, and 0.0207, respectively. The high correlations tell us that this variation among indices is small relative to the variation within each index, a consequence of the fact that even the returns on a broad stock market index such as the Wilshire 5000 are quite variable. □

Price-Weighted Indices

Another approach to computing a market index is to simply sum the prices $P_{1,t}, P_{2,t}, \ldots, P_{N,t}$ of the stocks represented in the index. Let

$$J_t = \frac{\sum_{j=1}^{N} P_{j,t}}{D_t}$$

where D_t is a divisor, with properties similar to those of a divisor for a cap-weighted index.

Suppose the divisor does not change from period t to period $t+1$. Then the return on the index J_t at time $t+1$ is

$$\frac{J_{t+1}}{J_t} - 1 = \frac{\sum_{j=1}^{N} P_{j,t+1}}{\sum_{j=1}^{N} P_{j,t}} - 1 = \frac{\sum_{j=1}^{N} (P_{j,t+1} - P_{j,t})}{\sum_{j=1}^{N} P_{j,t}}.$$

Writing $P_{j,t+1} - P_{j,t} = P_{j,t} R_{j,t+1}$,

$$\frac{J_{t+1}}{J_t} - 1 = \sum_{j=1}^{N} \frac{P_{j,t}}{\sum_{j=1}^{N} P_{j,t}} R_{j,t+1}$$

so that the return on J_t is equal to the return on a portfolio with asset weights

$$w_{j,t}^{p} = \frac{P_{j,t}}{\sum_{j=1}^{N} P_{j,t}}, \quad j = 1, 2, \ldots, N.$$

Therefore, an index of this type is said to be a *price-weighted* index.

The most well-known price-weighted index is the Dow Jones Industrial Average, which is based on the stocks of 30 large companies, in a variety of industries; the Yahoo Finance symbol for the Dow Jones Industrial Average is ^DJI.

Example 8.2 Suppose that the monthly excess returns on the Dow Jones Industrial Average for the period January 2010 to December 2014 have been calculated and are stored in the R variable `djia`.

```
> mean(djia)
[1] 0.00950
> sd(djia)
[1] 0.0347
```

Thus, the mean and standard deviation of the returns on the Dow Jones Industrial Average are close to, but slightly different than, those based on the returns on the cap-weighted indices considered earlier. Similarly, its returns are highly correlated with those of the cap-weighted indices, but the correlations are not as large as those among the four cap-weighted indices; of course, the Dow Jones is based on only 30 stocks, so we should not expect the same level of agreement seen earlier.

```
> cor(djia, indices)
     SP500 R1000 R3000 W5000
[1,] 0.977 0.971 0.968 0.967
```

□

8.3 The Model and Its Estimation

For $t = 1, 2, \ldots, T$, let $R_{i,t}$ denote the return on asset i at time t. Let $R_{m,t}$ denote the return on a market index at time t and let $R_{f,t}$ denote the return on the risk-free asset at time t. Recall that, although the return on the risk-free asset has zero variance, the rate of return itself varies over time.

Assume that the stochastic process given by $\{(R_{i,t} - R_{f,t}, R_{m,t} - R_{f,t})^T : t = 1, 2, \ldots\}$ is weakly stationary; weak stationarity of a pair of random variables holds if any real-valued linear function of the random variables is weakly stationary in the usual sense. Hence, under weak stationarity, the means, variances, and covariances of $R_{i,t}$ and $R_{m,t}$ do not depend on t.

The market model states that

$$R_{i,t} - R_{f,t} = \alpha_i + \beta_i(R_{m,t} - R_{f,t}) + \epsilon_{i,t}, \quad t = 1, 2, \ldots, T \tag{8.1}$$

where α_i, β_i are unknown parameters and $\epsilon_{i,1}, \epsilon_{i,2}, \ldots, \epsilon_{i,T}$ are unobserved random variables each with mean zero and variance $\sigma^2_{\epsilon,i}$. Furthermore, we assume that

$$\text{Cov}(\epsilon_{i,t}, R_{m,t}) = 0, \quad t = 1, 2, \ldots, T. \tag{8.2}$$

The random variables $\epsilon_{i,t}$, $t = 1, 2, \ldots, T$, are known as the *residual returns.* They may be interpreted as the component of the excess return on the asset that is uncorrelated with the market returns; see Corollary 7.1 for a similar random variable in the context of the CAPM.

Note that some analysts define the residual returns to be $\alpha_i + \epsilon_{i,t}$, $t = 1, 2, \ldots, T$ so that α_i represents the mean residual return, while here we define the residual returns to have mean zero; however, the basic idea is the same—residual returns represent that part of an asset's returns that remains after accounting for a linear relationship with the market return.

Note that assumption (8.2) is equivalent to the assumption that the parameter β_i in (8.1) can be expressed in terms of $\text{Cov}(R_{i,t}, R_{m,t})$ and $\text{Var}(R_{m,t})$:

$$\beta_i = \frac{\text{Cov}(R_{i,t}, R_{m,t})}{\text{Var}(R_{m,t})}. \tag{8.3}$$

To see this, first note that, if $\text{Cov}(\epsilon_{i,t}, R_{m,t}) = 0$, then, by (8.2),

$$\text{Cov}(R_{i,t}, R_{m,t}) = \beta_i \text{Var}(R_{m,t}),$$

which yields the expression (8.3) for β_i.

Conversely, according to (8.2),

$$\text{Cov}(R_{i,t}, R_{m,t}) = \beta_i \text{Var}(R_{m,t}) + \text{Cov}(\epsilon_{i,t}, R_{m,t}); \tag{8.4}$$

recall that $\text{Cov}(R_{m,t}, R_{f,t}) = 0$. Therefore, if (8.3) holds, then

$$\text{Cov}(R_{i,t}, R_{m,t}) = \text{Cov}(R_{i,t}, R_{m,t}) + \text{Cov}(\epsilon_{i,t}, R_{m,t})$$

so that $\text{Cov}(\epsilon_{i,t}, R_{m,t}) = 0$.

We also assume that the errors are uncorrelated,

$$\text{Cov}(\epsilon_{i,t}, \epsilon_{i,s}) = 0 \quad \text{for all} \quad t, s = 1, 2, \ldots, T, \quad t \neq s$$

and that

$$\text{Cov}(\epsilon_{i,t}, R_{m,s}) = 0 \quad \text{for all} \quad t, s = 1, 2, \ldots, T.$$

Therefore, the market model is a regression model with response variable

$$Y_t = R_{i,t} - R_{f,t}, \quad t = 1, 2, \ldots, T,$$

the excess returns of asset i, and predictor variable

$$X_t = R_{m,t} - R_{f,t}, \quad t = 1, 2, \ldots, T,$$

the excess returns on a market index

$$Y_t = \alpha + \beta X_t + \epsilon_t, \quad t = 1, 2, \ldots, T$$

where $\alpha = \alpha_i$, $\beta = \beta_i$, and $\epsilon_t = \epsilon_{i,t}$. Under the assumptions described here, the errors $\epsilon_1, \epsilon_2, \ldots, \epsilon_T$ have mean zero, constant variance, and are uncorrelated; furthermore, $\text{Cov}(\epsilon_t, X_t) = 0$. A model of this type for (Y_t, X_t), $t = 1, 2, \ldots, T$ is often described as a "simple linear regression model."

Relationship to the CAPM

The market model is similar to the CAPM, but there are important differences. The CAPM is a model for the relationship between the expected excess return on an asset and the expected excess return on a hypothetical market portfolio, which is assumed to achieve the maximum possible Sharpe ratio, while the market model is a statistical model for observed excess returns on an asset and the observed excess returns on a market index. Note that, even though the CAPM and the market model use the same basic notation for the returns on the market portfolio and the returns on a market index, these two sets of returns are not identical.

Taking expectations in (8.1), the market model implies that

$$\mu_i - \mu_f = \alpha_i + \beta_i(\mu_m - \mu_f),$$

where $\mu_i - \mu_f = \mathrm{E}(R_{i,t} - R_{f,t})$ and $\mu_m - \mu_f = \mathrm{E}(R_{m,t} - R_{f,t})$. Thus, if the return on the market index used in the market model may be viewed as the return on the market portfolio, the market model and the CAPM describe similar relationships. The most important difference between these models is that, under the CAPM, the efficiency of the market portfolio implies that the intercept parameter satisfies $\alpha_i = 0$; conversely, if $\alpha_i = 0$, the market model implies a form of the CAPM using the returns on a market index in place of the returns on the market portfolio.

Interpretation of β_i

As in the CAPM, the parameter β_i in the market model is a measure of the relationship between the excess returns on an asset and the excess returns on the market index and, in many respects, the interpretation of β_i follows from the interpretation of beta in the CAPM, as discussed in Chapter 7. For instance, it may be used to decompose the variance of an asset's returns into market and nonmarket components; this will be discussed in detail in Section 8.5.

An alternative interpretation of β_i is as a measure of the sensitivity of an asset's excess returns to the excess return on the market index. However, such an interpretation does not follow directly from the assumptions of the market model as given in this chapter. In particular, the interpretation of β_i as a measure of sensitivity is valid only if the relationship between $R_{i,t} - R_{f,t}$ and $R_{m,t} - R_{f,t}$ is a linear one.

That is, suppose that the condition that $\epsilon_{i,t}$ and $R_{m,t}$ are uncorrelated is strengthened to $\mathrm{E}(\epsilon_{i,t}|R_{m,t} - R_{f,t}) = 0$; then

$$\mathrm{E}(R_{i,t} - R_{f,t}|R_{m,t} - R_{f,t} = r) = \alpha_i + \beta_i r.$$

It follows that

$$\beta_i = \frac{d}{dr}\mathrm{E}(R_{i,t} - R_{f,t}|R_{m,t} - R_{f,t} = r)$$

and β_i may be interpreted as the measure of sensitivity described previously. However, if $\mathrm{E}(\epsilon_{i,t}|R_{m,t} - R_{f,t})$ is a nonzero function of $R_{m,t} - R_{f,t}$, then

$$\frac{d}{dr}\mathrm{E}(R_{i,t} - R_{f,t}|R_{m,t} - R_{f,t} = r)$$

might not be equal to β_i. Fortunately, it is generally reasonable to assume that $\mathrm{E}(\epsilon_{i,t}|R_{m,t} - R_{f,t}) = 0$ does hold and, hence, the interpretation of β_i as a measure of sensitivity is typically appropriate.

Estimation

We now consider estimation of the parameters of the market model. As discussed previously, the market model may be viewed as a simple linear regression model with response variable $Y_t = R_{i,t} - R_{f,t}$, where $R_{i,t}$ is the return on a specific asset in period t, and $R_{f,t}$ is the risk-free rate in period t, and predictor variable $X_t = R_{m,t} - R_{f,t}$, where $R_{m,t}$ is the return of a market index in period t.

Therefore, the parameters α_i and β_i may be estimated using ordinary least squares. The formulas for the estimators are

$$\hat{\beta}_i = \frac{\sum_{t=1}^{T}(Y_t - \bar{Y})(X_t - \bar{X})}{\sum_{t=1}^{T}(X_t - \bar{X})^2}$$

and

$$\hat{\alpha}_i = \bar{Y} - \hat{\beta}_i \bar{X}$$

where $\bar{Y}$ and $\bar{X}$ are the sample means of the Y_t and X_t, respectively; these expressions are sometimes useful for studying the properties of the estimators, but they are not needed for numerical work.

Thus, the remaining issue is selection of the data to be used in the analysis: the market index, the risk-free asset, the return interval, and the observation period.

As discussed in Section 8.2, the "market portfolio" is a hypothetical concept; hence, in estimating the parameters of the market model, we use a market index chosen to measure the general behavior of the equity market. The most commonly used index in this context is the S&P 500 index. Although it includes only 500 stocks, the return on the S&P 500 is generally believed to reflect the return on the entire market. There are a number of broader indices that can be used such as the Russell 3000 index and the Wilshire 5000 index. As shown in Example 8.1, the S&P 500 index, the Russell 1000 index, the Russell 3000 index, and the Wilshire 5000 index are generally highly correlated with each other; hence, the choice from among these indices has a relatively small impact on the estimates. Here we will use the return on the S&P 500 index as the return on the market portfolio.

For the risk-free rate to use in the analysis, we will use the return on a 3-month Treasury Bill, as discussed in Example 6.1. These are generally reported as annual percentage rates, which must be converted to proportional monthly rates. Let R_{fa} be an annual percentage rate; recall that this may be converted to a monthly rate by

$$R_f = (1 + R_{fa}/100)^{1/12} - 1.$$

For the return interval, we could use daily, weekly, monthly, quarterly, or yearly returns. The return interval should reflect the investment horizon of interest. For instance, if investment decisions are made on a monthly basis, it generally makes sense to use monthly returns. Here we will use monthly returns.

The observation period refers to the number of return intervals to use in the analysis; for example, for monthly data, we need to choose how many months of data to include. For a given return interval, a longer observation period clearly yields more data and smaller standard errors. However, in using a longer observation period, we are implicitly assuming that β is constant over that time. Over a short observation period, this may be reasonable, but such an assumption becomes questionable as the observation period increases due to changes in the firms under consideration or changes in economic conditions. Three to five years is commonly used. Here we will use five years of monthly data.

Example 8.3 Consider the monthly excess returns on IBM stock, which we assume have been calculated and stored in the variable `ibm`; as discussed in Example 8.1, the excess returns on the S&P 500 index are stored in the variable `sp500`. As with any statistical analysis, before estimating the parameters of the linear regression model relating `ibm` to `sp500`, it is a good idea to plot the data, such a plot is given in Figure 8.1. The plot indicates an approximate linear relationship among the variables that would be accurately described by the market model.

To estimate the parameters of the market model in R, we use the function `lm`. The syntax of the command to estimate the market model relating returns on IBM stock to returns on the S&P 500 index, as contained in the variables `ibm` and `sp500`, respectively, is

```
> lm(ibm~sp500)
```

The expression `ibm~sp500` is known as a *model formula* and may be read as "`ibm` is described by `sp500`." The screen output from the command contains

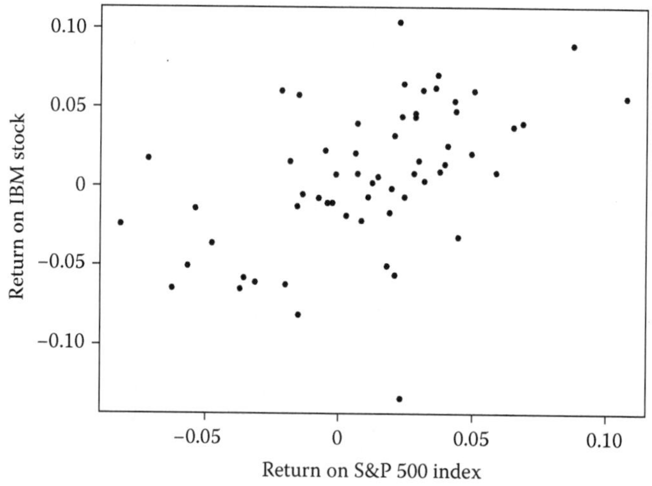

FIGURE 8.1
Plot of IBM monthly returns versus the returns on the S&P 500 index.

$\hat{\alpha}$ and $\hat{\beta}$, the least-squares estimates of the parameters α and β, respectively, in the market model for the returns on IBM stock:

```
> lm(ibm~sp500)
Coefficients:
(Intercept)        sp500
  -0.000707     0.618789
```

Therefore, for the data under consideration, $\hat{\beta} = 0.619$ and $\hat{\alpha} = -0.000707$.

However, much more information is available from the command, and it may be accessed by using certain *extractor* functions. Therefore, it is often useful to save the results of the `lm` function in a variable, which can be accessed as necessary:

```
> ibm.mm<-lm(ibm~sp500)
```

The variable `ibm.mm` now contains the results of the linear regression analysis relating the returns on IBM stock to the returns on the S&P 500 index. The `summary` command may be used to display a summary of the results:

```
> summary(ibm.mm)
Coefficients:
             Estimate Std. Error t value Pr(>|t|)
(Intercept) -0.000707   0.005358   -0.13      0.9
sp500        0.618789   0.138073    4.48  3.5e-05 ***
---
Signif. codes:  0 *** 0.001 ** 0.01 * 0.05 . 0.1   1

Residual standard error: 0.0398 on 58 degrees of freedom
Multiple R-squared:  0.257,     Adjusted R-squared:  0.244
F-statistic: 20.1 on 1 and 58 DF,  p-value: 3.54e-05
```

This output contains much useful information. For instance, the standard error of $\hat{\beta}$ is 0.138, so that an approximate 95% confidence interval for β is given by

$$0.619 \pm (1.96)(0.138) = (0.349, 0.889).$$

An estimate of the error standard deviation in the market model for returns on IBM stock, σ_ϵ, is given by the "Residual standard error" on the output. Hence, using σ_ϵ to denote the residual standard deviation in the market model for IBM stock, and using $\hat{\sigma}_\epsilon$ to denote the estimate of σ_ϵ based on least-squares regression, $\hat{\sigma}_\epsilon = 0.0398$.

It is also useful to know that this information may be accessed directly through the components of the result of the `summary` function. For instance, in the IBM example, `summary(ibm.mm)$coefficients` is a 2×4 matrix:

```
> summary(ibm.mm)$coefficients
             Estimate Std. Error t value Pr(>|t|)
(Intercept) -0.000707    0.00536  -0.132 8.96e-01
sp500        0.618789    0.13807   4.482 3.54e-05
```

For example, the estimate of β is given by `summary(ibm.mm)$coefficients [2,1]`:

```
> summary(ibm.mm)$coefficients[2,1]
[1] 0.619
```

Other useful components are `$sigma`, which contains $\hat{\sigma}_\epsilon$, and `$r.squared`, which contains R-squared for the regression:

```
> summary(ibm.mm)$sigma
[1] 0.0398
> summary(ibm.mm)$r.squared
[1] 0.257
```

The `lm` command may be used with a matrix argument as the response variable in order to provide results for several stocks at one time. Recall that in Chapter 6 we analyzed the variable `big8`, a matrix containing the monthly returns for the stocks of eight large companies; see Example 6.6 for details. To obtain the market model estimates of α_i, β_i for all the stocks represented in the data matrix `big8`, we use the command

```
> big8.mm<-lm(big8~sp500)
> big8.mm
Coefficients:
              AAPL       BAX        KO         CVS
(Intercept)  0.015361  -0.000353   0.004547   0.009559
sp500        0.920307   0.716442   0.485664   1.071935
              XOM        IBM        JNJ        DIS
(Intercept) -0.001335  -0.000707   0.005637   0.007799
sp500        0.878737   0.618789   0.540500   1.193193
```

Thus, the values of $\hat{\beta}$ for these stocks range from 0.486 for Coca-Cola to 1.193 for Disney. □

8.4 Testing the Hypothesis that an Asset Is Priced Correctly

Recall that, according to the CAPM, $\alpha_i \neq 0$ indicates that asset i is mispriced, with $\alpha_i > 0$ corresponding to an asset with a price that is too low and $\alpha_i < 0$ corresponding to an asset with a price that is too high; see Section 7.5.

Therefore, a test of the hypothesis $\alpha_i = 0$ in the market model may be used as a test of the hypothesis that the asset is priced correctly; rejection of this hypothesis suggests that the price of the asset is either too low or too high. It is important to keep in mind that such a conclusion is a statement about the market in periods $1, 2, \ldots, T$ and is not necessarily a statement about future prices.

The p-value for such a test is available in the result extracted using the `summary` function on the results from `lm`.

Example 8.4 Recall that, for IBM stock, the results of the `lm` function include the following:

```
Coefficients:
              Estimate Std. Error t value Pr(>|t|)
(Intercept) -0.000707   0.005358   -0.13       0.9
sp500        0.618789   0.138073    4.48   3.5e-05 ***
```

The p-value for testing $\alpha = 0$ is 0.9; therefore, we do not reject the hypothesis that the IBM stock is priced correctly.

To calculate the p-values for testing that $\alpha = 0$ for each of the stocks with returns included in the `big8` variable, we may use the `apply` function. Note that the `[1, 4]` element of the `$coefficients` component of the output from the `lm` function is the p-value for testing $\alpha = 0$

```
> summary(lm(ibm~sp500))$coefficients[1, 4]
[1] 0.8955
```

that is rounded to 0.9 in the output from `lm`.

Define a function `f.alphapval` that takes a vector of excess returns as an argument and returns this p-value:

```
> f.alphapval<-function(y)
+ {summary(lm(y~sp500))$coefficients[1, 4]}
> f.alphapval(ibm)
[1] 0.8955
```

The p-values for the eight stocks may then be calculated using `apply`

```
> apply(big8, 2, f.alphapval)
  AAPL    BAX     KO    CVS    XOM    IBM    JNJ    DIS
0.0883 0.9576 0.3675 0.0945 0.7591 0.8955 0.2110 0.1224
```

Therefore, for these eight stocks, the hypothesis that the stock is priced correctly is never rejected at the 0.05 level. Two stocks have a p-value less than 0.10—Apple, which has $\hat{\alpha} = 0.0154$ and a p-value of 0.088, and CVS, which has $\hat{\alpha} = 0.00956$ and a p-value of 0.095. □

Stock Screening and Multiple Testing

It is tempting to use tests of $\alpha_j = 0$ to screen a large number of stocks, hoping to find a few that are mispriced. However, when testing many hypotheses in this way, it is important to be aware of the *multiple testing problem.*

Suppose that we are testing m null hypotheses, each of the form $H_0 : \alpha_j = 0$. Recall that a p-value has the property that, if the null hypothesis is true, then the probability is approximately 0.05 that the p-value will be less than or equal to 0.05; more generally, the p-value is approximately distributed as a uniform random variable on the interval $(0, 1)$ when the null hypothesis is true.

Therefore, even if $\alpha_j = 0$ for all $j = 1, 2, \ldots, m$, we expect about 5% of the p-values to be less than 0.05. For instance, if we are testing $\alpha_j = 0$ for 100 stocks, even if all stocks are priced correctly, we expect about five significant p-values, defining significance in terms of a 0.05 level, that is, choosing each test to have a probability of Type I error of 0.05. Thus, if a few of the 100 p-values are significant, it may be inappropriate to conclude that those stocks are mispriced.

A simple way to deal with this issue is to modify the criterion for a significant p-value. Suppose that we take the null hypothesis to be the hypothesis that all α_j are 0; that is, consider the null hypothesis

$$\mathrm{H}_0 : \alpha_j = 0, \quad j = 1, 2, \ldots, m$$

or, equivalently,

$$\mathrm{H}_0 : \alpha_1 = \alpha_2 = \cdots = \alpha_m = 0.$$

For this testing problem, a Type I error corresponds to the event of rejecting $\alpha_j = 0$ for any j when, in fact, all α_j are 0. We can test this hypothesis using the p-values from the tests of the individual hypotheses $\alpha_j = 0$, which we denote by $q_1, q_2, \ldots, q_m$, respectively.

Suppose we want the level of our test of $H_0 : \alpha_1 = \alpha_2 = \cdots = \alpha_m = 0$ to be, at most, 0.05. If we reject $\alpha_j = 0$ when $q_j \leq c^*$, for some threshold c^*, then the probability of a Type I error is

$$\mathrm{P}\left(q_1 \leq c^* \ \cup\ q_2 \leq c^* \ \cup\ \cdots\ \cup\ q_m \leq c^*\right)$$

calculated under the assumption that

$$\alpha_1 = \alpha_2 = \cdots = \alpha_m = 0.$$

Exact calculation of this probability requires the joint distribution of $(q_1, q_2, \ldots, q_m)$; hence, it is difficult, if not impossible, without making strong assumptions.

However, it is generally possible to bound the probability. Recall that, for two events A and B,

$$\mathrm{P}(A \cup B) = \mathrm{P}(A) + \mathrm{P}(B) - \mathrm{P}(A \cap B)$$

and, hence,

$$\mathrm{P}(A \cup B) \leq \mathrm{P}(A) + \mathrm{P}(B).$$

An induction argument can be used to show that for events $A_1, A_2, \ldots, A_m$

$$\mathrm{P}(A_1 \cup A_2 \cup \cdots \cup A_m) \leq \sum_{j=1}^{m} \mathrm{P}(A_j),$$

a result known as the *Bonferroni inequality.*

Hence,

$$\mathrm{P}(q_1 \leq c^* \ \cup\ q_2 \leq c^* \ \cup \cdots\ \cup\ q_m \leq c^*) \leq \sum_{j=1}^{m} \mathrm{P}(q_j \leq c^*). \tag{8.5}$$

Using the fact that a p-value has a uniform distribution under the null hypothesis, $\mathrm{P}(q_j \leq c^*) = c^*$. It follows that

$$\mathrm{P}(q_1 \leq c^* \ \cup \ q_2 \leq c^* \ \cup \cdots \cup \ q_m \leq c^*) \leq mc^*.$$

Therefore, to guarantee that our test has a level less than or equal to 0.05, we can choose $c^* = 0.05/m$. Then the probability of concluding that any of the assets is mispriced when all are priced correctly is less than or equal to 0.05. Clearly, the same approach may be used for any desired level.

Hence, to address the multiple-testing problem, we modify the criterion for a significant p-value from 0.05 to $0.05/m$, where m is the number of hypotheses being tested; this is known as the *Bonferroni method.* An equivalent approach is to calculate "adjusted p-values," given by mq_j, $j = 1, 2, \ldots, m$; if $mq_j > 1$, we set the adjusted p-value to 1. The adjusted p-values can then be evaluated using the usual criteria; for instance, we can compare the adjusted p-values to 0.05 for a test with level 0.05.

Example 8.5 Consider stocks for firms represented in the S&P 100 index; stocks in the S&P 100 index are a subset of those in the S&P 500 index, representing a cross section of large U.S. companies. For each stock, five years of monthly returns were analyzed for the period ending December 31, 2014; only 96 of the 100 stocks had five years of monthly returns available.

For each of these 96 stocks, the p-value of the test of $\alpha_j = 0$ described earlier was calculated; the results are stored in the variable `sp96.pv`

```
> head(sp96.pv)
[1] 0.0883 0.2450 0.5338 0.9436 0.1488 0.0397
```

Thirteen of the p-values are less than 0.05, with the smallest at 0.0043.

```
> sort(sp96.pv)[1:15]
 [1] 0.00426 0.00548 0.00930 0.01299 0.01801 0.01960 0.02715
     0.02891 0.03139
[10] 0.03458 0.03966 0.04254 0.04811 0.05394 0.05474
```

For a test with level 0.05, the Bonferroni-corrected criterion is $0.05/96 = 0.00052$; all of the p-values exceed this threshold. Thus, although the p-values suggest that some of the stocks might be mispriced, after adjusting for multiple testing, we do not reject the hypothesis that all stocks are priced correctly.

Alternatively, if we compute the adjusted p-values, by multiplying the p-values by 96, we see that the smallest adjusted p-value is 0.41 (96 times 0.0043), leading to the same conclusion. □

False Discovery Rate

An important drawback of the Bonferroni method is that it is generally conservative, in the sense that the actual level of the test is less than 0.05; this is particularly true when m is large, as is often the case when analyzing stock return data. In the present context, this property means that there is a tendency for the procedure to conclude that all stocks are priced correctly even when one or more is mispriced.

An alternative approach to designing tests of many hypotheses is to control the *false discovery rate* (FDR) rather than to control the probability of a Type I error. Suppose we conduct a series of tests of the hypotheses that a stock is mispriced, that is, of the hypotheses of the form $\alpha_j = 0$, and that, based on the procedure used, we conclude that m_0 of the stocks are mispriced; that is, m_0 of the hypotheses that $\alpha_j = 0$ are rejected. Let m_1 denote the number of those rejected hypotheses for which α_j is actually 0.

We refer to a rejected hypothesis as a "discovery" and an incorrectly rejected hypothesis as a "false discovery." In the present context, a false discovery occurs if we conclude that a stock is mispriced when it is not. The false discovery proportion is defined as m_1/m_0 provided that $m_0 > 0$; if $m_0 = 0$, it is taken to be 0.

Note that the false discovery proportion is a random variable; the FDR is the expected value of this random variable. Therefore, the FDR is the expected proportion of rejected null hypotheses that were rejected incorrectly.

It is important to note that although the level of a test and its FDR are related, they are fundamentally different measures. The level of a test of $\alpha_1 = \alpha_2 = \cdots = \alpha_m = 0$ is the probability of rejecting this hypothesis, that is, of concluding that at least one α_j is nonzero when all are actually 0. The FDR measures the expected proportion of those cases in which $\alpha_j = 0$ is rejected for which α_j is actually 0. Hence, procedures that control the FDR do not control the level of the test. However, the FDR is an intuitively appealing concept in many applications, such as stock screening; furthermore, the procedures that control the FDR have higher power than those based on the Bonferroni correction, so that we are more likely to discover mispriced stocks.

Let q_j denote the p-value of the usual test of $\alpha_j = 0$, $j = 1, 2, \ldots, m$. To control the FDR at F, instead of comparing each q_j to a given threshold value, as in the Bonferroni method, we use the following procedure. First, order the p-values and let $q_{(1)}, q_{(2)}, \ldots, q_{(m)}$ denote the ordered values, so that $q_{(1)}$ is the smallest p-value, $q_{(2)}$ is the second smallest, and so on. Then, starting with $j = 1$ and moving through the list of p-values, we compare $q_{(j)}$ to $(j/m)F$.

If $q_{(j)} > (j/m)F$ for all $j = 1, 2, \ldots, m$, then we do not reject any of the hypotheses. Otherwise, find the largest j for which $q_{(j)} \leq (j/m)F$; denote this value by j^*. Then we reject the hypotheses corresponding to $q_{(1)}, q_{(2)}, \ldots, q_{(j^*)}$. Although this procedure is a bit complicated, fortunately, there is an R function that computes the corresponding adjusted p-values that can be compared to a given threshold in the usual way.

Although the conventional choice for the level of a test is 0.05, that is not necessarily the best choice for the FDR. For instance, an FDR of 0.10 or larger may be reasonable. In particular, if the tests of $\alpha_j = 0$ are used to screen stocks for further investigation, a threshold as large as 0.20 may be appropriate.

Example 8.6 Consider stocks for firms represented in the S&P 100 index analyzed in Example 8.5; consider testing $\alpha_j = 0$ for these stocks, controlling the FDR at 0.10.

The p-values for testing $\alpha_j = 0$ for each of the 96 stocks are stored in the variable `sp96.pv`. To compute the p-values adjusted for controlling the FDR, we use the following command:

```
> sp96.pv.fdr<-p.adjust(sp96.pv, method="fdr")
> head(sp96.pv)
[1] 0.0883 0.2450 0.5338 0.9436 0.1488 0.0397
> head(sp96.pv.fdr)
[1] 0.403 0.523 0.733 0.971 0.468 0.340
```

The function `p.adjust` can perform a number of different adjustments; using the argument `method="fdr"` specifies the adjustment to control the FDR, as described earlier.

The minimum adjusted p-value is given by

```
> min(sp96.pv.fdr)
[1] 0.26
```

Because this value exceeds 0.10, we conclude that all 96 stocks are priced correctly. If the minimum adjusted p-value had not exceeded 0.10, we would reject the hypothesis that $\alpha_j = 0$ for those assets with an adjusted p-value less than or equal to 0.10. □

8.5 Decomposition of Risk

When discussing the CAPM, it was shown that the variance of an asset's return may be expressed in terms of market and nonmarket components; see Section 7.3. Here we use the market model to estimate such components of variance.

Consider asset i for which the market model (8.1) holds. Since $\epsilon_{i,t}$ is uncorrelated with $R_{m,t}$,

$$\sigma_i^2 \equiv \text{Var}(R_{i,t}) = \text{Var}(\beta_i R_{m,t}) + \text{Var}(\epsilon_{i,t}) = \beta_i^2 \sigma_m^2 + \sigma_{\epsilon,i}^2,$$

where $\sigma_m^2 = \text{Var}(R_{m,t})$ and $\sigma_{\epsilon,i}^2 = \text{Var}(\epsilon_{i,t})$. Therefore, the risk of asset i, as measured by the variance, may be decomposed into two components, the market component, $\beta_i^2 \sigma_m^2$, and the nonmarket component, $\sigma_{\epsilon,i}^2$.

Estimates of β_i and $\sigma_{\epsilon,i}^2$ are available from the results of the linear regression analysis used to estimate the market model. To estimate σ_i^2 and σ_m^2, we interpret these as variances of excess returns:

$$\sigma_i^2 = \text{Var}(R_{i,t} - R_{f,t}) \quad \text{and} \quad \sigma_m^2 = \text{Var}(R_{m,t} - R_{f,t})$$

where $R_{f,t}$ denotes the return on the risk-free asset at time t.

Let S_i^2 denote the sample variance of $R_{i,t} - R_{f,t}$, $t = 1, 2, \ldots, T$, let S_m^2 denote the sample variance of $R_{m,t} - R_{f,t}$, $t = 1, 2, \ldots, T$, and let $\hat{\beta}_i$ and $\hat{\sigma}_{\epsilon,i}^2$

denote the estimators from least-squares regression, as discussed in Section 8.3. Then,

$$S_i^2 \doteq \hat{\beta}_i^2 S_m^2 + \hat{\sigma}_{\epsilon,i}^2; \tag{8.6}$$

note that the relationship does not hold exactly due to the different divisors used in the estimates: S_i^2 and S_m^2 use $T-1$, while $\hat{\sigma}_{\epsilon,i}^2$ uses $T-2$ to account for an additional degree of freedom lost when basing the estimator on the residuals from the regression. It follows that

$$S_i^2 = \hat{\beta}_i^2 S_m^2 + \frac{T-2}{T-1}\hat{\sigma}_{\epsilon,i}^2;$$

hence, except when T is very small, the relationship described in (8.6) holds to a high degree of approximation.

The proportion of the variance of $R_{i,t}$ explained by the return on the market index can be estimated by

$$\frac{\hat{\beta}_i^2 S_m^2}{S_i^2},$$

which is simply the R-squared value for the regression.

Example 8.7 Consider the market model for the returns on IBM stock; recall that the output from the linear regression analysis corresponding to the market model for IBM stock is stored in the variable `ibm.mm`. The R-squared value for the regression is available using the function `summary`.

```
Coefficients:
             Estimate Std. Error t value Pr(>|t|)
(Intercept) -0.000707   0.005358   -0.13      0.9
sp500        0.618789   0.138073    4.48  3.5e-05 ***

Residual standard error: 0.0398 on 58 degrees of freedom
Multiple R-squared:  0.257,     Adjusted R-squared:  0.244
F-statistic: 20.1 on 1 and 58 DF,  p-value: 3.54e-05
```

Therefore, 25.7% of the variability in IBM excess returns is attributable to the market. This result may be used to decompose the sample variance of IBM excess returns:

$$0.00210 = (0.257)(0.00210) + (1-0.257)(0.00210) = 0.00540 + 0.00156.$$

The R-squared value may also be obtained directly by accessing the `$r.squared` component of the result of `summary`:

```
> summary(ibm.mm)$r.squared
[1] 0.257
```

The `apply` function can be used to calculate the R-squared for all eight of the stocks represented in `big8`. Define a function `f.rsq` by

```
>f.rsq<-function(y){summary(lm(y~sp500))$r.squared}
```

Then the R-squared values may be calculated by

```
> apply(big8, 2, f.rsq)
 AAPL   BAX    KO    CVS   XOM   IBM   JNJ   DIS
0.219 0.234 0.196 0.485 0.516 0.257 0.276 0.599
```

Thus, for the eight stocks with return data in the `big8` variable, the R-squared values range from 0.196 for Coca-Cola to 0.599 for Disney. Therefore, nearly 60% of the variation in the returns on Disney stock, as measured by the return variance, can be explained by variation in the market return; on the other hand, for Coca-Cola stock, less than 20% of the variation in the returns can be explained by the market. □

8.6 Shrinkage Estimation and Adjusted Beta

Often, we are interested in estimating β for several assets. For instance, suppose we are analyzing N stocks and let $\beta_1, \beta_2, \ldots, \beta_N$ denote their respective parameters in the market model (8.1). In such cases, we may use shrinkage estimation, following the general approach described in Section 6.5.

Thus, in shrinkage estimation, we combine a simple estimator of β_i, such as the least-squares estimator, with an estimator based on assumptions regarding the parameters $\beta_1, \beta_2, \ldots, \beta_N$, by taking a weighted average of the two estimators.

Let $\hat{\beta}_1, \hat{\beta}_2, \ldots, \hat{\beta}_N$ denote the least-squares estimators of $\beta_1, \beta_2, \ldots, \beta_N$, respectively. For the assumption-based estimators, we may consider the assumption that $\beta_1 = \beta_2 = \cdots = \beta_N \equiv \beta$ for some β. To estimate β, we may use the average of the least-squares estimators:

$$\bar{\beta} = \frac{1}{N}\sum_{i=1}^{N} \hat{\beta}_i.$$

Then a shrinkage estimator of β_i is given by a weighted average of $\hat{\beta}_i$ and $\bar{\beta}$.

When $\hat{\beta}_1, \hat{\beta}_2, \ldots, \hat{\beta}_N$ are approximately equal, we give more weight to $\bar{\beta}$. When $\hat{\beta}_1, \hat{\beta}_2, \ldots, \hat{\beta}_N$ do not follow the assumption of equal beta, in the sense that there is a great deal of variability in $\hat{\beta}_1, \hat{\beta}_2, \ldots, \hat{\beta}_N$, then more weight is given to $\hat{\beta}_i$.

In order to choose the weights given to $\hat{\beta}_i$ and $\bar{\beta}$, we adapt the procedure used in Section 6.5 when estimating mean returns. Let $\text{SE}(\hat{\beta}_i)$ denote the standard error of $\hat{\beta}_i$, as given in the output of the `lm` function, and let

$$\overline{\text{SE}^2} = \frac{1}{N}\sum_{i=1}^{N} \text{SE}(\hat{\beta}_i)^2.$$

Then a shrinkage estimator of β_i is given by

$$\psi\bar{\beta} + (1 - \psi)\hat{\beta}_i$$

where

$$\psi = \frac{\overline{\text{SE}^2}}{\overline{\text{SE}^2} + \tau_\beta^2}$$

and

$$\tau_\beta^2 = \frac{1}{N}\sum_{ij=1}^{N}(\hat{\beta}_i - \bar{\beta})^2.$$

Example 8.8 One use for shrinkage estimation is in estimating the values of β for a number of assets for which it is reasonable to expect similar relationships with the market index. For example, here we consider the four airline stocks, American Airlines Group, Inc. (symbol AAL), Delta Air Lines, Inc. (DAL), Southwestern Airline Company (LUV), and United Continental Holdings, Inc. (UAL).

Five years of monthly returns for the period ending December 31, 2014, were computed for these stocks. The results are stored in variables with the name matching the stock symbol; for example, the returns on American Airlines stock are stored in the variable `aal`. The returns for all the four stocks are stored as a matrix in the variable `air`, which is similar to the variable `big8`. Estimates of beta for each of the four stocks may be computed as follows:

```
> air.mm<-lm(air~sp500)
> air.beta<-air.mm$coefficients[2,]
> air.beta
  AAL   DAL   LUV   UAL
0.610 0.825 1.016 0.679
```

Therefore, $\bar{\beta}$ and τ_β^2 may be calculated by

```
> beta.bar<-mean(air.beta)
> beta.bar
[1] 0.782
> tausq.beta<-mean((air.beta-beta.bar)^2)
> tausq.beta
[1] 0.0242
```

To compute $\text{SE}(\hat{\beta}_i)$ for each stock, we may use the `apply` function. Note that the `[2, 2]` element of the component `$coefficients` of `summary` applied to the output from the `lm` function yields the standard error of $\hat{\beta}$.

Define a function `f.betase` by

```
> f.betase<-function(y){ summary(lm(y~sp500))$coefficients[2,2]}
```

Then the vector of standard errors of $\hat{\beta}_i$ can be computed by

```
> air.betase<-apply(air, 2, f.betase)
> air.betase
   AAL    DAL    LUV    UAL
0.5628 0.3405 0.2400 0.3879
```

and $\overline{\text{SE}^2}$ is given by

```
> sesq.bar<-mean(air.betase^2)
> sesq.bar
[1] 0.160
```

It follows that the weight ψ used in the shrinkage estimator is given by

```
> air.psi<-sesq.bar/(sesq.bar+tausq.beta)
> air.psi
[1] 0.869
```

and the shrinkage estimates of beta for the four airline stocks are given by

```
> air.psi*beta.bar + (1-air.psi)*air.beta
  AAL   DAL   LUV   UAL
0.760 0.788 0.813 0.769
```

Recall that the least-squares estimates are given by

```
  AAL   DAL   LUV   UAL
0.610 0.825 1.016 0.679
```

Note that the variation of $\hat{\beta}_1, \hat{\beta}_2, \hat{\beta}_3, \hat{\beta}_4$ is small relative to the standard errors of the $\hat{\beta}_j$ so that the shrinkage estimates are all relatively close to the average of the $\hat{\beta}_j$. □

An important feature of the shrinkage estimator described previously is that the same weight ψ is used for each asset. However, in many cases, the standard error of $\hat{\beta}_i$ varies considerably for the different assets. For assets in which beta is estimated accurately, in the sense that $\text{SE}(\hat{\beta}_i)$ is relatively small, we may want to give more weight to $\hat{\beta}_i$; on the other hand, for assets for which $\text{SE}(\hat{\beta}_i)$ is relatively large, it may be preferable to give relatively little weight to $\hat{\beta}_i$.

Thus, it may be desirable to use asset-specific values of ψ, $\psi_1, \psi_2, \ldots, \psi_N$, particularly when $\text{SE}(\hat{\beta}_i)$, $i = 1, 2, \ldots, N$ exhibit large variation. Let

$$\psi_i = \frac{\text{SE}(\hat{\beta}_i)^2}{\text{SE}(\hat{\beta}_i)^2 + \tau_\beta^2}.$$

Then an alternative shrinkage estimator of β_i is given by

$$\psi_i\bar{\beta} + (1 - \psi_i)\hat{\beta}_i.$$

Example 8.9 Consider the four airline stocks analyzed in Example 8.8. Recall that the excess return data for the four stocks are stored in the variables `aal`, `dal`, `luv`, and `ual`; the data matrix for these four assets is stored in the variable `air`. The variable `sp500` contains the excess returns on the S&P 500 index.

Consider estimation of β for American Airlines stock. Using the results of the `lm` function applied to the market model for American Airlines,

```
> summary(lm(aal~sp500))
Coefficients:
            Estimate Std. Error t value Pr(>|t|)
(Intercept)   0.0459     0.0218    2.10     0.04 *
sp500         0.6095     0.5628    1.08     0.28

Residual standard error: 0.162 on 58 degrees of freedom
Multiple R-squared:  0.0198,    Adjusted R-squared:  0.00292
F-statistic: 1.17 on 1 and 58 DF,  p-value: 0.283
```

we have that $SE(\hat{\beta}_i) = 0.5628$; recall that here τ_β^2 is given by 0.0242. Hence, the weight for AAL is

$$\frac{0.5628^2}{((0.5828)^2 + 0.0242)} = 0.9290.$$

It follows that the shrinkage estimate of beta for AAL is given by

$$(0.929)\bar{\beta} + (1 - 0.929)(0.6095) = (0.929)(0.7822) + (1 - 0.929)(0.6095) = 0.7699.$$

To calculate the shrinkage estimates of β for all four stocks, recall that the variable `air.betase` contains the standard errors of $\hat{\beta}_i$ for the four stocks

```
> air.betase
   AAL    DAL    LUV    UAL
0.5628 0.3405 0.2400 0.3879
```

The vector of weights used in this procedure along with the vector of shrinkage estimates of β for all four stocks may now be computed as follows:

```
> psi.air<-(air.betase^2)/(tausq.beta + air.betase^2)
> psi.air
  AAL   DAL   LUV   UAL
0.929 0.827 0.704 0.861
> psi.air*beta.bar + (1-psi.air)*air.beta
  AAL   DAL   LUV   UAL
0.770 0.790 0.851 0.768
```

Recall that the shrinkage estimates based on a global value for ψ are given by

```
  AAL   DAL   LUV   UAL
0.760 0.788 0.813 0.769
```

The two sets of estimates are very similar. The greatest difference occurs for LUV; note that LUV has the largest value of $\hat{\beta}_i$ and the smallest value of the standard error of $\hat{\beta}_i$. □

Adjusted Beta

When estimating β for a large number of assets, a simpler type of shrinkage estimator is sometimes used. Since the value of beta for the entire market is, by definition, equal to 1, it is often reasonable to use 1 in place of $\bar{\beta}$. Using global values for the weights given to one and $\hat{\beta}_i$ leads to an estimator of the form

$$(1-k)+k\hat{\beta}_i \tag{8.7}$$

for some constant k. Often $k = 2/3$ is used, yielding the estimator of β_i given by

$$\hat{\beta}_{i,\text{adj}} = \frac{1}{3} + \frac{2}{3}\hat{\beta}_i, \tag{8.8}$$

which is known as *adjusted beta.* It is sometimes attributed to analysts at the brokerage firm Merrill Lynch, Pierce, Fenner & Smith, Inc. (Vasicek 1973); this type of adjusted beta is used most notably by the financial data firm Bloomberg L. P. (www.bloomberg.com) so it is sometimes referred to as "Bloomberg adjusted beta."

This estimator has the advantage of requiring only $\hat{\beta}_i$ in order to estimate β_i. However, it has the drawbacks of always shrinking the estimates toward 1 and of always using the weights 1/3 and 2/3; these choices may not be appropriate in all cases.

Example 8.10 Stocks for firms represented in the S&P 100 index were considered; these data were also analyzed in Example 8.5. For each of the 96 stocks, five years of monthly returns were analyzed for the period ending December 31, 2014.

For each stock, the least-squares estimate $\hat{\beta}_i$ was calculated along with the adjusted beta for stock i, $\hat{\beta}_{i,\text{adj}}$. To measure the accuracy in these estimates as predictions of future beta values, they were compared with the least-squares estimates of β based on the 12 monthly returns in 2015, which we denote by $\hat{\beta}_i^*$, $i = 1, 2, \ldots, 96$.

The average error in the least-squares estimates is given by

$$\frac{1}{96}\sum_{i=1}^{96} |\hat{\beta}_i - \hat{\beta}_i^*| = 0.369;$$

for adjusted beta, the average error is

$$\frac{1}{96}\sum_{i=1}^{96} |\hat{\beta}_{i,\text{adj}} - \hat{\beta}_i^*| = 0.321.$$

Thus, use of adjusted beta reduces the error in predicting the estimates of β for 2015 by about 13%.

For comparison, the shrinkage estimates of the β_i were also calculated. The average error in these estimates in predicting $\hat{\beta}_i^*$ is 0.347 using a global value of ψ and 0.344 using asset-specific values of ψ. Thus, at least for this example, adjusted beta appears to be at least as successful as the shrinkage estimators in predicting future beta values.

It is worth noting that none of the estimators are particularly accurate in predicting future beta estimates. In evaluating these results, it is important to keep in mind that the future beta values used for comparison are estimates, with their own sampling variability, in addition to the sampling variability of the least-squares, adjusted, and shrinkage estimates of beta. □

8.7 Applying the Market Model to Portfolios

Although the discussion in this chapter has focused on the analysis of individual securities, the same approach may be applied to a portfolio.

Consider a portfolio based on N assets, with returns $R_{1,t}, R_{2,t}, \ldots, R_{N,t}$ in period t. Let $w_1, w_2, \ldots, w_N$ denote the portfolio weights so that the return on the portfolio in period t is given by

$$R_{p,t} = \sum_{i=1}^{N} w_i R_{i,t}, \quad t = 1, 2, \ldots, T.$$

Suppose that the market model holds for each asset so that, for each $i = 1, 2, \ldots, N$,

$$R_{i,t} - R_{f,t} = \alpha_i + \beta_i(R_{m,t} - R_{f,t}) + \epsilon_{i,t}, \quad t = 1, 2, \ldots, T \tag{8.9}$$

where $\epsilon_{i,t}$ has mean 0 and is uncorrelated with the market return $R_{m,t}$. Then

$$\begin{aligned} R_{p,t} - R_{f,t} &= \sum_{i=1}^{N} w_i(R_{i,t} - R_{f,t}) \\ &= \sum_{i=1}^{N} w_i(\alpha_i + \beta_i(R_{m,t} - R_{f,t}) + \epsilon_{i,t}) \\ &= \sum_{i=1}^{N} w_i\alpha_i + \left(\sum_{i=1}^{N} w_i\beta_i\right)(R_{m,t} - R_{f,t}) + \sum_{i=1}^{N} w_i\epsilon_{i,t}, \quad t = 1, 2, \ldots, T. \end{aligned}$$

Note that

$$\mathrm{E}\left(\sum_{i=1}^{N} w_i\epsilon_{i,t}\right) = \sum_{i=1}^{N} w_i\mathrm{E}(\epsilon_{i,t}) = 0$$

and, because $\mathrm{Cov}(\epsilon_{i,t}, R_{m,t}) = 0$ for all $i = 1, 2, \ldots, N$,

$$\mathrm{Cov}\left(\sum_{i=1}^{N} w_i\epsilon_{i,t}, R_{m,t}\right) = \sum_{i=1}^{N} w_i\mathrm{Cov}(\epsilon_{i,t}, R_{m,t}) = 0, \quad t = 1, 2, \ldots, T.$$

It follows that the market model holds for the portfolio with parameters

$$\alpha_p = \sum_{i=1}^{N} w_i \alpha_i \quad \text{and} \quad \beta_p = \sum_{i=1}^{N} w_i \beta_i.$$

Furthermore, the least-squares estimators of α_p and β_p follow these relationships as well. That is, if $\hat{\alpha}_i$ and $\hat{\beta}_i$ are the least-squares estimators of α_i and β_i, then $\hat{\alpha}_p$ and $\hat{\beta}_p$, the least-squares estimators of α_p and β_p, respectively, are given by

$$\hat{\alpha}_p = \sum_{i=1}^{N} w_i \hat{\alpha}_i \quad \text{and} \quad \hat{\beta}_p = \sum_{i=1}^{N} w_i \hat{\beta}_i.$$

To see why this holds, consider the $N = 2$ case. For $j = 1, 2$, let $Y_{j,t} = R_{j,t} - R_{f,t}$, and let $X_t = R_{m,t} - R_{f,t}$. Then, as discussed in Section 8.3, for $j = 1, 2$,

$$\hat{\beta}_j = \frac{\sum_{t=1}^{T} (Y_{j,t} - \bar{Y}_j)(X_t - \bar{X})}{\sum_{t=1}^{T} (X_t - \bar{X})^2}$$

where $\bar{Y}_j$ is the sample mean of the $Y_{j,t}$.

Now consider a portfolio with return $R_{p,t} = wR_{1,t} + (1 - w)R_{2,t}$ at time t. The least-squares estimator of beta for the portfolio may be written

$$\hat{\beta}_p = \frac{\sum_{t=1}^{T} (Y_{p,t} - \bar{Y}_p)(X_t - \bar{X})}{\sum_{t=1}^{T} (X_t - \bar{X})^2}$$

where $Y_{p,t} = R_{p,t} - R_{f,t}$ and $\bar{Y}_p$ is the sample mean of the $Y_{p,t}$. Because

$$Y_{p,t} = wY_{1,t} + (1 - w)Y_{2,t}$$

it follows that

$$\begin{aligned}
\hat{\beta}_p &= \frac{\sum_{t=1}^{T} (wY_{1,t} + (1 - w)Y_{2,t} - w\bar{Y}_1 - (1 - w)\bar{Y}_2)(X_t - \bar{X})}{\sum_{t=1}^{T} (X_t - \bar{X})^2} \\
&= w\frac{\sum_{t=1}^{T} (Y_{1,t} - \bar{Y}_1)(X_t - \bar{X})}{\sum_{t=1}^{T} (X_t - \bar{X})^2} + (1 - w)\frac{\sum_{t=1}^{T} (Y_{2,t} - \bar{Y}_2)(X_t - \bar{X})}{\sum_{t=1}^{T} (X_t - \bar{X})^2} \\
&= w\hat{\beta}_1 + (1 - w)\hat{\beta}_2.
\end{aligned}$$

The argument for $\hat{\alpha}_p$ is similar.

Example 8.11 Consider the eight stocks with returns stored in the variable `big8`. Recall that the results from the market model regression on all eight stocks are stored in the variable `big8.mm`. The estimated regression coefficients from the eight regression analyses are available in the component `coefficients` of `big8.mm`:

```
> big8.mm$coefficients
                AAPL       BAX     KO    CVS      XOM      IBM    JNJ
(Intercept) 0.015 -0.00035 0.0045 0.0096 -0.0013 -0.00071 0.0056
sp500       0.920  0.71644 0.4857 1.0719  0.8787  0.61879 0.5405
               DIS
(Intercept) 0.0078
sp500       1.1932
```

These estimates form a 2×8 matrix; therefore, the second row of this vector contains the eight estimates of β:

```
> big8.mm$coefficients[2,]
  AAPL    BAX     KO    CVS    XOM    IBM    JNJ    DIS
0.9203 0.7164 0.4857 1.0719 0.8787 0.6188 0.5405 1.1932
```

Consider the equally weighted portfolio of the eight stocks; the returns on this portfolio may be calculated as

```
> big8.port<-apply(big8, 1, mean)
```

yielding a vector consisting of the average return in each time period. Alternatively, we can perform the calculation using matrix multiplication

```
> big8.port<-big8%*%rep(1/8, 8)
```

Here, `rep(1/8, 8)` is a vector of length 8 of the form $(1/8, 1/8, \ldots, 1/8)$.

The estimates of α_p and β_p may be calculated directly from the returns on the portfolio.

```
> lm(big8.port~sp500)
Coefficients:
(Intercept)        sp500
    0.00506      0.80320
```

These estimates can also be obtained as the averages of the coefficient estimates from the analyses on the eight individual stocks.

```
> apply(big8.mm$coefficients, 1, mean)
(Intercept)       sp500
    0.00506     0.80320
```

□

Time-Dependent Portfolio Weights

Note that the analysis thus far in this section is based on the assumption that the portfolio weights do not depend on t. However, in some cases, the portfolio weights may change over time; for instance, the holdings in a mutual fund may be modified to account for changing economic conditions or changing beliefs about the future returns of certain stocks.

Let $w_{1,t}, w_{2,t}, \ldots, w_{N,t}$ denote the weights at time t so that

$$\sum_{i=1}^{N} w_{i,t} = 1 \quad \text{for} \quad t = 1, 2, \ldots, T.$$

If the market model holds for each asset, that is, if (8.9) holds, then, for each $t = 1, 2, \ldots, T$,

$$\begin{aligned} R_{p,t} - R_{f,t} &= \sum_{i=1}^{N} w_{i,t}(R_{i,t} - R_{f,t}) \\ &= \sum_{i=1}^{N} w_{i,t}\alpha_i + \sum_{i=1}^{N} w_{i,t}\beta_i(R_{m,t} - R_{f,t}) + \sum_{i=1}^{N} w_{i,t}\epsilon_{i,t} \\ &= \alpha_{p,t} + \beta_{p,t}(R_{m,t} - R_{f,t}) + \epsilon_{p,t} \end{aligned}$$

where

$$\alpha_{p,t} = \sum_{i=1}^{N} w_{i,t}\alpha_i \quad \text{and} \quad \beta_{p,t} = \sum_{i=1}^{N} w_{i,t}\beta_i.$$

Note that the conditions that $\mathrm{E}(\epsilon_{p,t}) = 0$ and $\mathrm{Cov}(\epsilon_{p,t}, R_{m,t}) = 0$ for each $t = 1, 2, \ldots, T$ continue to hold so that the market model holds for the portfolio in each specific time period; however, the values of α and β for the portfolio now depend on t.

If the weights are approximately constant over time, for example, if the portfolio corresponds to a specific investment strategy with periodic minor adjustments, then it may be reasonable to assume that $\alpha_{p,t}$ and $\beta_{p,t}$ are approximately constant over time and, hence, the market model is appropriate for the portfolio. On the other hand, if major changes are made regularly to the portfolio so that, in effect, there is a different portfolio in each time period, then the market model assumption of constant α_p and β_p is inappropriate.

8.8 Diversification and the Market Model

In Chapter 4, we saw that it is generally possible to reduce the risk of a portfolio by diversification. In this section, we look at the implications of the market model on diversification.

Consider two assets, with returns $R_{i,t}$ and $R_{j,t}$, respectively, at time t. Let $\sigma_i^2 = \mathrm{Var}(R_{i,t}), \sigma_j^2 = \mathrm{Var}(R_{j,t})$, and let ρ_{ij} denote the correlation of $R_{i,t}, R_{j,t}$. Suppose that the market model holds for each asset so that

$$R_{i,t} - R_{f,t} = \alpha_i + \beta_i(R_{m,t} - R_{f,t}) + \epsilon_{i,t}, \quad t = 1, 2, \ldots, T$$

and

$$R_{j,t} - R_{f,t} = \alpha_j + \beta_j(R_{m,t} - R_{f,t}) + \epsilon_{j,t} \quad t = 1, 2, \ldots, T.$$

Now consider a portfolio of assets i and j, with return of the form

$$R_{p,t} = wR_{i,t} + (1-w)R_{j,t},$$

where $0 < w < 1$. Let $\sigma_p^2 = \text{Var}(R_{p,t})$; then

$$\sigma_p^2 = w^2\sigma_i^2 + (1-w)^2\sigma_j^2 + 2w(1-w)\rho_{ij}\sigma_i\sigma_j.$$

Suppose, for simplicity, that $\sigma_i^2 = \sigma_j^2$. Then

$$\begin{aligned}\sigma_p^2 &= \left(w^2 + (1-w)^2 + 2w(1-w)\rho_{ij}\right)\sigma_i^2 \\ &= \left(1 - 2(1-\rho_{ij})w(1-w)\right)\sigma_j^2.\end{aligned}$$

Because $w(1-w) > 0$ for $0 < w < 1$ and $\rho_{ij} < 1$, it follows that $\sigma_p^2 < \sigma_j^2$ for any $0 < w < 1$; see Section 4.2 for further details. That is, for any $0 < w < 1$, the risk of the portfolio is less than that of either of the two assets used to form it. However, diversification has very different effects on the market and nonmarket components of risk.

As discussed in the previous section, the market model holds for the portfolio:

$$R_{p,t} - R_{f,t} = \alpha_p + \beta_p(R_{m,t} - R_{f,t}) + \epsilon_{p,t}, \quad t = 1, 2, \ldots, T$$

where

$$\alpha_p = w\alpha_i + (1-w)\alpha_j \quad \text{and} \quad \beta_p = w\beta_i + (1-w)\beta_j.$$

Then the market component of the variance of the portfolio return is given by $\beta_p^2\sigma_m^2$.

Suppose that, without loss of generality, $\beta_i \leq \beta_j$. Then

$$\beta_i \leq \beta_p \leq \beta_j.$$

Therefore, provided that $\beta_i > 0$, as is typically the case,

$$\beta_i^2\sigma_m^2 \leq \beta_p^2\sigma_m^2 \leq \beta_j^2\sigma_m^2$$

so that the market component of variance for the portfolio return lies between the market components of return variance for the two assets. Thus, the market component of return variance for the portfolio cannot be reduced below the smaller of the market components of return variance for the two assets.

In particular, if $\beta_i = \beta_j$ then $\beta_p = \beta_i$ and, hence, the market component of the variance of $R_{p,t}$ is $\beta_i^2\sigma_m^2$, the same as the market components of variance for returns on each of assets i and j. That is, in this case, diversification does not reduce the market component of risk. Therefore, the reduction in the variance of the portfolio return, as compared to the return variances of assets i and j, is entirely because of a reduction in the nonmarket component of variance.

Example 8.12 Consider two assets, asset i and asset j; assume that $\sigma_i = \sigma_j = 0.5$, $\beta_i = \beta_j = 0.8$, and ρ_{ij}, the correlation of the returns of the two assets, is 0.2. Suppose that the market portfolio has return standard deviation of 0.4. Then the equally weighted portfolio of these assets has return variance

$$\left(\frac{1}{2}\right)^2 \sigma_i^2 + \left(\frac{1}{2}\right)^2 \sigma_j^2 + 2\frac{1}{2}\frac{1}{2}\rho_{ij}\sigma_i\sigma_j = 0.15.$$

That is, the risk of 0.5 for each asset is reduced to $\sqrt{0.15} \doteq 0.39$ for the portfolio.

The market component of the variance of each of the two asset returns is

$$(0.8)^2(0.4)^2 = 0.1024,$$

and for each asset, the nonmarket component of the return variance is

$$0.25 - 0.1024 = 0.1476.$$

The equally weighted portfolio has $\beta_p = 0.8$; hence, its market component of return variance is also 0.1024. It follows that the nonmarket component of the return variance for the portfolio is

$$0.15 - 0.1024 = 0.0476.$$

Thus, each asset has a return variance of 0.25, with a market component of 0.1024. The return variance of the portfolio is 0.15, but the market component of that variance is the same as the market component of the return variance for the two assets, 0.1024. The nonmarket component of return variance for each of the two assets is 0.1476, while the nonmarket component of return variance for the portfolio is only 0.0476. □

Recall that, according to the CAPM, it is the market component of risk that is rewarded by a higher expected return, as discussed in Section 7.3. Thus, not only does diversification tend to reduce risk, the reduction is greater for the nonmarket component of risk, the component of risk that is not rewarded with a higher expected return.

Portfolios of Several Assets

Similar considerations apply to a portfolio of N assets. Let $R_{1,t}, R_{2,t}, \ldots, R_{N,t}$ denote the asset returns at time t and suppose that the market model holds for each asset, that is, for each $i = 1, 2, \ldots, N$,

$$R_{i,t} - R_{f,t} = \alpha_i + \beta_i(R_{m,t} - R_{f,t}) + \epsilon_{i,t}, \quad t = 1, 2, \ldots, T.$$

Let

$$\boldsymbol{\alpha} = \begin{pmatrix} \alpha_1 \\ \alpha_2 \\ \vdots \\ \alpha_N \end{pmatrix}, \quad \boldsymbol{\beta} = \begin{pmatrix} \beta_1 \\ \beta_2 \\ \vdots \\ \beta_N \end{pmatrix}, \quad \text{and} \quad \boldsymbol{\epsilon}_t = \begin{pmatrix} \epsilon_{1,t} \\ \epsilon_{2,t} \\ \vdots \\ \epsilon_{N,t} \end{pmatrix}$$

denote $N \times 1$ vectors. Consider a portfolio based on a weight vector $\boldsymbol{w} \in \Re^N$, with return $R_{p,t}$ at time t. Then

$$R_{p,t} - R_{f,t} = \alpha_p + \beta_p(R_{m,t} - R_{f,t}) + \epsilon_{p,t}$$

where $\alpha_p = \boldsymbol{w}^T\boldsymbol{\alpha}, \beta_p = \boldsymbol{w}^T\boldsymbol{\beta}$, and $\epsilon_{p,t} = \boldsymbol{w}^T\boldsymbol{\epsilon}_t$.

Let $\boldsymbol{\Sigma}_\epsilon$ denote the covariance matrix of $\boldsymbol{\epsilon}_t$. Then the market component of $\text{Var}(R_{p,t})$ is $\beta_p^2\sigma_m^2$ and the nonmarket component is

$$\sigma_{\epsilon,p}^2 \equiv \text{Var}(\epsilon_{p,t}) = \text{Var}(\boldsymbol{w}^T\boldsymbol{\epsilon}_t) = \boldsymbol{w}^T\boldsymbol{\Sigma}_\epsilon\boldsymbol{w}.$$

Because of the benefits of diversification, the variance of the return on the portfolio tends to be small relative to the variances of the returns on the individual assets; as in the two-asset case, this is generally because of a reduction in the nonmarket components of return variance.

Example 8.13 Consider the equally weighted portfolio of eight stocks analyzed in Example 8.11, with returns in `big8`. For the eight individual stocks, the standard deviations are given by

```
> apply(big8, 2, sd)
  AAPL    BAX     KO    CVS    XOM    IBM    JNJ    DIS
0.0739 0.0556 0.0412 0.0578 0.0459 0.0458 0.0386 0.0579
```

Now consider estimation of the nonmarket component of risk for each of the stocks. Note that $\hat{\sigma}_{\epsilon,i}$ for asset i is available from the results of the `lm` function by extracting the `sigma` component of the results from the `summary` function; for example,

```
> summary(lm(ibm~sp500))$sigma
[1] 0.0398
```

Define a function `f.sighat` by

```
> f.sighat<-function(y){summary(lm(y~sp500))$sigma}
```

Note that

```
> f.sighat(ibm)
[1] 0.0398
```

The nonmarket components of risk—that is, the standard deviations corresponding to the nonmarket components of variance—for the eight stocks may now be calculated using the `apply` function in the usual way as follows:

```
> apply(big8, 2, f.sighat)
  AAPL    BAX     KO    CVS    XOM    IBM    JNJ    DIS
0.0659 0.0490 0.0372 0.0418 0.0322 0.0398 0.0331 0.0370
```

Now consider the equally weighted portfolio of the stocks represented in `big8`; recall that the returns for this portfolio are stored in the variable `big8.port` and that the estimate of β for the portfolio is $\hat{\beta}_p = 0.803$. The total risk and the nonmarket risk for this portfolio may be calculated using the functions `sd` and `f.sighat`, respectively.

```
> sd(big8.port)
[1] 0.0340
>  f.sighat(big8.port)
[1] 0.0158
```

Note that the standard deviation of the portfolio return is about two-thirds as large as the average return standard deviation for the eight stocks, given by

```
> mean(apply(big8, 2, sd))
[1] 0.0521
```

However, the standard deviation corresponding to the nonmarket component of variance for the portfolio is only about one-third as large as the average nonmarket component of return standard deviation for the eight stocks, given by

```
> mean(apply(big8, 2, f.sighat))
[1] 0.042
```

Recall that, for the observed returns on the S&P 500 index, $S_m^2 = 0.00141$; it follows that the market component of risk for the "big8" portfolio is

$$\hat{\beta}_p S_m = (0.803)\sqrt{0.00141} = 0.0302.$$

This value is similar to the market components of risk for the eight individual stocks:

```
> big8.mm$coefficients[2,]*sd(sp500)
  AAPL    BAX     KO    CVS    XOM    IBM    JNJ    DIS
0.0346 0.0269 0.0182 0.0403 0.0330 0.0232 0.0203 0.0448
```

Therefore, the total risk of the portfolio, 0.0340, is smaller than the total risks of the individuals stocks, which range from 0.0386 to 0.0739; this difference is attributable primarily to a decrease in the nonmarket risk. □

Because the total variance of the portfolio return consists of the market component, which is similar to the market components of return variance for the individual assets in the portfolio, and the nonmarket component, which tends to be much less than the individual nonmarket components of return variance, the proportion of return variance explained by the market return tends to be higher for the portfolio than for the individual assets. That is, R-squared for the portfolio tends to be larger than R-squared for the individual stocks.

Example 8.14 Consider the returns for the eight stocks stored in the variable `big8` and analyzed in the previous example. Recall that the R-squared values for these stocks are given by

```
> apply(big8, 2, f.rsq)
 AAPL   BAX    KO   CVS   XOM   IBM   JNJ   DIS
0.219 0.234 0.196 0.485 0.516 0.257 0.276 0.599
```

The R-squared value for the equally weighted portfolio of the eight stocks may be calculated using `f.rsq` as well:

```
> f.rsq(big8.port)
[1] 0.787
```

Therefore, R-squared for the portfolio is considerably larger than R-squared for the individual stocks. □

Note that a relatively large value of R-squared for a portfolio indicates that it is well diversified, in the sense that most of its risk is because of its relationship with the market portfolio which, by definition, is diversified.

Some Further Results on Portfolio Risk

The properties of the portfolios in Examples 8.13 and 8.14 hold in general, at least to some degree. As noted previously, let $R_{1,t}, R_{2,t}, \ldots, R_{N,t}$ denote the asset returns at time t and suppose that the market model holds for each asset so that

$$R_{i,t} - R_{f,t} = \alpha_{i,t} + \beta_i(R_{m,t} - R_{f,t}) + \epsilon_{i,t}, \quad t = 1, 2, \ldots, T$$

for $i = 1, 2, \ldots, N$. Let $\boldsymbol{\epsilon}_t$ denote the vector $(\epsilon_{1,t}, \epsilon_{2,t}, \ldots, \epsilon_{N,t})^T$ and let $\boldsymbol{\Sigma}_\epsilon$ denote the covariance matrix of $\boldsymbol{\epsilon}_t$.

Suppose that for any i and j, the residual returns $\epsilon_{i,t}$ and $\epsilon_{j,t}$ are uncorrelated. Then

$$\boldsymbol{\Sigma}_\epsilon = \begin{pmatrix} \sigma^2_{\epsilon,1} & 0 & \cdots & 0 \\ 0 & \ddots & \ddots & \vdots \\ \vdots & \ddots & \ddots & 0 \\ 0 & \ldots & 0 & \sigma^2_{\epsilon,N} \end{pmatrix} \tag{8.10}$$

where $\sigma^2_{\epsilon,j} = \text{Var}(\epsilon_{j,t})$. The assumption that $\boldsymbol{\Sigma}_\epsilon$ is a diagonal matrix is a strong assumption that leads to the *single-index model* for the returns

$R_{1,t}, R_{2,t}, \ldots, R_{N,t}$. The single-index model and its implications for portfolio theory are covered in detail in Chapter 9.

Consider the equally weighted portfolio with weight vector $\boldsymbol{w} = (1/N)\mathbf{1}$. Then

$$\sigma_{\epsilon,p}^2 = \frac{1}{N^2}\mathbf{1}^T\boldsymbol{\Sigma}_\epsilon\mathbf{1} = \frac{1}{N}\left(\frac{1}{N}\sum_{j=1}^{N}\sigma_{\epsilon,j}^2\right) = \frac{1}{N}\bar{\sigma}_\epsilon^2$$

where

$$\bar{\sigma}_\epsilon^2 = \frac{1}{N}\sum_{j=1}^{N}\sigma_{\epsilon,j}^2$$

is the average nonmarket component of the return variance of the assets. Therefore, if N is large and the residual returns for different assets are uncorrelated, then the nonmarket component of return variance for an equally weighted portfolio tends to be small.

It is important to note that such a conclusion depends on the assumption that $\boldsymbol{\Sigma}_\epsilon$ is a diagonal matrix, that is, on the assumption that $\epsilon_{i,t}$ and $\epsilon_{j,t}$ are uncorrelated for $i \neq j$. For a general covariance matrix $\boldsymbol{\Sigma}_\epsilon$ the minimum possible nonmarket return variance can be found by choosing $\boldsymbol{w}$ to minimize $\boldsymbol{w}^T\boldsymbol{\Sigma}_\epsilon\boldsymbol{w}$ subject to $\boldsymbol{w}^T\mathbf{1} = 1$; note that this is the same as finding the weight vector for the minimum variance portfolio, but with $\boldsymbol{\Sigma}_\epsilon$ replacing $\boldsymbol{\Sigma}$, the covariance matrix of the returns.

Therefore, according to Proposition 5.3, $\boldsymbol{w}^T\boldsymbol{\Sigma}_\epsilon\boldsymbol{w}$ is minimized by

$$\tilde{\boldsymbol{w}} = \frac{\boldsymbol{\Sigma}_\epsilon^{-1}\mathbf{1}}{\mathbf{1}^T\boldsymbol{\Sigma}_\epsilon^{-1}\mathbf{1}}$$

and the minimum nonmarket component of return variance is given by

$$\tilde{\boldsymbol{w}}^T\boldsymbol{\Sigma}_\epsilon\tilde{\boldsymbol{w}} = \frac{1}{\mathbf{1}^T\boldsymbol{\Sigma}_\epsilon^{-1}\mathbf{1}}.$$

For instance, suppose that $\boldsymbol{\Sigma}_\epsilon$ is of the form $\sigma_\epsilon^2\boldsymbol{M}_\rho$ where

$$\boldsymbol{M}_\rho = \begin{pmatrix} 1 & \rho & \cdots & \rho \\ \rho & \ddots & \ddots & \vdots \\ \vdots & \ddots & \ddots & \rho \\ \rho & \ldots & \rho & 1 \end{pmatrix} \tag{8.11}$$

and $\sigma_\epsilon^2 > 0$; see Example 5.7. Then all residual returns have standard deviation σ_ϵ and any pair of residual returns for different assets in the same time period has correlation ρ. In Example 5.7, it was shown that the equally weighted portfolio is the minimum variance portfolio in this case and it has variance

$$\frac{\sigma_\epsilon^2}{N^2}\mathbf{1}^T\boldsymbol{M}_\rho\mathbf{1}\sigma_\epsilon^2 = \frac{N + N(N-1)\rho}{N^2} = \left(\rho + \frac{1-\rho}{N}\right)\sigma_\epsilon^2.$$

Therefore, the nonmarket component of return variance when $\boldsymbol{\Sigma}_\epsilon = \sigma_\epsilon^2\boldsymbol{M}_\rho$ is never less than $\rho\sigma_\epsilon^2$. Thus, even for a large number of assets and a diversified portfolio, the nonmarket component of the variance is not negligible unless $\rho\sigma_\epsilon$ is close to 0.

It is worth noting that, even when it is not negligible, the nonmarket component of risk for the portfolio, which is approximately $\sqrt{\rho}\sigma_\epsilon$ for large N, is still less than that of the individual assets, which is given by σ_ϵ.

8.9 Measuring Portfolio Performance

A primary goal of portfolio theory and methodology is the construction of portfolios that yield large returns. However, as we have seen, large returns are generally associated with large risk. Therefore, in assessing portfolio performance, we need to consider both the average return on the portfolio and the variability of the returns as measured by the standard deviation or some similar measure.

One such measure of portfolio performance that we have used is the Sharpe ratio, given by $(\mu_p - \mu_f)/\sigma_p$, where μ_p and σ_p denote the mean and standard deviation, respectively, of the return on a given portfolio and μ_f denotes the return on the risk-free asset. The Sharpe ratio is estimated by $\widehat{\text{SR}} = (\bar{R}_p - \bar{R}_f)/S_p$, where $\bar{R}_p - \bar{R}_f$ and S_p are the sample mean and sample standard deviation, respectively, of the observed excess returns.

In this section, we consider some alternative measures of portfolio performance based on the estimates of the market model parameters for a portfolio.

Treynor Ratio

The Sharpe ratio considers the mean excess return of a portfolio relative to the portfolio risk, as measured by standard deviation of the returns. An important aspect of this measure is that it uses total risk, including both the market and the nonmarket components.

According to the CAPM, the expected excess return on an asset depends on the market component of its risk, as measured by $\beta_p\sigma_m$, where β_p denotes the value of beta for the portfolio, assumed to be positive, and σ_m is the standard deviation of the return on the market portfolio. Therefore, portfolios with large values of β_p are expected to have larger returns than portfolios with values of β_p close to 0.

A version of the Sharpe ratio with total risk replaced by the market component of risk is given by $(\mu_p - \mu_f)/(\beta_p\sigma_m)$. Note that the market risk, as measured by σ_m, is the same for each portfolio considered.

The *Treynor ratio* measures the excess return of a portfolio relative to its value of beta:

$$\text{TR} = \frac{\mu_p - \mu_f}{\beta_p}.$$

Here we can interpret β_p as the portfolio's value of beta in the market model.

Recall that we may write

$$\beta_p = \rho_p \frac{\sigma_p}{\sigma_m}$$

where ρ_p denotes the correlation between the portfolio's return and the return on the market index. It follows that the Treynor ratio and the Sharpe ratio of a portfolio are related by

$$\text{TR} = \frac{\sigma_m}{\rho_p}(\text{SR}).$$

Therefore, the Treynor ratio rewards portfolios that have a large Sharpe ratio while having returns that have low correlation with the returns on the market index.

To estimate the Treynor ratio, we may use the least-squares estimator $\hat{\beta}_p$ of the parameter β_p in the market model applied to the portfolio returns. Then an estimator of the Treynor ratio is given by

$$\widehat{\text{TR}} = \frac{\bar{R}_p - \bar{R}_f}{\hat{\beta}_p}.$$

Example 8.15 Although the methods described here may be applied to any portfolio of assets, investors often choose to invest in mutual funds, which are essentially professionally managed, regulated portfolios. Therefore, the performance measures described in this chapter will be illustrated based on mutual funds. A mutual fund combines the capital of a number of investors for the purpose of investing it on their behalf. The investments made by a fund form a type of portfolio and may include investments in stocks, bonds, and cash. A given mutual fund has a specific investment strategy, as described in its prospectus, but the exact investments generally vary over time; in that respect, a mutual fund differs from the portfolios we have been considering. In spite of this variation over time, we will analyze the returns on a mutual fund as if they are the returns on a given portfolio.

We will consider the returns of four mutual funds: Vanguard U.S. Growth Portfolio Fund (symbol VWUSX), T. Rowe Price New Horizons Fund (PRNHX), Fidelity Select Utilities Portfolio (FSUTX), and BlackRock Natural Resources Trust (MDGRX). These funds differ in their investment objectives. The Vanguard U.S. Growth Portfolio Fund focuses on stocks exhibiting long-term capital appreciation; T. Rowe Price New Horizons Fund also looks for long-term growth but focuses on the stocks of small companies, before they are widely recognized. The other two funds are more specialized. Fidelity Select Utilities Portfolio invests in companies in the utilities industry and BlackRock Natural Resources Trust invests in companies with substantial assets in natural resources.

The value of a share in a mutual fund changes over time, like the price of a stock, and these share values may be downloaded from Yahoo Finance, using the same procedure we used to download stock price information. Furthermore, the returns on the mutual funds are calculated in the same way. The data matrix `funds` contains five years of monthly excess returns on these four funds for the period ending December 31, 2014; thus, `funds` plays the same role that `big8` played in the analysis of the stock returns of eight large companies. Note that, although these are excess returns, we will generally refer to them simply as "returns."

```
> head(funds)
        VWUSX   PRNHX   FSUTX   MDGRX
[1,] -0.06323 -0.0325 -0.0439 -0.0574
[2,]  0.03687  0.0452 -0.0120  0.0323
[3,]  0.06054  0.0830  0.0375  0.0238
[4,]  0.00576  0.0398  0.0342  0.0299
[5,] -0.08571 -0.0646 -0.0559 -0.0929
[6,] -0.05843 -0.0643 -0.0135 -0.0513
```

The mean and standard deviation of the fund returns may be calculated using the `apply` function.

```
> funds.mean<-apply(funds, 2, mean)
> funds.mean
  VWUSX   PRNHX   FSUTX   MDGRX
0.01262 0.01736 0.01185 0.00341
> funds.sd<-apply(funds, 2, sd)
> funds.sd
 VWUSX  PRNHX  FSUTX  MDGRX
0.0440 0.0474 0.0331 0.0626
```

Estimates of α_p and β_p for the four funds can be estimated using the same approach used to estimate the parameters of the market model for the "big8" stocks in Example 8.3.

```
> funds.mm<-lm(funds~sp500)
> funds.alpha<-funds.mm$coefficients[1,]
> funds.beta<-funds.mm$coefficients[2,]
> funds.alpha
    VWUSX     PRNHX     FSUTX     MDGRX
 0.000377  0.005150  0.006691 -0.010968
> funds.beta
VWUSX PRNHX FSUTX MDGRX
1.123 1.120 0.473 1.319
```

Using the least-squares estimates, it is straightforward to estimate the Treynor ratio for the funds.

```
> funds.treynor<-funds.mean/funds.beta
> funds.treynor
  VWUSX    PRNHX    FSUTX    MDGRX
0.01124 0.01550 0.02505 0.00259
```

Therefore, the Utilities Portfolio has the largest estimated Treynor ratio. Note that the estimate of beta for this fund is much lower than those of the other funds and its estimated correlation with the return on the S&P 500 index is much lower than those of the other funds.

```
> cor(sp500, funds)
      VWUSX PRNHX FSUTX MDGRX
SP500 0.959 0.888 0.537 0.791
```

For comparison, we can estimate the Sharpe ratio for each fund.

```
> funds.sharpe<-funds.mean/funds.sd
> funds.sharpe
 VWUSX  PRNHX  FSUTX  MDGRX
0.2870 0.3665 0.3578 0.0545
```

Hence, the New Horizons Fund has the largest estimated Sharpe ratio and the Natural Resources Trust has the smallest.

In interpreting these results, it is important to keep in mind that the Treynor and Sharpe ratios calculated here are estimates of the underlying "true" ratios corresponding to the funds considered. In particular, in evaluating and comparing funds, it is important to take into account the sampling variability of the estimates, as measured by their standard errors. Such issues will be considered in detail in the following section. □

Jensen's Alpha

According to the CAPM, the expected return on a portfolio depends on its value of beta:

$$\mu_p = \mu_f + \beta_p(\mu_m - \mu_f).$$

Thus, a portfolio with $\mu_p > \mu_f + \beta_p(\mu_m - \mu_f)$ has a larger expected return than predicted by the CAPM for its value of beta. A similar interpretation holds if $\mu_p < \mu_f + \beta_p(\mu_m - \mu_f)$.

Therefore, one way to evaluate the performance of a portfolio is to compare its mean return to what is predicted by the CAPM. Let

$$\alpha_p = \mu_p - \mu_f - \beta_p(\mu_m - \mu_f);$$

α_p is known as *Jensen's alpha* for the portfolio.

When $\alpha_p > 0$, the mean portfolio return is greater than expected for the portfolio's value of β; when $\alpha_p < 0$, the mean return is less than expected. Jensen's alpha has the advantage of being easy to interpret because it is measured on the same scale as returns. Recall that α_p may also be interpreted in terms of a portfolio that is over- or underpriced, as discussed in Section 8.4.

Jensen's alpha may be estimated by $\hat{\alpha}_p$, the least-squares estimator of the intercept in the market model regression for the portfolio.

Example 8.16 For the four funds represented in the data matrix `funds`, the output from estimation of the market model is stored in the variable `funds.mm` and the estimates of alpha are stored in the variable `funds.alpha`.

```
> funds.alpha
    VWUSX      PRNHX      FSUTX      MDGRX
 0.000377  0.005150  0.006691 -0.010968
```

Hence, the U.S. Growth Portfolio, the New Horizons Fund, and the Utilities Portfolio all have an estimated mean return greater than that predicted by the CAPM, with the largest difference for the Utilities Portfolio; the estimated mean return of the Natural Resources Trust is less than what is predicted by the CAPM.

As discussed in the previous example, it is important to keep in mind that the values reported here are only estimates of the true parameter values. □

Appraisal Ratio

One drawback of Jensen's alpha is that it does not explicitly incorporate portfolio risk. A portfolio with a large value of α_p, but with large risk, may be less desirable than a less-risky portfolio that has a smaller value of α_p.

The market component of return variance for a portfolio, $\beta_p^2\sigma_m^2$, is directly tied to its predicted mean return according to the CAPM through the parameter β_p. Thus, in evaluating the value of α_p for a portfolio, we compare it to the portfolio's nonmarket component of risk.

The *appraisal ratio* is a risk-adjusted form of Jensen's alpha given by

$$\frac{\alpha_p}{\sigma_{\epsilon,p}},$$

where $\sigma_{\epsilon,p}$ is the error standard deviation in the market model for the portfolio.

Thus, the appraisal ratio is the difference between the expected return on the portfolio and its predicted expected return according to the CAPM based on the value of β_p, relative to the nonmarket component of risk for the portfolio. Recall that, according to the CAPM, nonmarket risk is not compensated by a larger expected return. Thus, a portfolio with a large appraisal ratio is apparently realizing some reward for its nonmarket risk.

There is an alternative interpretation of the appraisal ratio as the increase in the Sharpe ratio that may be achieved by combining the portfolio under consideration with the market portfolio; this result will be discussed in detail in Section 9.6.

Example 8.17 Consider the four mutual funds with return data stored in the variable `funds`. The estimates $\hat{\sigma}_{\epsilon,p}$ for these assets may be calculated using the `apply` function with the function `f.sighat` defined in Example 8.13:

```
> funds.s<-apply(funds, 2, f.sighat)
> funds.s
 VWUSX  PRNHX  FSUTX  MDGRX
0.0125 0.0220 0.0282 0.0386
```

Using these results, the appraisal ratios are easily calculated.

```
> funds.appraisal<-funds.alpha/funds.s
> funds.appraisal
  VWUSX   PRNHX   FSUTX   MDGRX
 0.0301  0.2344  0.2374 -0.2843
```

Hence, the New Horizons Fund and the Utilities Portfolio have the largest estimated appraisal ratios for the four funds considered; as with the other performance measures, these values are only estimates of the funds' true appraisal ratios. □

8.10 Standard Errors of Estimated Performance Measures

The estimated portfolio performance measures we have discussed are just estimates of a portfolio's "true" performance measures. Therefore, when interpreting such results, it is important to assess the uncertainty in these estimates by calculating their standard errors.

For some statistics, such as a sample mean or an estimated regression parameter, calculating the standard error is straightforward. For instance, for the average mean return $\bar{R}_p$, the standard error is given by $S_p/\sqrt{T}$, where S_p is the sample standard deviation of the returns and T is the sample size. In some cases, the standard error of a statistic is given as part of the R output of the function used to calculate the statistic. For instance, the standard error of Jensen's alpha, the estimated intercept parameter in the market model, is available from the regression output from estimating the market model using the `lm` function.

One role of the standard error of an estimate is in calculating an approximate confidence interval for a parameter; for instance, an approximate 95% confidence interval for μ_p is given by

$$\bar{R}_p \pm 1.96\frac{S_p}{\sqrt{T}}.$$

Example 8.18 Consider the returns on the Vanguard U.S. Growth Portfolio, which are stored in the variable `vwusx`. Output from estimating the market model using the `lm` function includes the table

```
Coefficients:
             Estimate Std. Error t value Pr(>|t|)
(Intercept) 0.000377   0.001687    0.22     0.82
sp500       1.122536   0.043467   25.83   <2e-16 ***
```

Therefore, the estimate of Jensen's alpha for this fund is 0.000377 and the standard error is 0.001687, leading to an approximate 95% confidence interval of

$$0.000377 \pm 1.96(0.001687) = (-0.00293, 0.00368). \quad \square$$

For more general statistics, such as an estimated Sharpe ratio, simple expressions for the standard error are not available. Hence, here we consider a general method of computing a standard error based on Monte Carlo simulation.

Monte Carlo simulation was considered in Section 6.7. The approach used in that section was based on assumptions regarding the distribution of the data. One drawback of such an approach is that the results depend on the assumptions used; here we use the observed data as the basis for the simulated data so that no such distributional assumptions are required.

We begin with a simple example to illustrate the mechanics of the Monte Carlo procedure. Let $R_{j,1}, R_{j,2}, \ldots, R_{j,T}$ denote the returns on a given asset and consider calculating the standard error of the sample mean return $\bar{R}_j$. Of course, in this case, we know that this standard error is given by $S_j/\sqrt{T}$ where S_j is the sample standard deviation of the returns. However, suppose that such a formula for this standard error is not available.

The standard error of $\bar{R}_j$ is an estimate of the standard deviation of the sampling distribution of $\bar{R}_j$. This sampling distribution is based on the assumption that $R_{j,1}, R_{j,2}, \ldots, R_{j,T}$ is a random sample from some population, in this case, the (hypothetical) population of all possible returns on the asset under consideration. Hence, here we assume that the $R_{j,t}$ are independent, identically distributed random variables.

Suppose that the population values are known; denote them by $\tilde{R}_{j,k}$, $k = 1, 2, \ldots, K$, so that the population of return values may be written

$$\mathcal{P} = \{\tilde{R}_{j,1}, \tilde{R}_{j,2}, \ldots, \tilde{R}_{j,K}\}.$$

Then we can calculate the standard deviation of $\bar{R}_j$ by drawing repeated samples of size T from $\mathcal{P}$, calculating the sample mean of each sample and then calculating the standard deviation of these simulated sample means of returns.

That is, for each $i = 1, 2, \ldots, I$, let

$$R_{j,1}^{(i)}, R_{j,2}^{(i)}, \ldots, R_{j,T}^{(i)}$$

be a sample with replacement drawn from $\mathcal{P}$, and let $\bar{R}_j^{(i)} = \sum_{t=1}^{T} R_j^{(i)}/T$ denote the corresponding sample mean. Then

$$\bar{R}_j^{(1)}, \bar{R}_j^{(2)}, \ldots, \bar{R}_j^{(I)}$$

is a sample from the distribution of $\bar{R}_j$.

The standard error of $\bar{R}_j$ may now be calculated as the sample standard deviation of the values $\bar{R}_j^{(i)}$, $i = 1, 2, \ldots, I$. Provided that I is sufficiently large, that is, provided that we draw a sufficiently large number of samples from $\mathcal{P}$, the standard error calculated in this way will be an accurate estimator of the standard deviation of the sampling distribution of $\bar{R}_j$.

The flaw in this approach, of course, is that we do not have the population $\mathcal{P}$. In Section 6.7, we dealt with this issue by making some assumptions regarding the distribution of the data and used those assumptions to draw the Monte Carlo samples. Here we use the information regarding $\mathcal{P}$ provided by the observed values $R_{j,1}, R_{j,2}, \ldots, R_{j,T}$ themselves.

Let $\mathcal{P}_O = \{R_{j,1}, R_{j,2}, \ldots, R_{j,T}\}$ denote the set of observed return values. Then we may estimate the standard error of $\bar{R}_j$ by replacing the hypothetical population $\mathcal{P}$ of return values by the observed set of return values, $\mathcal{P}_O$. This procedure is known as the *bootstrap* because we are apparently estimating the standard error by "pulling ourselves up by our bootstraps"; that is, we estimate the standard error without any assistance in the form of distributional assumptions.

Example 8.19 Consider estimating the mean return on the New Horizons Fund; recall that five years of monthly returns on this mutual fund are stored in the variable prnhx. For illustration, we will use only the first five of these values, which are stored in prnhx5:

```
> prnhx5<-prnhx[1:5]
> prnhx5
[1] -0.0325  0.0452  0.0830  0.0398 -0.0646
```

The sample mean of these five values is 0.0142 with a standard error of 0.0272:

```
> mean(prnhx5)
[1] 0.0142
> sd(prnhx5)/(5^.5)
[1] 0.0272
```

Now consider the simulation-based approach to estimating this standard error. The function sample may be used to draw a random sample from a set of integers. Specifically, sample(5, replace=T) draws a random sample with replacement from the set $\{1, 2, 3, 4, 5\}$:

```
> samp<-sample(5, replace=T)
> samp
[1] 2 1 1 5 3
```

Using the sampled integers as the indices of the vector prnhx5 yields a random sample with replacement from the set of returns values in prnhx5:

```
> prnhx5[samp]
[1]  0.0452 -0.0325 -0.0325 -0.0646  0.0830
```

The sample mean of prnhx5[samp] yields a simulated value of the sample mean return $\bar{R}_j$ for this asset:

```
> mean(prnhx5[samp])
[1] -0.0003
```

This procedure may be repeated multiple times; for example,

```
> mean(prnhx5[sample(5, replace=T)])
[1] 0.0286
> mean(prnhx5[sample(5, replace=T)])
[1] 0.0142
```

and so on.

Note that each time mean(prnhx5[sample(5, replace=T)]) is calculated, a new set of random numbers is drawn. Suppose we perform this procedure 1000 times, storing the sample means in the variable prnhx5.boot, these values represent a type of random sample drawn from the distribution of the sample mean of five returns on the New Horizons Fund. Here are the first eight values.

```
> prnhx5.boot[1:8]
[1] -0.0003  0.02860  0.01420  0.05270 -0.15400  0.03400
[7] 0.01523  0.00447
```

The sample standard deviation of prnhx5.boot yields an estimate of the standard error of $\bar{R}_j$ for this fund.

```
> sd(prnhx5.boot)
[1] 0.0247
```

Note that the bootstrap standard error is close to, but not exactly the same as, the value given by the usual formula for the standard error of the sample mean, 0.0272. There are two reasons for the difference. One is that the sample standard deviation uses a divisor of $T-1$; it may be shown that the estimate of the standard deviation implicity used by the bootstrap method is equivalent to the one with a divisor of T. The effect of these different divisors is highlighted in the example because in that case $T = 5$. In more realistic settings, such as the analysis of five years of monthly data, $T = 60$ and $\sqrt{60/59} = 1.0084$ so that the difference is unlikely to be important.

The other reason for the difference between the bootstrap standard error and the usual value is that the bootstrap method is based on a random sample. If the method is repeated, a different standard error will be obtained. For instance, the bootstrap method was repeated three times, with results

```
> sd(prnhx5.boot1)
[1] 0.0237
> sd(prnhx5.boot2)
[1] 0.0245
> sd(prnhx5.boot3)
[1] 0.0246
```

If a very large bootstrap sample size is used, we expect that the result will be closer to that obtained by the usual method. For example, `prnhx5.boot10k` contains a random sample of size 10,000 from the distribution of $\bar{R}_j$ for the New Horizons Fund.

```
> sd(prnhx.boot10k)
[1] 0.0241
```

Hence, note that $\sqrt{5/4} \doteq 1.118$ so that, after accounting for the difference in divisors, the result is nearly identical to the 0.0272 obtained from the usual formula. □

Obviously, the bootstrap method is not needed to calculate the standard error of a sample mean. However, it is extremely useful for calculating the standard error of more complicated statistics for which a simple formula is not available. Although it is possible to carry out the calculations for any statistic by following the procedure described earlier for the sample mean, fortunately, there are convenient R functions available for that purpose.

Here we will use the function `boot` in the package `boot` (Canty and Ripley 2015). This function takes three arguments (there are other, optional arguments for which we will use the default values). The most important of these is a function calculating the statistic of interest for a given set of data.

Example 8.20 Consider estimation of the Sharpe ratio based on a sequence of excess returns; we will write a function `Sharpe` to compute the Sharpe ratio.

A function to be used in `boot` must take two arguments: the data, in the form of a vector or matrix, and the indices of the data values to be used in the computation, similar to the way the indices vector was used in the aforementioned sample mean example.

Consider the function

```
> Sharpe<-function(x, ind){mean(x[ind])/sd(x[ind])}
```

This function takes the values in `x` corresponding to the indices in the vector `ind` and uses those values to compute the Sharpe ratio. For example, consider the excess return data in the vector `prnhx`. To use all the data, we set `ind` to `1:60`, the vector of integers from 1 to 60.

```
> Sharpe(prnhx, 1:60)
[1] 0.367
```

This yields the same result as computing the Sharpe ratio directly for the data in prnhx:

```
> mean(prnhx)/sd(prnhx)
[1] 0.367
```

If 1:60 is replaced by 1:5, the result is the Sharpe ratio based on the first five values; recall that these are stored in the variable prnhx5.

```
> Sharpe(prnhx, 1:5)
[1] 0.233
> mean(prnhx5)/sd(prnhx5)
[1] 0.233
```

□

The other arguments to boot are data, the data used in calculating the statistic of interest, and R, the number of bootstrap replications to be used.

Example 8.21 Consider estimation of the Sharpe ratio for the New Horizons Fund. To calculate the standard error of the estimated Sharpe ratio for the data in prnhx based on a bootstrap sample size of 1000, we use the command

```
>library(boot)
> boot(prnhx, Sharpe, 1000)
ORDINAR{Y} NONPARAMETRIC BOOTSTRAP
Bootstrap Statistics :
    original  bias    std. error
t1*    0.367 0.00582        0.138
```

The output gives the value of the estimate, under the heading "original"; hence, the estimated Sharpe ratio for these data is 0.367, as calculated previously. The standard error, given under the "std. error" heading, is 0.138.

□

The output of the boot function includes an estimate of the bias of the estimator; recall that the bias is the expected value of the estimator minus the true value of the parameter.

A *bias-corrected* estimate may be formed by subtracting the bias from the estimate; for example, a bias-corrected estimate of the Sharpe ratio for the New Horizons Fund is $0.367 - 0.006 = 0.361$. Whenever the bias is small relative to the standard error, the impact of the bias correction is small and, hence, it may be ignored. A simple rule of thumb is that the estimated bias may be ignored when it is less than one-fourth of the standard error; of course, such a guideline will not be appropriate in all cases.

The usefulness of the bootstrap method arises from the fact that it can be applied to a wide range of statistics, by modifying the function used as the argument to boot. For instance, it may be applied when the statistic under consideration depends on the returns of more than one asset; this is illustrated in the following example.

Example 8.22 Consider calculation of the standard error of the estimated Treynor ratio for the New Horizons Fund.

The first step in using the bootstrap method is to construct a function that calculates the estimated Treynor ratio based on a set of returns, together with a vector of indices. The complication in this example is that the Treynor ratio depends on two sets of returns—the returns on the asset under consideration and the returns on the market index, used to calculate $\hat{\beta}_p$. Hence, we take the data for estimation procedure to be a matrix, in which the first column is the asset (excess) returns and the second column is the (excess) returns on market index, in this case, the S&P 500 index. The indices variable will then select the rows of the matrix to be included in the estimate. This approach is implemented in the following function:

```
> Treynor<-function(rmat, ind){
+ ret<-rmat[ind, 1]
+ mkt<-rmat[ind, 2]
+ beta<-lm(ret~mkt)$coefficients[2]
+ mean(ret)/beta}
```

In this function, the return data are input in the matrix `rmat`, which is assumed to have two columns, the first with the return data for the asset and the second with the return data for the market index. The variable `ind` contains the indices of the returns to be used in the estimation of the Treynor ratio. The first two lines of the function extract the relevant return data using `ind` and places them in two variables `ret` and `mkt`, which contain the return data for the asset and for the market index, respectively, corresponding to `ind`. The third line obtains the estimate of beta for the data in `ret` and `mkt`, and the final line returns the estimate of the Treynor ratio.

For example, using the function with the first argument taken to be `cbind(prnhx, sp500)`, the matrix formed by combining `prnhx` and `sp500` as column vectors, and taking the second argument to be the sequence of integers `1:60`, yields the estimated Treynor ratio for the New Horizons Fund

```
> Treynor(cbind(prnhx, sp500), 1:60)
  0.01550
```

in agreement with what was obtained in Example 8.15.

We are now in position to apply the `boot` function. Suppose we are interested in estimating the standard error of $\widehat{\mathrm{TR}}$ for the data in the variable `prnhx`. This is obtained from the command

```
> boot(cbind(prnhx, sp500), Treynor, 10000)
ORDINAR{Y} NONPARAMETRIC BOOTSTRAP
Bootstrap Statistics :
    original     bias    std. error
t1*   0.0155 -1.21e-05     0.00534
```

Therefore, the standard error for the estimated Treynor ratio for the New Horizons Fund is 0.00534; the estimated bias is very small relative to the standard error and, hence, it may be ignored. □

Comparison of Portfolios

A common goal in calculating measures of portfolio performance is to compare portfolios. Hence, we may be interested in estimating the difference between measures of performance for two portfolios. Estimation of such a difference is straightforward. We may estimate the difference in performance measures by the difference of corresponding estimates. To calculate the standard error of such a difference, we may again use the bootstrap procedure, as implemented in the `boot` function, by defining the inputs to `boot` appropriately. This is illustrated in the following example.

Example 8.23 Suppose that we are interested in comparing the Sharpe ratios of the U.S. Growth Portfolio and the New Horizons Fund, the estimated Sharpe ratios are 0.287 for the U.S. Growth Portfolio and 0.367 for the New Horizons Fund, suggesting that the Sharpe ratio for the New Horizons Fund is larger. However, these are only estimates and it is of interest to take into account the sampling variability in evaluating the difference in the estimates.

First, consider calculation of the standard error for each of these individual estimates. The return data for the U.S. Growth Portfolio are stored in the variable `vwusx` and the returns for the New Horizons Fund are stored in the variable `prnhx`. Recall that `Sharpe` is a function that calculates the Sharpe ratio of an asset, which can be used in the function `boot`.

The standard errors for the individual Sharpe ratios are given by

```
> boot(vwusx, Sharpe, 10000)
ORDINAR{Y} NONPARAMETRIC BOOTSTRAP
Bootstrap Statistics :
    original  bias    std. error
t1*    0.287 0.00786       0.141

> boot(prnhx, Sharpe, 10000)
ORDINAR{Y} NONPARAMETRIC BOOTSTRAP


Bootstrap Statistics :
    original  bias    std. error
t1*    0.367 0.00751       0.137
```

Thus, the standard error of the estimated Sharpe ratio for the U.S. Growth Portfolio is 0.141 and the standard error of the estimated Sharpe ratio for the New Horizons Fund is 0.137.

Now consider calculation of the standard error for the difference in two estimated Sharpe ratios. Note that we cannot use standard error based on the individual standard errors because the two estimates are likely to be correlated. Hence, we use an approach similar to that used when calculating the standard error for the difference of means for matched-pair data.

Define a function `Sharpe_diff` by

```
> Sharpe_diff<-function(rets, ind){
+ Sharpe(rets[,1], ind)-Sharpe(rets[,2], ind)
+ }
```

This function takes the return data in the matrix `rets`, with the returns for the first asset in column 1 and the returns for the second asset in column 2, and computes the difference in the Sharpe ratios corresponding to the indices in `ind`.

To form a matrix with columns given by the variables `vwusx` and `prhnx`, we use the `cbind` function. Therefore, standard error of the difference in Sharpe ratios is given by

```
> boot(cbind(vwusx, prnhx), Sharpe_diff, 10000)
ORDINAR{Y} NONPARAMETRIC BOOTSTRAP
Bootstrap Statistics :
    original  bias     std. error
t1*  -0.0795 0.00208       0.0589
```

Note that the estimated difference, -0.0795, agrees with the difference in the estimated Sharpe ratios calculated in Example 8.15, $0.2870 - 0.3665$. The standard error of the difference is 0.0589 and, hence, the difference is not statistically significant at the 5% level; that is, a 95% confidence interval for the difference includes zero. The estimated bias is small relative to the standard error and may be ignored.

It is worth noting that the standard error of the difference of the estimates based on the standard errors of the individual estimates along with the assumption that the estimates are uncorrelated,

$$\left((0.141)^2 + (0.137)^2\right)^{\frac{1}{2}} = 0.197$$

is much larger than the estimate given previously. This is because the method used to obtain the value 0.197 ignores the fact that the returns on the two funds are correlated. □

It is important to keep in mind that the results based on the bootstrap method are based on the random numbers generated by the `boot` function. Hence, if the procedure is repeated, the results will vary. It is generally a good idea to repeat the standard error calculation in order to assess this variation; if it is large enough to affect the conclusions of the analysis, then the bootstrap sample size should be increased.

Example 8.24 Consider calculation of the standard error of the difference of the Sharpe ratios for the U.S. Growth Portfolio and the New Horizons Fund. Recall that in Example 8.23, the standard error was found to be 0.0589.

Repeating the calculation twice yields

```
> boot(cbind(vwusx, prnhx), Sharpe_diff, 10000)
ORDINAR{Y} NONPARAMETRIC BOOTSTRAP
Bootstrap Statistics :
    original    bias    std. error
t1*  -0.0795 0.000631       0.0596

> boot(cbind(vwusx, prnhx), Sharpe_diff, 10000)
ORDINAR{Y} NONPARAMETRIC BOOTSTRAP
Bootstrap Statistics :
    original    bias    std. error
t1*  -0.0795 0.000612       0.0591
```

Clearly, the variation in the estimated standard error is fairly small and the conclusion that the estimated difference of the Sharpe ratios is not statistically significant is not affected. Although the estimated biases obtained indicate some variation, it clearly does not affect the conclusion that the bias is negligible.

Given these additional results, it is reasonable to include them in our calculation of the standard error. For instance, to estimate the standard error, we could average the values $0.0589, 0.0596$, and 0.0591, leading to a new estimate of 0.0592. □

8.11 Suggestions for Further Reading

Although many factors affect the return on a security, the return on a market portfolio is generally considered to be the most important and, hence, the market model presented in this chapter is discussed in many books on financial modeling; see, for example, Benninga (2008, Chapter 2), Campbell et al. (1997, Chapter 4), and Fama (1976, Chapter 4). Bradfield (2003) presents a detailed discussion of estimating the parameters of the market model addressing a number of practical issues; see also Damodaran (1999). The relationship between assumptions on the error term in the model and the interpretation of beta is discussed by Severini (2016).

Standard & Poors has published a technical report providing many details of stock market indices; it is available at

http://www.spindices.com/documents/methodologies/
methodology-index-math.pdf

Many authors define the market model in terms of regular returns instead of excess returns; that is the case, for example, with Campbell et al. (1997) and Fama (1976). The interpretation of β is the same in both cases, although the value of the estimate generally changes slightly. The interpretation of the intercept, on the other hand, depends on the type of return used; however, it is a simple matter to translate the parameter for the model based on regular returns to the parameter based on excess returns and vice versa.

As discussed in Section 8.3, the market model may be viewed as a simple linear regression model and, hence, standard least-squares-based methods may be used for inference for the parameters of the model; Newbold et al. (2013, Chapter 11) offer a good introduction to these methods and include a discussion of their application to the market model in their Section 11.8. Like least-squares estimators in general, the least-squares estimator of beta is sensitive to outliers; see Martin and Simin (2003) for discussion of an outlier-resistant estimator.

Further details on shrinkage estimation of β are given by Elton et al. (2007, Chapter 7), Francis and Kim (2013, Section 17.4), and Vasicek (1973). Multiple testing and the Bonferroni correction are discussed by Tamhane and Dunlop (2000, Section 6.3.9), who also present other useful information on hypothesis tests and their properties. The method of controlling the expected FDR is known as the Benjamini-Hochberg method. Foulkes (2009, Chapter 4) presents a good introduction to this and other methods of multiple testing; although the subject of this book is statistical genetics, a field in which multiple testing is routinely used, it is not difficult to see how the methods described there may also be applied to problems in financial statistics.

Modigliani and Pogue (1974) offer a detailed discussion of the implications of the market model for understanding risk. An extension of the market model allows beta to change over time; see Ruppert (2004, Section 7.10) for an introduction to such models.

Good general discussions of measures of portfolio performance are available from Elton et al. (2007, Chapter 25) and Francis and Kim (2013, Chapter 18). The bootstrap is an important method in statistics that can be used to calculate standard errors and confidence intervals for a wide range of statistics; see, for example, Efron and Tibshirani (1993) and Davison and Hinkley (1997). In particular, the `boot` package used here is attributed to Davison and Hinkley (1997).

8.12 Exercises

1. Calculate five years of monthly excess returns on Bed Bath & Beyond Inc., stock (symbol BBBY) and five years of monthly excess returns on the S&P 500 index (symbol ^GSPC) for the period ending December 31, 2015. For the risk-free rate, use the return on the three-month Treasury Bill, available on the Federal Reserve website.

   Using these data, estimate the parameters α and β of the market model for Bed Bath & Beyond stock, using the S&P 500 as the market index; see Example 8.3. Give the standard error of each estimate. Based on these results, does it appear that Bed Bath & Beyond stock is priced correctly?

2. Repeat Exercise 1 using the Wilshire 5000 index (symbol ^W5000) as the market index. Compare the results to those obtained in Exercise 1.

3. Calculate five years of monthly excess returns for the period ending December 31, 2015, for five stocks, Papa John's International, Inc. (symbol PZZA), Bed Bath & Beyond, Inc. (BBBY), Netflix, Inc. (NFLX), Time Warner, Inc. (TWX), and Verizon Communications, Inc. (VZ); for the risk-free rate, use the return on the three-month Treasury Bill, available on the Federal Reserve website. Using the S&P 500 index (symbol ^GSPC) as the market index, calculate the estimate of beta for the market model for each stock; see Example 8.3.

   Which stock is most sensitive to the market? Which stock is least sensitive?

4. Consider the return data for the five stocks given in Exercise 3 and take the market index to be the S&P 500 index.

   a. Using the Bonferroni method, test the hypothesis that all five stocks are priced correctly; see Example 8.5. Using a significance level of 0.05, what do you conclude?
   b. Repeat the analysis controlling the expected FDR at the level 0.10; see Example 8.6. Do your conclusions change?

5. For each of the five stocks listed in Exercise 3, estimate $\sigma_{\epsilon,i}$, the standard deviation corresponding to the nonmarket component of the return variance, and give the value of R^2 for the market model regression; see Example 8.7. Use the S&P 500 as the market index.

   For which stock is the proportion of the return variance explained by the market the greatest? For which stock is it the smallest?

6. For the return data on the five stocks given in Exercise 3, use the shrinkage method to estimate the values of beta in the market model with the S&P 500 index as the market index. Compute the estimates using a global value of the weight ψ, as in Example 8.8, and then repeat the analysis using an asset-specific value of ψ, as in Example 8.9.

   Which shrinkage method appears to be more appropriate here? Why?

7. For the return data on the five stocks given in Exercise 3, calculate the value of adjusted beta for each stock. Use the S&P 500 index as the market index.

8. Consider the five stocks given in Exercise 3. Suppose that we construct a portfolio of these stocks placing weight 0.1 on PZZA, 0.2 on BBBY, 0.3 on NFLX, 0.3 on TWX, and 0.1 on VZ.

   Using the S&P 500 index as the market index, find the estimate of beta for the portfolio based on five years of monthly excess returns for the period ending December 31, 2015; see Example 8.11.

9. For the portfolio considered in Exercise 8, estimate the return standard deviation along with the standard deviations corresponding to the market and nonmarket components of return variance; see Example 8.11. Estimate the proportion of its return variance that is explained by the market.

10. Consider three mutual funds, Copley (symbol COPLX), Edgewood Growth Retail (EGFFX), and Fidelity Contrafund (FCNTX). Using five years of monthly excess returns for the period ending December 31, 2015, and the S&P 500 index as the market index, determine which of these three funds is the most diversified. Justify your answer.

11. Consider the returns on the three mutual funds described in Exercise 10. Estimate the values of alpha and beta in the market model for an equally weighted portfolio of the three funds using the S&P 500 as the market index. Is there evidence that the portfolio is priced incorrectly? Why or why not? Estimate the proportion of the portfolio return variance that is explained by the market.

12. Consider the returns on three mutual funds described in Exercise 10. For each fund, estimate the Treynor ratio using the market model with the S&P 500 index as the market index; see Example 8.15. According to these estimates, rank the performances of funds. Compare the results to the ranking based on the estimated Sharpe ratios.

13. Consider two assets. Let T_j denote the Treynor ratio and let β_j denote the value of beta in the market model for asset j, $j = 1, 2$. Suppose we construct a portfolio of these two assets, giving weight w to asset 1 and weight $1 - w$ to asset 2. Let μ_j denote the mean return on asset j, $j = 1, 2$ and let μ_f denote the mean return on the risk-free asset. Assume that $\mu_j > \mu_f$, $j = 1, 2$.

    a. Find the Treynor ratio of the portfolio, as a function of $T_1, T_2, \beta_1, \beta_2$, and w.
    b. Find the value of w, $0 \leq w \leq 1$, that maximizes the Treynor ratio of the portfolio.

14. Consider the returns on the three mutual funds described in Exercise 10. For each fund, estimate the appraisal ratio using the market model with the S&P 500 index as the market index;

see Example 8.17. According to these results, rank the performances of the funds. Compare the results to the ranking based on the estimated Sharpe ratios.

15. Consider the returns on the three mutual funds described in Exercise 10. For each fund, use the bootstrap method as described in Example 8.22 to calculate the standard error of the estimated Treynor ratio. Use the estimates from the market model with the S&P 500 index as the market index. Calculate an approximate 95% confidence interval for the true Treynor ratio of each fund. Take the bootstrap sample size to be 10,000.

16. Consider the returns on the three mutual funds described in Exercise 10. For each fund, use the bootstrap method to calculate the bias of the estimated appraisal ratios and calculate the bias-corrected estimate. Use the market model with the S&P 500 index as the market index and a bootstrap sample size of 10,000.

17. Consider the returns on the three mutual funds described in Exercise 10. Estimate the difference in the Sharpe ratios for the Copley and Edgewood funds and use the bootstrap method to calculate the standard error of the estimate; see example 8.23. Use the market model with the S&P 500 index as the market index and a bootstrap sample size of 10,000.

    Repeat the procedure for the difference in the Sharpe ratios of the Copley and Fidelity funds and for the difference in the Sharpe ratios of the Edgewood and Fidelity funds.

    Based on these results, what do you conclude regarding the differences in the Sharpe ratios of the three funds?

9

The Single-Index Model

9.1 Introduction

In the previous chapter, the market model, which relates the return on an asset to the return on a market index, was presented. According to the market model, the excess return on an asset may be written in terms of two random variables, the excess return on a market index and the residual return, which is uncorrelated with the return on the market index.

The decomposition given by the market model is useful for understanding the properties of the expected return and risk of an asset. However, the market model applies only to a single asset and, hence, it is not useful in explaining the relationships among the returns on several assets, as required in portfolio theory.

In this chapter, we consider the *single-index model*, which is a type of extension of the market model to a set of N assets. In particular, the single-index model leads to a simple model for the covariance matrix of an asset return vector.

9.2 The Model

Consider the market model for asset i, with returns $R_{i,1}, R_{i,2}, \ldots, R_{i,T}$:

$$R_{i,t} - R_{f,t} = \alpha_i + \beta_i(R_{m,t} - R_{f,t}) + \epsilon_{i,t}, \quad t = 1, 2, \ldots, T \tag{9.1}$$

where $R_{m,t}$ denotes the return on a market index at time t, $R_{f,t}$ denotes the risk-free rate at time t, $\epsilon_{i,t}$ is an error term that has mean zero and is uncorrelated with $R_{m,t}$, as discussed in Section 8.3, and α_i and β_i are parameters.

As we have seen, the market model is useful for understanding the relationship between the return on an asset and the return on the market, as reflected in a market index; in particular, it gives a decomposition of the variance of an asset's returns into market and nonmarket components. The *single-index model* uses the same approach to describe the covariance structure of a set of assets; thus, the single-index model is a model for the returns on a set of assets.

Consider a set of N assets with returns $R_{1,t}, R_{2,t}, \ldots, R_{N,t}$, in period t. Suppose that, for each asset, the market model (9.1) holds; then we have N equations describing the asset returns in terms of the returns on the market index: for $t = 1, 2, \ldots, T$,

$$\begin{aligned}
R_{1,t} - R_{f,t} &= \alpha_1 + \beta_1(R_{m,t} - R_{f,t}) + \epsilon_{1,t} \\
R_{2,t} - R_{f,t} &= \alpha_2 + \beta_2(R_{m,t} - R_{f,t}) + \epsilon_{2,t} \\
\vdots \quad & \quad \vdots \\
R_{N,t} - R_{f,t} &= \alpha_N + \beta_N(R_{m,t} - R_{f,t}) + \epsilon_{N,t}.
\end{aligned}$$

Writing these equations in matrix form, we have

$$\boldsymbol{R}_t - R_{f,t}\mathbf{1} = \boldsymbol{\alpha} + (R_{m,t} - R_{f,t})\boldsymbol{\beta} + \boldsymbol{\epsilon}_t \tag{9.2}$$

where

$$\boldsymbol{R}_t = \begin{pmatrix} R_{1,t} \\ R_{2,t} \\ \vdots \\ R_{N,t} \end{pmatrix}, \quad \boldsymbol{\alpha} = \begin{pmatrix} \alpha_1 \\ \alpha_2 \\ \vdots \\ \alpha_N \end{pmatrix}, \quad \boldsymbol{\beta} = \begin{pmatrix} \beta_1 \\ \beta_2 \\ \vdots \\ \beta_N \end{pmatrix},$$

and

$$\boldsymbol{\epsilon}_t = \begin{pmatrix} \epsilon_{1,t} \\ \epsilon_{2,t} \\ \vdots \\ \epsilon_{N,t} \end{pmatrix}.$$

As with the market model, we assume that the stochastic process

$$\{((\boldsymbol{R}_t - R_{f,t}\mathbf{1})^T, R_{m,t} - R_{f,t})^T : t = 1, 2, \ldots\}$$

is weakly stationary; we may interpret this condition as requiring that, for any $\boldsymbol{b} \in \Re^N$ and any scalar c, the real-valued stochastic process

$$\boldsymbol{b}^T(\boldsymbol{R}_t - R_{f,t}\mathbf{1}) + c(R_{m,t} - R_{f,t}), \quad t = 1, 2, \ldots$$

satisfies the conditions of weak stationarity given in Section 2.4. In particular, $\boldsymbol{\epsilon}_1, \boldsymbol{\epsilon}_2, \ldots$ is a weakly stationary process. Furthermore, we assume that $\boldsymbol{\epsilon}_t$ and $R_{m,t}$ are uncorrelated, in the sense that for any $\boldsymbol{b} \in \Re^N$, $\boldsymbol{b}^T\boldsymbol{\epsilon}_t$ and $R_{m,t}$ are uncorrelated; we may write this condition as

$$\text{Cov}(\boldsymbol{\epsilon}_t, R_{m,t}) = \mathbf{0}.$$

Also, we assume that

$$\text{Cov}(\boldsymbol{\epsilon}_t, \boldsymbol{\epsilon}_s) = \mathbf{0}$$

for any t, s, $t \neq s$, which we interpret to mean that for any $\boldsymbol{a}, \boldsymbol{b} \in \Re^N$, $\boldsymbol{a}^T\boldsymbol{\epsilon}_t$ and $\boldsymbol{b}^T\boldsymbol{\epsilon}_s$ are uncorrelated; in particular, $\epsilon_{i,t}$ and $\epsilon_{j,s}$ are uncorrelated for any $i, j = 1, \ldots, N$ and any $t \neq s$.

At this point, all we have done is rewrite the market models for the N assets using matrix notation. The single-index model goes further, by assuming that the covariance matrix of $\boldsymbol{\epsilon}_t$ is a diagonal matrix,

$$\boldsymbol{\Sigma}_\epsilon = \begin{pmatrix} \sigma_{\epsilon,1}^2 & 0 & \cdots & 0 \\ 0 & \sigma_{\epsilon,2}^2 & \ddots & \vdots \\ \vdots & \ddots & \ddots & 0 \\ 0 & \cdots & 0 & \sigma_{\epsilon,N}^2 \end{pmatrix}.$$

That is, the single-index model is an extension of the market model to a set of asset returns in which the market model holds for each asset and the residual returns for different assets are uncorrelated.

We will say that *the single-index model holds for* $\boldsymbol{R}_t$ whenever $\boldsymbol{R}_t$ follows the model given in (9.2) and the preceding assumptions are satisfied.

9.3 Covariance Structure of Returns under the Single-Index Model

The single-index model yields a simple expression for the covariance matrix of $\boldsymbol{R}_t$, as described in the following proposition.

Proposition 9.1. *For each $t = 1, 2, \ldots, T$, let $\boldsymbol{R}_t$ denote an $N \times 1$ vector of asset returns and suppose that the single-index model holds for $\boldsymbol{R}_t$.*

Let $\boldsymbol{\Sigma}$ denote the covariance matrix of $\boldsymbol{R}_t$ and let $\boldsymbol{\Sigma}_\epsilon$ denote the covariance matrix of $\boldsymbol{\epsilon}_t$. Then

$$\boldsymbol{\Sigma} = \sigma_m^2 \boldsymbol{\beta}\boldsymbol{\beta}^T + \boldsymbol{\Sigma}_\epsilon \tag{9.3}$$

where $\sigma_m^2 = Var(R_{m,t})$.

Proof. Let $\boldsymbol{a}$ and $\boldsymbol{b}$ be elements of $\Re^N$ and consider $\text{Cov}(\boldsymbol{a}^T\boldsymbol{R}_t, \boldsymbol{b}^T\boldsymbol{R}_t)$. According to the single-index model,

$$\begin{aligned} \text{Cov}(\boldsymbol{a}^T\boldsymbol{R}_t, \boldsymbol{b}^T\boldsymbol{R}_t) &= \text{Cov}(\boldsymbol{a}^T\boldsymbol{\beta}R_{m,t} + \boldsymbol{a}^T\boldsymbol{\epsilon}_t, \boldsymbol{b}^T\boldsymbol{\beta}R_{m,t} + \boldsymbol{b}^T\boldsymbol{\epsilon}_t) \\ &= \text{Cov}(\boldsymbol{a}^T\boldsymbol{\beta}R_{m,t}, \boldsymbol{b}^T\boldsymbol{\beta}R_{m,t}) + \text{Cov}(\boldsymbol{a}^T\boldsymbol{\beta}R_{m,t}, \boldsymbol{b}^T\boldsymbol{\epsilon}_t) \\ &\quad + \text{Cov}(\boldsymbol{a}^T\boldsymbol{\epsilon}_t, \boldsymbol{b}^T\boldsymbol{\beta}R_{m,t}) + \text{Cov}(\boldsymbol{a}^T\boldsymbol{\epsilon}_t, \boldsymbol{b}^T\boldsymbol{\epsilon}_t). \end{aligned}$$

Note that $\boldsymbol{a}^T\boldsymbol{\beta}R_{m,t} = (\boldsymbol{a}^T\boldsymbol{\beta})R_{m,t}$ where $(\boldsymbol{a}^T\boldsymbol{\beta})$ is a scalar; it follows that

$$\text{Cov}(\boldsymbol{a}^T\boldsymbol{\beta}R_{m,t}, \boldsymbol{b}^T\boldsymbol{\beta}R_{m,t}) = (\boldsymbol{a}^T\boldsymbol{\beta})(\boldsymbol{b}^T\boldsymbol{\beta})\sigma_m^2 = (\boldsymbol{a}^T\boldsymbol{\beta})(\boldsymbol{\beta}^T\boldsymbol{b})\sigma_m^2 = \boldsymbol{a}^T\boldsymbol{\beta}\boldsymbol{\beta}^T\boldsymbol{b}\sigma_m^2.$$

By the properties of covariance matrices,

$$\text{Cov}(\boldsymbol{a}^T\boldsymbol{\epsilon}_t, \boldsymbol{b}^T\boldsymbol{\epsilon}_t) = \boldsymbol{a}^T\boldsymbol{\Sigma}_\epsilon\boldsymbol{b}$$

and, by the assumptions of the single-index model,

$$\text{Cov}(\boldsymbol{a}^T \boldsymbol{\beta} R_{m,t}, \boldsymbol{b}^T \boldsymbol{\epsilon}_t) = \text{Cov}(\boldsymbol{a}^T \boldsymbol{\epsilon}_t, \boldsymbol{b}^T \boldsymbol{\beta} R_{m,t}) = 0.$$

It follows that

$$\begin{aligned}\text{Cov}(\boldsymbol{a}^T \boldsymbol{R}_t, \boldsymbol{b}^T \boldsymbol{R}_t) &= (\boldsymbol{a}^T \boldsymbol{\beta})(\boldsymbol{b}^T \boldsymbol{\beta})\sigma_m^2 + \boldsymbol{a}^T \boldsymbol{\Sigma}_\epsilon \boldsymbol{b} \\ &= (\boldsymbol{a}^T \boldsymbol{\beta})(\boldsymbol{\beta}^T \boldsymbol{b})\sigma_m^2 + \boldsymbol{a}^T \boldsymbol{\Sigma}_\epsilon \boldsymbol{b} \\ &= \boldsymbol{a}^T \left(\boldsymbol{\beta}\boldsymbol{\beta}^T \sigma_m^2 + \boldsymbol{\Sigma}_\epsilon\right) \boldsymbol{b}.\end{aligned}$$

Since this holds for any vectors $\boldsymbol{a}, \boldsymbol{b}$, it follows that $\boldsymbol{\beta}\boldsymbol{\beta}^T \sigma_m^2 + \boldsymbol{\Sigma}_\epsilon$ must be the covariance matrix of $\boldsymbol{R}_t$. □

Under the assumption that $\boldsymbol{\Sigma}_\epsilon$ is a diagonal matrix, the correlation between the returns of any two assets is attributable entirely to the fact that the returns are related to the market index. That is, the correlation structure of $\boldsymbol{R}_t$ is attributable to a "single index," the return on the portfolio corresponding to $R_{m,t}$.

The primary assumption underlying the single-index model, that the covariance matrix of $\boldsymbol{\epsilon}_t$ is a diagonal matrix, is a strong one, but it greatly simplifies the structure of the covariance matrix of $\boldsymbol{R}_t$. In general, the covariance matrix of $\boldsymbol{R}_t$ has $N(N+1)/2$ parameters, N variances plus $(N^2 - N)/2$ covariances. Under the single-index model, (9.3) has $2N+1$ parameters, N elements of $\boldsymbol{\beta}$, N diagonal elements of $\boldsymbol{\Sigma}_\epsilon$, and σ_m^2. For example, for $N = 50$ assets, in general, the covariance matrix of $\boldsymbol{R}_t$ has 1225 unknown parameters, while under the single-index model it has only 101.

Furthermore, the parameters in (9.3) can be estimated using the regression output from applying the market model to each of the N assets together with the sample variance of the excess returns on the market index.

Correlation of Asset Returns under the Single-Index Model

To better understand the implications of the single-index model, consider the properties of two assets, i and j. Let σ_i^2 and σ_j^2 denote the variances of $R_{i,t}$ and $R_{j,t}$, respectively. Then, under either the market model or the single-index model, σ_i^2 and σ_j^2 can be decomposed into market and nonmarket components:

$$\sigma_i^2 = \beta_i^2 \sigma_m^2 + \sigma_{\epsilon,i}^2 \quad \text{and} \quad \sigma_j^2 = \beta_j^2 \sigma_m^2 + \sigma_{\epsilon,j}^2.$$

Let ρ_i and ρ_j denote the correlations of $R_{i,t}$ and $R_{j,t}$, respectively, with $R_{m,t}$. Under either the market model or the single-index model,

$$\text{Cov}(R_{i,t} R_{m,t}) = \beta_i \sigma_m^2 \quad \text{and} \quad \text{Cov}(R_{j,t}, R_{m,t}) = \beta_j \sigma_m^2;$$

it follows that

$$\rho_i = \frac{\beta_i \sigma_m^2}{\sigma_m \sqrt{\beta_i^2 \sigma_m^2 + \sigma_{\epsilon,i}^2}} = \frac{\beta_i \sigma_m}{\sqrt{\beta_i^2 \sigma_m^2 + \sigma_{\epsilon,i}^2}}$$

and

$$\rho_j = \frac{\beta_j \sigma_m}{\sqrt{\beta_j^2 \sigma_m^2 + \sigma_{\epsilon,j}^2}}.$$

Now consider ρ_{ij}, the correlation of $R_{i,t}$ and $R_{j,t}$. Under the single-index model, the covariance of $R_{i,t}$ and $R_{j,t}$ is the (i,j)th element of

$$\sigma_m^2 \boldsymbol{\beta}\boldsymbol{\beta}^T + \boldsymbol{\Sigma}_\epsilon;$$

hence,

$$\text{Cov}(R_{i,t}, R_{j,t}) = \beta_i \beta_j \sigma_m^2.$$

It follows that

$$\rho_{ij} = \frac{\beta_i \beta_j \sigma_m^2}{\sqrt{\beta_i^2 \sigma_m^2 + \sigma_{\epsilon,i}^2}\sqrt{\beta_j^2 \sigma_m^2 + \sigma_{\epsilon,j}^2}} = \rho_i \rho_j.$$

Thus, we have proven the following result.

Corollary 9.1. *For each $t = 1, 2, \ldots, T$, let $\boldsymbol{R}_t$ denote an $N \times 1$ vector of asset returns and suppose that $\boldsymbol{\Sigma}$, the covariance matrix of $\boldsymbol{R}_t$ is of the form*

$$\boldsymbol{\Sigma} = \sigma_m^2 \boldsymbol{\beta}\boldsymbol{\beta}^T + \boldsymbol{\Sigma}_\epsilon \tag{9.4}$$

where $\boldsymbol{\Sigma}_\epsilon$ is a diagonal matrix.

Let ρ_{ij} denote the correlation of $R_{i,t}$ and $R_{j,t}$, let ρ_i denote the correlation of $R_{i,t}$ and $R_{m,t}$, and let ρ_j denote the correlation of $R_{j,t}$ and $R_{m,t}$. Then for $i, j = 1, 2, \ldots, N$, $i \neq j$,

$$\rho_{ij} = \rho_i \rho_j.$$

Thus, under the single-index model, the correlation of the returns on any two assets is equal to the product of the correlations of each assets' returns with the returns on the market index.

Example 9.1 Suppose that for a set of three assets, the single-index model holds with $\boldsymbol{\beta} = (0.8, 0.5, 1.1)^T$, $\sigma_{\epsilon,1} = 0.20$, $\sigma_{\epsilon,2} = 0.25$, $\sigma_{\epsilon,3} = 0.10$, and $\sigma_m = 0.05$. Then the covariance matrix of the assets is given by

```
> beta<-c(0.8, 0.5, 1.1)
> Sig1<-(0.05^2)*(beta%*%t(beta)) + diag(c(0.2,0.25, 0.1)^2)
> Sig1
        [,1]    [,2]    [,3]
[1,] 0.0416 0.00100 0.00220
[2,] 0.0010 0.06313 0.00138
[3,] 0.0022 0.00138 0.01303
```

The corresponding correlation matrix can be obtained using the function cov2cor, which converts a covariance matrix into a correlation matrix.

```
> cov2cor(Sig1)
        [,1]   [,2]   [,3]
[1,] 1.0000 0.0195 0.0945
[2,] 0.0195 1.0000 0.0480
[3,] 0.0945 0.0480 1.0000
```

Recall that, for a given asset, $\beta_i = \rho_i(\sigma_i/\sigma_m)$ where $\sigma_i^2 = \text{Var}(R_{i,t})$. Therefore, the vector of correlations $(\rho_1, \rho_2, \rho_3)^T$ of each asset's returns with the returns on the market index is given by

```
> cor_vec<-beta*(0.05/(diag(Sig1)^.5))
> cor_vec
[1] 0.1961 0.0995 0.4819
```

Products of the form $\rho_i\rho_j$ for $i \neq j$ may be easily obtained from the off-diagonal elements of cor_vec%*%t(cor_vec):

```
> cor_vec%*%t(cor_vec)
        [,1]   [,2]   [,3]
[1,] 0.0385 0.0195 0.0945
[2,] 0.0195 0.0099 0.0480
[3,] 0.0945 0.0480 0.2322
```

Note that the off-diagonal elements of the matrix cor_vec%*%t(cor_vec) are identical to the off-diagonal elements of cov2cor(Sig1). □

Partial Correlation

This property of the correlation between the returns of any two assets in a given time period may be described in terms of their *partial correlation*. If, as in the single-index model, the correlation between $R_{i,t}$ and $R_{j,t}$ is attributable entirely to the fact that both assets' returns are linearly related to the market return, then $\rho_{ij} = \rho_i\rho_j$. It follows that

$$\rho_{ij} - \rho_i\rho_j$$

represents the correlation of $R_{i,t}$ and $R_{j,t}$ relative to the value of the correlation that would be obtained if the correlation between $R_{i,t}$ and $R_{j,t}$ is because of the relationships of the assets' returns to the market return.

The *partial correlation coefficient* of $R_{i,t}$ and $R_{j,t}$ given $R_{m,t}$ is a scaled version of this difference:

$$\rho_{ij\cdot m} = \frac{\rho_{ij} - \rho_i\rho_j}{\sqrt{(1-\rho_i^2)(1-\rho_j^2)}}. \tag{9.5}$$

This partial correlation coefficient describes the extent to which $R_{i,t}$ and $R_{j,t}$ are linearly related, after removing the effect of the market return on this relationship; this is often expressed by saying that $\rho_{ij \cdot m}$ gives the correlation between $R_{i,t}$ and $R_{j,t}$ "controlling for" the market return. Like the usual correlation coefficient, it is based on the assumption that the relationships among $R_{i,t}, R_{j,t}$, and $R_{m,t}$ are all linear ones. If the single-index model holds, then $\rho_{ij \cdot m} = 0$ for all $i, j = 1, 2, \ldots, N$, $i \neq j$.

Example 9.2 In this chapter, we apply the single-index model to the returns on the stocks of five companies, Cablevision Systems Corp. (symbol CVC), Edison International (EIX), Expedia, Inc. (EXPE), Humana, Inc. (HUM), and Wal-Mart Stores, Inc. (WMT). Five years of monthly returns for the period ending December 31, 2014, were calculated for each stock and stored in variables with the same name as the stock symbol; for example, `cvc` contains the excess returns for Cablevision.

The matrix of excess returns for all five stocks is stored in the variable `stks`, which is analogous to the matrix stored in the variable `big8` used in Example 6.6, as well as other examples in Chapters 6 through 8. The variable `sp500` contains similar excess returns on the Standard & Poors (S&P) 500 index.

An estimate of the partial correlation coefficient is given by replacing the correlation coefficients in (9.5) by the sample correlation coefficients. Although such an estimate is easily calculated using results from the `cor` function, it is convenient to use the `pcor.test` function in package `ppcor`, which also includes a test of $\rho_{ij \cdot m} = 0$.

For instance, consider the estimated partial correlation of the returns on Edison stock and Wal-Mart stock given the returns on the S&P 500 index.

```
> library(ppcor)
> cor(eix, wmt)
       V1
V1 0.2769
> pcor.test(eix, wmt, sp500)
  estimate p.value statistic  n gp  Method
1   0.1574   0.229     1.203 60  1 pearson
```

Therefore, the sample correlation coefficient for the returns on Edison and Wal-Mart stock is 0.277 and the estimated partial correlation coefficient is 0.157. The test of the null hypothesis that the partial correlation coefficient for Edison and Wal-Mart returns, controlling for the returns on the S&P 500 index, is zero has p-value 0.229. Hence, there is no evidence to reject the null hypothesis, and it appears that the correlation between Edison and Wal-Mart stock returns is attributable entirely to their relationships with the market return, as measured by the return on the S&P 500 index. That is, there is no evidence to reject the hypothesis that the single-index model holds for Edison and Wal-Mart stock.

To calculate the partial correlation coefficients for all pairs of assets, we can use nested loops that calculate the partial correlation and the corresponding p-value for the returns on each pair of stocks.

```
> pcor<-pvalue<-matrix(0, 5, 5)
>  for (i in 1:4){
+      for (j in (i+1):5){
+         res<-pcor.test(stks[,i], stks[, j], sp500)
+         pcor[i, j]<-res[1, 1]
+         pvalue[i, j]<-res[1, 2]
+       }
+  }
```

The matrix pcor will contain the estimated partial correlation coefficients and the matrix pvalue will contain the associated p-values. Note, although there are five columns in the return matrix stks, i runs from 1 to 4 and j runs from $i+1$ to 5 to avoid computing each estimate and p-value twice and to avoid computing the partial correlation coefficient of a vector of returns with itself.

It is easier to read the results if we add column names and row names to the result matrices:

```
> rownames(pcor)<-colnames(stks)
> colnames(pcor)<-colnames(stks)
> rownames(pvalue)<-colnames(stks)
> colnames(pvalue)<-colnames(stks)
> pcor
     CVC     EIX    EXPE     HUM     WMT
CVC    0 -0.0856 -0.0354 -0.0318 -0.1018
EIX    0  0.0000 -0.0396 -0.0285  0.1574
EXPE   0  0.0000  0.0000 -0.2486  0.1526
HUM    0  0.0000  0.0000  0.0000 -0.0667
WMT    0  0.0000  0.0000  0.0000  0.0000
> pvalue
     CVC   EIX  EXPE    HUM   WMT
CVC    0 0.517 0.789 0.8101 0.440
EIX    0 0.000 0.765 0.8297 0.229
EXPE   0 0.000 0.000 0.0527 0.244
HUM    0 0.000 0.000 0.0000 0.614
WMT    0 0.000 0.000 0.0000 0.000
```

Because all p-values are relatively large, there is no evidence that the single-index model is inappropriate.

When there are many zeros in a table, it is often convenient to replace them by a different symbol; this is particularly true when, as in the present case, the zeros do not provide any information—they are simply placeholders. This can be achieved by converting the matrix to a "table" and then using the function `print`, which includes an argument for the symbol used for zeros. For example,

```
> print(as.table(pvalue), zero.print=".")
        CVC     EIX    EXPE     HUM     WMT
CVC       . 0.5167 0.7892 0.8101 0.4398
EIX       .      . 0.7649 0.8297 0.2290
EXPE      .      .      . 0.0527 0.2436
HUM       .      .      .      . 0.6138
WMT       .      .      .      .      .
```

Using nested loops in this way generally works well when analyzing a small or moderate number of assets. However, when analyzing a large number of assets, it may be preferable to use one of the vector-based functions available in R, such as `outer`, which is often more efficient in such cases. □

Note that when testing a large number of hypotheses, as in the previous example, it is important to be aware of the multiple testing issue, as discussed in Section 8.4. That is, when interpreting the p-values for tests of $\rho_{ij\cdot m} = 0$ for a large number of stocks, we expect a few small p-values even if the single-index model holds for all stocks. As discussed in Section 8.4, the Bonferroni method may be used in such cases to calculate an adjusted p-value that is valid for testing the hypothesis that the single-index model holds for all stocks considered.

9.4 Estimation

As in the previous section, consider a set of N assets, with returns $R_{1,t}, R_{2,t}, \ldots, R_{N,t}$ at time t, $t = 1, 2, \ldots, T$, and suppose the single-index model (9.2) holds. In this section, we consider estimation of the parameters of the model: the vector $\boldsymbol{\beta}$, $\boldsymbol{\beta} = (\beta_1, \beta_2, \ldots, \beta_N)^T$, the vector $\boldsymbol{\alpha} = (\alpha_1, \alpha_2, \ldots, \alpha_N)^T$, and the standard deviations of the residual returns, $\sigma_{\epsilon,1}, \sigma_{\epsilon,2}, \ldots, \sigma_{\epsilon,N}$.

Note that, under the assumptions of the single-index model, the parameter estimates for each asset may be obtained by estimating the parameters of the market model for that asset. Hence, parameter estimation for the single-index model uses the methods discussed in Section 8.3.

Example 9.3 Consider the stock returns for the five companies listed in Example 9.2. Using the same procedure used in Chapter 8 for the data in `big8`, the results for the market model applied to the returns in `stks` are stored in the variable `stks.mm`. The estimated regression coefficients are therefore in the `$coefficients` component of `stks.mm`.

```
> stks.mm<-lm(stks~sp500)
> stks.mm$coefficients
                   CVC     EIX   EXPE    HUM    WMT
(Intercept) -9.75e-05 0.00914 0.0108 0.0162 0.0059
sp500        1.07e+00 0.47407 0.8902 0.6516 0.4568
```

Therefore, the estimates $\hat{\boldsymbol{\alpha}}$ and $\hat{\boldsymbol{\beta}}$ may be obtained from `stks.mm $coefficients`:

```
> stks.alpha<-stks.mm$coefficients[1,]
> stks.alpha
      CVC        EIX       EXPE        HUM        WMT
-9.75e-05  9.14e-03  1.08e-02  1.62e-02  5.90e-03
> stks.beta<-stks.mm$coefficients[2,]
> stks.beta
  CVC   EIX  EXPE   HUM   WMT
1.073 0.474 0.890 0.652 0.457
```

The estimates $\hat{\sigma}_{\epsilon,j}$, $j = 1, 2, \ldots, 5$ may be calculated using the procedure described in Example 8.13, in which the `apply` function is used with the function `f.sighat` we have defined; the results are stored in the variable `stks.s`. Therefore, the residual standard deviations are given by

```
> stks.s<-apply(stks, 2, f.sighat)
> stks.s
   CVC    EIX   EXPE    HUM    WMT
0.0822 0.0457 0.1055 0.0695 0.0410
```

Thus, the parameters of the single-index model may be estimated using the methods used to estimate the parameters of the market model. □

Under the assumptions of the single-index model, the covariance matrix of the asset returns is of the form

$$\boldsymbol{\Sigma} = \sigma_m^2 \boldsymbol{\beta}\boldsymbol{\beta}^T + \boldsymbol{\Sigma}_\epsilon \tag{9.6}$$

where

$$\boldsymbol{\Sigma}_\epsilon = \begin{pmatrix} \sigma_{\epsilon,1}^2 & 0 & \ldots & 0 \\ 0 & \sigma_{\epsilon,2}^2 & \ddots & \vdots \\ \vdots & \ddots & \ddots & 0 \\ 0 & \ldots & 0 & \sigma_{\epsilon,N}^2 \end{pmatrix}.$$

Let

$$\widehat{\boldsymbol{\Sigma}}_\epsilon = \begin{pmatrix} \hat{\sigma}_{\epsilon,1}^2 & 0 & \ldots & 0 \\ 0 & \hat{\sigma}_{\epsilon,2}^2 & \ddots & \vdots \\ \vdots & \ddots & \ddots & 0 \\ 0 & \ldots & 0 & \hat{\sigma}_{\epsilon,N}^2 \end{pmatrix}.$$

Then an estimate of $\boldsymbol{\Sigma}$ based on the single-index model is given by

$$\widehat{\boldsymbol{\Sigma}} = S_m^2 \hat{\boldsymbol{\beta}}\hat{\boldsymbol{\beta}}^T + \widehat{\boldsymbol{\Sigma}}_\epsilon \tag{9.7}$$

where S_m^2 is the sample variance of $R_{m,1} - R_{f,1}, R_{m,2} - R_{f,2}, \ldots, R_{m,T} - R_{f,T}$.

Example 9.4 Consider the data analyzed in Example 9.3. The matrix $\widehat{\mathbf{\Sigma}}_\epsilon$ is the diagonal matrix with diagonal elements $\hat{\sigma}^2_{\epsilon,1}, \ldots, \hat{\sigma}^2_{\epsilon,5}$. Hence, for this example, it may be formed from the estimates in `stks.s` using the `diag` function, which forms a diagonal matrix from a vector of diagonal elements.

```
> stks.Sigeps<-diag(stks.s^2)
> rownames(stks.Sigeps)<-colnames(stks.Sigeps)<-c(labels(stks.s))
> print(as.table(stks.Sigeps), zero.print=".")

         CVC     EIX    EXPE     HUM     WMT
CVC  0.00676       .       .       .       .
EIX        . 0.00209       .       .       .
EXPE       .       . 0.01114       .       .
HUM        .       .       . 0.00483       .
WMT        .       .       .       . 0.00168
```

The command

```
> rownames(stks.Sigeps)<-colnames(stks.Sigeps)<-c(labels(stks.s))
```

assigns the labels from the vector `stks.s` to the row and column names of the matrix `stks.Sigeps`. The `print` command is used with `as.table` in order to print the zeros in the matrix as dots.

The matrix $\hat{\boldsymbol{\beta}}\hat{\boldsymbol{\beta}}^T$ may be obtained using matrix multiplication with the vector `stks.beta`. Recall that the matrix multiplication operator is `%*%` and `t` is the transpose function.

```
> stks.beta%*%t(stks.beta)
       CVC   EIX  EXPE   HUM   WMT
[1,] 1.152 0.509 0.955 0.699 0.490
[2,] 0.509 0.225 0.422 0.309 0.217
[3,] 0.955 0.422 0.792 0.580 0.407
[4,] 0.699 0.309 0.580 0.425 0.298
[5,] 0.490 0.217 0.407 0.298 0.209
```

The estimate $\widehat{\mathbf{\Sigma}}$ defined in (9.7) may now be obtained by combining `diag(stks.s^2)` with `stks.beta%*%t(stks.beta)` and the sample variance of the returns on the S&P 500 index.

```
> stks.Sig<-var(c(sp500))*(stks.beta%*%t(stks.beta)) +
+ diag(stks.s^2)
> stks.Sig
          CVC      EIX     EXPE      HUM      WMT
[1,] 0.008388 0.000718 0.001348 0.000987 0.000692
[2,] 0.000718 0.002410 0.000595 0.000436 0.000305
[3,] 0.001348 0.000595 0.012255 0.000818 0.000574
[4,] 0.000987 0.000436 0.000818 0.005428 0.000420
[5,] 0.000692 0.000305 0.000574 0.000420 0.001979
```

Note that the variance of the returns on the S&P 500 index is calculated by `var(c(sp500))` rather than by `var(sp500)` since `sp500` is a 60×1 matrix rather than a vector and, hence, `var(sp500)` returns a 1×1 matrix, rather than a scalar; `c(sp500)` converts `sp500` to a vector.

The estimate `stks.Sig` may be compared to the sample covariance matrix of the returns in `stks`.

```
> cov(stks)
          CVC      EIX      EXPE       HUM      WMT
CVC  0.008273 0.000401  0.001046  0.000808 0.000354
EIX  0.000401 0.002374  0.000407  0.000347 0.000596
EXPE 0.001046 0.000407  0.012067 -0.000974 0.001223
HUM  0.000808 0.000347 -0.000974  0.005346 0.000233
WMT  0.000354 0.000596  0.001223  0.000233 0.001950
```

The two estimates appear to be generally similar, although it is difficult to compare covariance matrices in this way. A useful alternative is to look at the standard deviations of the asset returns along with the correlation matrix of the return vector.

Recall that the `diag` function may also be used to extract the diagonal from a matrix. Therefore, the estimated standard deviations based on the single-index model may be compared to the sample standard deviations by comparing the values in `diag(stks.Sig)^.5` to those in `diag(cov(stks))^.5`.

Because

$$\mathrm{Var}(R_{j,t}) = \beta_j^2\sigma_m^2 + \sigma_{\epsilon,j}^2,$$

the two estimates of standard deviation should be very close, with the difference due to the slightly different divisors used in the estimates.

```
> diag(stks.Sig)^.5
[1] 0.0916 0.0491 0.1107 0.0737 0.0445
> diag(cov(stks))^.5
   CVC    EIX   EXPE    HUM    WMT
0.0910 0.0487 0.1098 0.0731 0.0442
```

That is, the estimates of the asset return standard deviations based on the single-index model are essentially the same as the return sample standard deviations.

The correlation matrix corresponding to `stks.Sig` is given by

```
> cov2cor(stks.Sig)
       CVC   EIX  EXPE   HUM   WMT
[1,] 1.000 0.160 0.133 0.146 0.170
[2,] 0.160 1.000 0.110 0.120 0.140
[3,] 0.133 0.110 1.000 0.100 0.116
[4,] 0.146 0.120 0.100 1.000 0.128
[5,] 0.170 0.140 0.116 0.128 1.000
```

this may be compared to the sample correlation matrix of the excess returns,

```
> cor(stks)
        CVC    EIX    EXPE     HUM    WMT
CVC  1.0000 0.0905  0.1047  0.1215 0.0881
EIX  0.0905 1.0000  0.0761  0.0973 0.2769
EXPE 0.1047 0.0761  1.0000 -0.1212 0.2522
HUM  0.1215 0.0973 -0.1212  1.0000 0.0721
WMT  0.0881 0.2769  0.2522  0.0721 1.0000
```

In looking for discrepancies between the two matrices, it is often useful to compute their difference.

```
> cov2cor(stks.Sig)-cor(stks)
        CVC     EIX    EXPE    HUM     WMT
[1,] 0.0000  0.0691  0.0283 0.0247  0.0817
[2,] 0.0691  0.0000  0.0334 0.0232 -0.1370
[3,] 0.0283  0.0334  0.0000 0.2216 -0.1357
[4,] 0.0247  0.0232  0.2216 0.0000  0.0560
[5,] 0.0817 -0.1370 -0.1357 0.0560  0.0000
```

Another approach to estimating the covariance matrix of the asset returns is to use a shrinkage estimate, as discussed in Section 6.5. For instance, the shrinkage estimate of the covariance matrix of the returns on the stocks represented in the data matrix `stks`, based on the target matrix taken to be a matrix of the form $\sigma^2 \boldsymbol{I}$, may be obtained using the function `shrinkcovmat.equal` in the package `ShrinkCovMat` (Touloumis 2015). The estimates of the asset standard deviations are given by

```
> stks.shrink<-shrinkcovmat.equal(t(stks))$Sigmahat
> diag(stks.shrink)^.5
   CVC    EIX   EXPE    HUM    WMT
0.0878 0.0571 0.1029 0.0742 0.0542
```

These estimates are similar to those based on the single-index model, with some differences, particularly for EIX and WMT. The shrinkage estimate of the correlation matrix is given by

```
> cov2cor(stks.shrink)
        CVC    EIX    EXPE     HUM    WMT
CVC  1.0000 0.0604  0.0874  0.0936 0.0561
EIX  0.0604 1.0000  0.0524  0.0618 0.1452
EXPE 0.0874 0.0524  1.0000 -0.0963 0.1655
HUM  0.0936 0.0618 -0.0963  1.0000 0.0437
WMT  0.0561 0.1452  0.1655  0.0437 1.0000
```

Although the three estimates of the return correlation matrix are similar, there are some differences. The most important is the correlation between

Humana and Expedia stocks. Note that, according to the sample covariance matrix and the shrinkage estimate of the covariance matrix, the returns on these stocks are negatively correlated. However, both stocks have positive estimates of beta, 0.890 for Expedia and 0.652 for Humana. Therefore, according to the single-index model, both stocks are positively correlated with the market index and, hence, they are positively correlated with each other. Specifically, the estimate of the correlation based on the single-index model is 0.100, while the sample covariance is −0.121 and the shrinkage estimate is −0.096.

Because the shrinkage estimate is a weighted average of the sample covariance matrix and a scaled identity matrix, it is not surprising that the shrinkage correlation estimates are closer to zero than are the sample correlations. The shrinkage estimates also tend to be closer to zero than are the correlation estimates based on the single-index model. □

9.5 Applications to Portfolio Analysis

In Proposition 9.1, it was shown that the single-index model implies a simple form for the covariance matrix of asset returns:

$$\boldsymbol{\Sigma} = \sigma_m^2 \boldsymbol{\beta}\boldsymbol{\beta}^T + \boldsymbol{\Sigma}_\epsilon \tag{9.8}$$

where $\boldsymbol{\Sigma}_\epsilon$ is an $N \times N$ diagonal matrix, $\boldsymbol{\beta}$ denotes the vector of betas with respect to the return on a market index, and σ_m^2 is the variance of the return on the market index. In this section, the implications of this form of the covariance matrix for portfolio theory are considered.

Recall that the inverse covariance matrix, $\boldsymbol{\Sigma}^{-1}$, plays a central role in calculating the weight vectors of several types of optimal portfolios. For instance, the weight vector of a "risk-averse" portfolio, as derived in Section 5.6, is of the form

$$\boldsymbol{w}_{mv} + \frac{1}{\lambda}\boldsymbol{\Sigma}^{-1}(\boldsymbol{\mu} - \mu_{mv}\mathbf{1}) \tag{9.9}$$

where $\boldsymbol{w}_{mv}$ is the weight vector of the minimum-variance portfolio, given by

$$\boldsymbol{w}_{mv} = \frac{\boldsymbol{\Sigma}^{-1}\mathbf{1}}{\mathbf{1}^T\boldsymbol{\Sigma}^{-1}\mathbf{1}},$$

λ is a scalar constant, and $\boldsymbol{\mu}$ is the vector of asset return means. The weight vector of the tangency portfolio is given by

$$\boldsymbol{w}_T = \frac{1}{\mathbf{1}^T\boldsymbol{\Sigma}^{-1}(\boldsymbol{\mu} - \mu_f\mathbf{1})}\boldsymbol{\Sigma}^{-1}(\boldsymbol{\mu} - \mu_f\mathbf{1})$$

where μ_f denotes the risk-free rate; see Proposition 5.7.

When the single-index model holds,

$$\boldsymbol{\mu} - \mu_f\mathbf{1} = \boldsymbol{\alpha} + \boldsymbol{\beta}(\mu_m - \mu_f)$$

and

$$\boldsymbol{\Sigma} = \sigma_m^2 \boldsymbol{\beta}\boldsymbol{\beta}^T + \boldsymbol{\Sigma}_\epsilon \tag{9.10}$$

where $\sigma_m^2 = \text{Var}(R_m)$ and $\boldsymbol{\Sigma}_\epsilon$ is a diagonal matrix.

To estimate the weight vector of a given portfolio under the single-index model, we can simply use the single-index-model estimates of $\boldsymbol{\Sigma}$ and $\boldsymbol{\mu} - \mu_f \mathbf{1}$ in the expression for the portfolio weights. The following example illustrates this for the case of the tangency portfolio.

Example 9.5 Consider the example of the returns on the stocks of the five companies discussed in the examples in this chapter. Recall that `stks.Sig` is an estimate of the return covariance matrix under the assumption that the single-index model holds and `stks.alpha` and `stks.beta` are estimates of $\boldsymbol{\alpha}$ and $\boldsymbol{\beta}$, respectively. Therefore, an estimate of $\boldsymbol{w}_T$ is given by

```
> wgt<-solve(stks.Sig, stks.alpha + stks.beta*mean(sp500))
> w_T_si<-wgt/sum(wgt)
> w_T_si
  CVC    EIX  EXPE    HUM    WMT
0.009 0.353 0.080 0.269 0.289
```

Note that `sp500` contains the excess returns on the S&P 500 index.

The weights calculated here may be compared to the estimated weights calculated without assuming that the single-index model holds.

```
> w_1T<-solve(cov(stks), apply(stks, 2, mean))
> w_T<-w_1T/sum(w_1T)
> w_T
  CVC    EIX  EXPE    HUM    WMT
0.035 0.331 0.119 0.314 0.201
```

The two sets of weights are similar but there are differences; this is not surprising because the single-index model is an important assumption regarding the relationship among the returns of the different stocks.

Note that the differences in the estimates are a consequence of the differences in the estimates of $\boldsymbol{\Sigma}$. Because of the properties of least-squares estimates, the estimate of $\boldsymbol{\mu} - \mu_f \mathbf{1}$ under the single-index model agrees with the estimate based on the sample means of the excess returns:

```
> apply(stks, 2, mean)
   CVC     EIX    EXPE     HUM     WMT
0.0116 0.0143 0.0205 0.0233 0.0109
> stks.alpha + stks.beta*mean(sp500)
   CVC     EIX    EXPE     HUM     WMT
0.0116 0.0143 0.0205 0.0233 0.0109
```

The estimates of the tangency portfolio weights calculated earlier can also be compared to the estimates based on a shrinkage estimate of the covariance matrix. Using the shrinkage estimate based on a target matrix of the form $\sigma^2 \boldsymbol{I}$, which is stored in the variable `stks.shrink` (see Example 9.4), the estimate of tangency weight vector is given by

```
> w.sh1<-solve(stks.shrink, apply(stks, 2, mean))
> w.sh1/sum(w.sh1)
  CVC    EIX  EXPE    HUM    WMT
0.061 0.278 0.149 0.331 0.180
```

The three sets of estimates are generally similar, with some differences; as noted previously, such differences are not unexpected. □

Note that many of the portfolio weight vectors we have considered depend on the return covariance matrix $\boldsymbol{\Sigma}$ through its inverse $\boldsymbol{\Sigma}^{-1}$. Under the form of $\boldsymbol{\Sigma}$ using the single-index model, it is possible to derive a relatively simple expression for this inverse. Although such an expression is not needed for numerical work, it is useful for understanding how such portfolio weights are related to the parameters of the single-index model.

Hence, we now consider such an expression for $\boldsymbol{\Sigma}^{-1}$; we begin with some useful results on matrix inverses.

Some Preliminary Results on Matrix Inverses

We first consider the inverse of a matrix of the form $\boldsymbol{I} + c\boldsymbol{d}\boldsymbol{d}^T$ where $\boldsymbol{d}$ is an $N \times 1$ vector, $\boldsymbol{I}$ is the $N \times N$ identity matrix, and c is a scalar. This result will then be used to derive the inverse of a matrix of the form (9.8).

Lemma 9.1. *For any scalar c and any vector $\boldsymbol{d} \in \Re^N$,*

$$(\boldsymbol{I} + c\,\boldsymbol{d}\boldsymbol{d}^T)^{-1} = \boldsymbol{I} - \frac{c}{1 + c\,\boldsymbol{d}^T\boldsymbol{d}}\boldsymbol{d}\boldsymbol{d}^T.$$

Proof. The result may be established by verifying that both

$$(\boldsymbol{I} + c\,\boldsymbol{d}\boldsymbol{d}^T)\left(\boldsymbol{I} - \frac{c}{1 + c\,\boldsymbol{d}^T\boldsymbol{d}}\boldsymbol{d}\boldsymbol{d}^T\right) \tag{9.11}$$

and

$$\left(\boldsymbol{I} - \frac{c}{1 + c\,\boldsymbol{d}^T\boldsymbol{d}}\boldsymbol{d}\boldsymbol{d}^T\right)(\boldsymbol{I} + c\,\boldsymbol{d}\boldsymbol{d}^T) \tag{9.12}$$

are equal to the identity matrix.

Consider (9.11).

$$(\boldsymbol{I} + c\,\boldsymbol{d}\boldsymbol{d}^T)\left(\boldsymbol{I} - \frac{c}{1 + c\,\boldsymbol{d}^T\boldsymbol{d}}\boldsymbol{d}\boldsymbol{d}^T\right) = \boldsymbol{I} + c\,\boldsymbol{d}\boldsymbol{d}^T - \frac{c}{1 + c\boldsymbol{d}^T\boldsymbol{d}}\boldsymbol{d}\boldsymbol{d}^T - \frac{c^2}{1 + c\,\boldsymbol{d}^T\boldsymbol{d}}\boldsymbol{d}\boldsymbol{d}^T\boldsymbol{d}\boldsymbol{d}^T \tag{9.13}$$

Note that, because $\boldsymbol{d}^T\boldsymbol{d}$ is a scalar, we may write

$$\boldsymbol{d}\boldsymbol{d}^T\boldsymbol{d}\boldsymbol{d}^T = \boldsymbol{d}(\boldsymbol{d}^T\boldsymbol{d})\boldsymbol{d}^T = (\boldsymbol{d}^T\boldsymbol{d})\boldsymbol{d}\boldsymbol{d}^T$$

and, hence,

$$\begin{aligned} c\, \boldsymbol{d}\boldsymbol{d}^T &- \frac{c}{1+c\,\boldsymbol{d}^T\boldsymbol{d}}\boldsymbol{d}\boldsymbol{d}^T - \frac{c^2}{1+c\,\boldsymbol{d}^T\boldsymbol{d}}\boldsymbol{d}\boldsymbol{d}^T\boldsymbol{d}\boldsymbol{d}^T \\ &= c\left(1 - \frac{1}{1+c\,\boldsymbol{d}^T\boldsymbol{d}} - \frac{c\,\boldsymbol{d}^T\boldsymbol{d}}{1+c\,\boldsymbol{d}^T\boldsymbol{d}}\right)\boldsymbol{d}\boldsymbol{d}^T \\ &= c\left(1 - \frac{1+c\,\boldsymbol{d}^T\boldsymbol{d}}{1+c\,\boldsymbol{d}^T\boldsymbol{d}}\right)\boldsymbol{d}\boldsymbol{d}^T = 0. \end{aligned}$$

It follows that the right-hand side of (9.13) is equal to $\boldsymbol{I}$. A similar argument can be used to show (9.12). □

The result in Lemma 9.1 may now be used to determine the inverse of a matrix of the form $\sigma_m^2\boldsymbol{\beta}\boldsymbol{\beta}^T + \boldsymbol{\Sigma}_\epsilon$.

Lemma 9.2. *Let $\boldsymbol{\Sigma}_\epsilon$ be an $N \times N$ positive-definite matrix and let $\boldsymbol{\beta}$ be an element of $\Re^N$. Then, for all $\sigma_m^2 \geq 0$,*

$$\left(\sigma_m^2\boldsymbol{\beta}\boldsymbol{\beta}^T + \boldsymbol{\Sigma}_\epsilon\right)^{-1} = \boldsymbol{\Sigma}_\epsilon^{-1} - \frac{\sigma_m^2}{1+\sigma_m^2\boldsymbol{\beta}^T\boldsymbol{\Sigma}_\epsilon^{-1}\boldsymbol{\beta}}\boldsymbol{\Sigma}_\epsilon^{-1}\boldsymbol{\beta}\boldsymbol{\beta}^T\boldsymbol{\Sigma}_\epsilon^{-1}. \tag{9.14}$$

Proof. Note that we may write

$$\sigma_m^2\boldsymbol{\beta}\boldsymbol{\beta}^T + \boldsymbol{\Sigma}_\epsilon = \boldsymbol{\Sigma}_\epsilon + \sigma_m^2\boldsymbol{\beta}\boldsymbol{\beta}^T = \boldsymbol{\Sigma}_\epsilon^{\frac{1}{2}}\left(\boldsymbol{I} + \sigma_m^2(\boldsymbol{\Sigma}_\epsilon^{-\frac{1}{2}}\boldsymbol{\beta})(\boldsymbol{\Sigma}_\epsilon^{-\frac{1}{2}}\boldsymbol{\beta})^T\right)\boldsymbol{\Sigma}_\epsilon^{\frac{1}{2}}.$$

Hence,

$$(\boldsymbol{\Sigma}_\epsilon + \sigma_m^2\boldsymbol{\beta}\boldsymbol{\beta}^T)^{-1} = \boldsymbol{\Sigma}_\epsilon^{-\frac{1}{2}}\left(\boldsymbol{I}_N + \sigma_m^2(\boldsymbol{\Sigma}_\epsilon^{-\frac{1}{2}}\boldsymbol{\beta})(\boldsymbol{\Sigma}_\epsilon^{-\frac{1}{2}}\boldsymbol{\beta})^T\right)^{-1}\boldsymbol{\Sigma}_\epsilon^{-\frac{1}{2}}. \tag{9.15}$$

Applying Lemma 9.1 with $c = \sigma_m^2$ and $\boldsymbol{d} = \boldsymbol{\Sigma}_\epsilon^{-\frac{1}{2}}\boldsymbol{\beta}$, it follows that

$$(\boldsymbol{I}_N + \sigma_m^2(\boldsymbol{\Sigma}_\epsilon^{-\frac{1}{2}}\boldsymbol{\beta})(\boldsymbol{\Sigma}_\epsilon^{-\frac{1}{2}}\boldsymbol{\beta})^T)^{-1} = \boldsymbol{I}_N - \frac{c}{1+c(\boldsymbol{\Sigma}_\epsilon^{-\frac{1}{2}}\boldsymbol{\beta})^T(\boldsymbol{\Sigma}_\epsilon^{-\frac{1}{2}}\boldsymbol{\beta})}(\boldsymbol{\Sigma}_\epsilon^{-\frac{1}{2}}\boldsymbol{\beta})(\boldsymbol{\Sigma}_\epsilon^{-\frac{1}{2}}\boldsymbol{\beta})^T.$$

Using this result in (9.15) yields

$$(\sigma_m^2\boldsymbol{\beta}\boldsymbol{\beta}^T + \boldsymbol{\Sigma}_\epsilon)^{-1} = (\boldsymbol{\Sigma}_\epsilon + \sigma_m^2\boldsymbol{\beta}\boldsymbol{\beta}^T)^{-1} = \boldsymbol{\Sigma}_\epsilon^{-1} - \frac{\sigma_m^2}{1+\sigma_m^2\boldsymbol{\beta}^T\boldsymbol{\Sigma}_\epsilon^{-1}\boldsymbol{\beta}}\boldsymbol{\Sigma}_\epsilon^{-1}\boldsymbol{\beta}\boldsymbol{\beta}^T\boldsymbol{\Sigma}_\epsilon^{-1} \tag{9.16}$$

as given in the statement of the lemma. □

Note that Lemma 9.1 does not require $\boldsymbol{\Sigma}_\epsilon$ to be a diagonal matrix; however, when it is a diagonal matrix, a simple expression for $\boldsymbol{\Sigma}_\epsilon^{-1}$ is available. Lemma 9.1 also shows that the covariance matrix $\sigma_m^2\boldsymbol{\beta}\boldsymbol{\beta}^T + \boldsymbol{\Sigma}_\epsilon$ is invertible provided that $\boldsymbol{\Sigma}_\epsilon$ is invertible. The same argument shows that the estimated covariance matrix $S_m^2\hat{\boldsymbol{\beta}}\hat{\boldsymbol{\beta}}^T + \widehat{\boldsymbol{\Sigma}}_\epsilon$ is invertible under the weak condition that $\hat{\sigma}_{\epsilon,j}^2 > 0$ for $j = 1, 2, \ldots, N$.

The explicit expression for the inverse of $\boldsymbol{\Sigma}$ in this setting can be used to derive explicit expressions for the weight vectors of some of the optimal portfolios we have considered in terms of $\boldsymbol{\beta}$ and $\boldsymbol{\Sigma}_\epsilon$.

Weight Vector of the Tangency Portfolio under the Single-Index Model

The explicit expression for the matrix inverse in Lemma 9.2 may be used to obtain an explicit expression for the weights of the tangency portfolio. As noted previously, such an expression is not needed for numerical work, but it is useful for understanding how the tangency portfolio weights depend on the parameters $\boldsymbol{\alpha}$, $\boldsymbol{\beta}$, and $\boldsymbol{\Sigma}_\epsilon$.

Proposition 9.2. *Consider a set of N assets and let $\boldsymbol{R}_t$ denote the corresponding vector of asset returns at time t. Assume that $\boldsymbol{R}_t$ follows the single-index model so that $\boldsymbol{\mu}$, the mean vector of $\boldsymbol{R}_t$, is of the form*

$$\boldsymbol{\mu} = \mu_f \mathbf{1} + \boldsymbol{\alpha} + (\mu_m - \mu_f)\boldsymbol{\beta}$$

and $\boldsymbol{\Sigma}$, the covariance matrix of $\boldsymbol{R}_t$, is of the form

$$\boldsymbol{\Sigma} = \sigma_m^2 \boldsymbol{\beta}\boldsymbol{\beta}^T + \boldsymbol{\Sigma}_\epsilon$$

where $\boldsymbol{\Sigma}_\epsilon$ is a diagonal matrix with diagonal elements $\sigma_{\epsilon,1}^2, \sigma_{\epsilon,2}^2, \ldots, \sigma_{\epsilon,N}^2$.

Let $\boldsymbol{w}_T$ denote the weight vector of the tangency portfolio based on $\boldsymbol{R}_t$. Then

$$\boldsymbol{w}_T = \frac{\boldsymbol{v}}{\mathbf{1}^T \boldsymbol{v}}$$

where

$$\boldsymbol{v} = \boldsymbol{\Sigma}_\epsilon^{-1}\left(\boldsymbol{\alpha} + (\mu_m - \mu_f - \gamma)\boldsymbol{\beta}\right)$$

and

$$\begin{aligned}\gamma &= \frac{\sigma_m^2}{1 + \sigma_m^2 \boldsymbol{\beta}^T \boldsymbol{\Sigma}_\epsilon^{-1} \boldsymbol{\beta}} \boldsymbol{\beta}^T \boldsymbol{\Sigma}_\epsilon^{-1} (\boldsymbol{\mu} - \mu_f \mathbf{1}) \\ &= \frac{\sigma_m^2}{1 + \sigma_m^2 \boldsymbol{\beta}^T \boldsymbol{\Sigma}_\epsilon^{-1} \boldsymbol{\beta}} \boldsymbol{\beta}^T \boldsymbol{\Sigma}_\epsilon^{-1} \left(\boldsymbol{\alpha} - (\mu_m - \mu_f)\boldsymbol{\beta}\right).\end{aligned}$$

Proof. For a given value of the mean vector $\boldsymbol{\mu}$, consider calculation of the weight vector of the tangency portfolio when $\boldsymbol{\Sigma}$ is given by $\sigma_m^2 \boldsymbol{\beta}\boldsymbol{\beta}^T + \boldsymbol{\Sigma}_\epsilon$. Using (9.14), this weight vector is proportional to

$$\left(\boldsymbol{\Sigma}_\epsilon^{-1} - \frac{\sigma_m^2}{1 + \sigma_m^2 \boldsymbol{\beta}^T \boldsymbol{\Sigma}_\epsilon^{-1} \boldsymbol{\beta}} \boldsymbol{\Sigma}_\epsilon^{-1} \boldsymbol{\beta}\boldsymbol{\beta}^T \boldsymbol{\Sigma}_\epsilon^{-1}\right) (\boldsymbol{\mu} - \mu_f \mathbf{1}), \tag{9.17}$$

which may be written as

$$\boldsymbol{\Sigma}_\epsilon^{-1}(\boldsymbol{\mu} - \mu_f \mathbf{1}) - \frac{\sigma_m^2}{1 + \sigma_m^2 \boldsymbol{\beta}^T \boldsymbol{\Sigma}_\epsilon^{-1} \boldsymbol{\beta}} \boldsymbol{\Sigma}_\epsilon^{-1} \boldsymbol{\beta} (\boldsymbol{\beta}^T \boldsymbol{\Sigma}_\epsilon^{-1} (\boldsymbol{\mu} - \mu_f \mathbf{1})).$$

Because $\boldsymbol{\beta}^T \boldsymbol{\Sigma}_\epsilon^{-1}(\boldsymbol{\mu} - \mu_f \mathbf{1})$ is a scalar,

$$\frac{\sigma_m^2}{1 + \sigma_m^2 \boldsymbol{\beta}^T \boldsymbol{\Sigma}_\epsilon^{-1} \boldsymbol{\beta}} \boldsymbol{\Sigma}_\epsilon^{-1} \boldsymbol{\beta} (\boldsymbol{\beta}^T \boldsymbol{\Sigma}_\epsilon^{-1} (\boldsymbol{\mu} - \mu_f \mathbf{1})) = \gamma \boldsymbol{\Sigma}_\epsilon^{-1} \boldsymbol{\beta}$$

where

$$\gamma = \frac{\sigma_m^2}{1+\sigma_m^2\boldsymbol{\beta}^T\boldsymbol{\Sigma}_\epsilon^{-1}\boldsymbol{\beta}}\boldsymbol{\beta}^T\boldsymbol{\Sigma}_\epsilon^{-1}(\boldsymbol{\mu}-\mu_f\mathbf{1})\frac{\sigma_m^2}{1+\sigma_m^2\boldsymbol{\beta}^T\boldsymbol{\Sigma}_\epsilon^{-1}\boldsymbol{\beta}}\boldsymbol{\beta}^T\boldsymbol{\Sigma}_\epsilon^{-1}\left(\boldsymbol{\alpha}-(\mu_m-\mu_f)\boldsymbol{\beta}\right).$$

Therefore, (9.17) may be written as

$$\boldsymbol{\Sigma}_\epsilon^{-1}(\boldsymbol{\mu}-\mu_f\mathbf{1})-\gamma\boldsymbol{\Sigma}_\epsilon^{-1}\boldsymbol{\beta};$$

substituting $\boldsymbol{\alpha}+(\mu_m-\mu_f)\boldsymbol{\beta}$ for $\boldsymbol{\mu}-\mu_f\mathbf{1}$ shows that, under the single-index model, $\boldsymbol{w}_T$ is proportional to $\boldsymbol{v}$, as defined in the statement of the proposition. The result now follows from noting that the tangency weights must sum to 1. □

This result is useful for gaining some insight into the relationship between weights of the tangency portfolio and the parameters $(\alpha_j, \beta_j, \sigma_{\epsilon,j}^2)$, $j = 1, 2, \ldots, N$. Note that v_j, the jth element of $\boldsymbol{v}$, may be written as

$$v_j = \alpha_j/\sigma_{\epsilon,j}^2 + \delta\beta_j/\sigma_{\epsilon,j}^2, \quad j = 1, 2, \ldots, N$$

where $\delta = \mu_m - \mu_f - \gamma$. That is, the tangency weight for asset j is proportional to a linear function of $\alpha_j/\sigma_{\epsilon,j}^2$ and $\beta_j/\sigma_{\epsilon,j}^2$.

If $\boldsymbol{\alpha} = \mathbf{0}$, as would be the case if the capital asset pricing model (CAPM) holds, then

$$\boldsymbol{v} = (\mu_m - \mu_f - \gamma_0)\boldsymbol{\Sigma}^{-1}\boldsymbol{\beta}$$

where

$$\gamma_0 = \frac{\sigma_m^2}{1+\sigma_m^2\boldsymbol{\beta}^T\boldsymbol{\Sigma}_\epsilon^{-1}\boldsymbol{\beta}}\boldsymbol{\beta}^T\boldsymbol{\Sigma}_\epsilon^{-1}\boldsymbol{\beta}(\mu_m - \mu_f).$$

That is, $\boldsymbol{v}$ is proportional to $\boldsymbol{\Sigma}_\epsilon^{-1}\boldsymbol{\beta}$ so that the weight given to asset j in the tangency portfolio is proportional to $\beta_j/\sigma_{\epsilon,j}^2$. Such weights may be viewed as a simple approximation to the tangency weights that holds when the $|\alpha_j|$ are small. According to this approximation, asset j receives a large weight in the tangency portfolio when β_j is large relative to the residual variance for asset j.

Example 9.6 For the stocks analyzed in Example 9.5, consider the approximation to the tangency portfolio weights based on $\beta_j/\sigma_{\epsilon,j}^2$, $j = 1, \ldots, 5$. Recall that the estimates of β_j for these stocks are stored in the variable `stks.beta` and the estimates of $\sigma_{\epsilon,j}$ are stored in the variable `stks.s`. Then the approximations to the tangency portfolio weights are given by

```
> stks.beta/(stks.s^2)/sum(stks.beta/(stks.s^2))
  CVC   EIX  EXPE   HUM   WMT
0.182 0.260 0.092 0.155 0.311
```

These weights may be compared to the estimated tangency portfolio weights based on the single-index model

```
> w_T_si
  CVC   EIX  EXPE   HUM   WMT
0.009 0.353 0.080 0.269 0.289
```

□

9.6 Active Portfolio Management and the Treynor–Black Method

According to the assumptions used to derive the CAPM, the market portfolio is the optimal choice for investors; see Section 7.1. In practice, such a market portfolio is unobservable; hence, we use a suitably chosen market index as a substitute. If we accept the optimality of the portfolio corresponding to the market index, then the portfolio problem is solved—simply invest in a portfolio that tracks the market index along with the risk-free asset. This is sometimes referred to as a *passive* approach to investing because the investor does not need to choose the individual assets or their weights in constructing a portfolio. However, if the portfolio corresponding to the market index is not efficient, then it may be possible to modify it in order to improve portfolio performance, a process known as *active portfolio management.*

Active portfolio management is based on a *benchmark portfolio*, assumed to have desirable properties—high expected return and low risk—but not assumed to satisfy any optimality criteria. The goal in active portfolio management is to construct a portfolio with a return that is highly correlated with the return on the benchmark portfolio but that improves on it by having a higher expected return and/or lower risk.

The benchmark portfolio is generally taken to be the portfolio based on a market index, such as the S&P 500 index. In this section, we will take the benchmark portfolio to be the portfolio corresponding to the market index used in the single-index model; we will refer to this portfolio as the market portfolio. However, it is important to keep in mind that we do not assume that the market portfolio is efficient or nearly efficient; on the contrary, it is exactly because of the inefficiency of the market portfolio that active portfolio management is considered to be a worthwhile activity.

Consider a set of N assets and let $\boldsymbol{R}_t$ be the corresponding vector of asset returns at time t, $t = 1, 2, \ldots, T$. We assume that the single-index model holds for $\boldsymbol{R}_t$ so that, as described in Section 9.2,

$$\mathrm{E}(\boldsymbol{R}_t) = \boldsymbol{\alpha} + (\mu_m - \mu_f)\boldsymbol{\beta}$$

and $\boldsymbol{\Sigma}$, the covariance matrix of $\boldsymbol{R}_t$, is of the form

$$\boldsymbol{\Sigma} = \sigma_m^2 \boldsymbol{\beta}\boldsymbol{\beta}^T + \boldsymbol{\Sigma}_\epsilon,$$

where $\boldsymbol{\Sigma}_\epsilon$ is a diagonal matrix.

Let $\boldsymbol{w}_p$ denote the weight vector of a given portfolio and let $R_{p,t} = \boldsymbol{w}_p^T R_t$ denote the corresponding return variable, $t = 1, 2, \ldots, T$. Then

$$\mathrm{E}(R_{p,t}) = \boldsymbol{w}_p^T \boldsymbol{\alpha} + \boldsymbol{w}_p^T \boldsymbol{\beta}(\mu_m - \mu_f) = \alpha_p + \beta_p(\mu_m - \mu_f),$$

where $\alpha_p = \boldsymbol{w}_p^T \boldsymbol{\alpha}$ and $\beta_p = \boldsymbol{w}_p^T \boldsymbol{\beta}$ are the values of alpha and beta, respectively, for the portfolio. The variance of the portfolio return is given by

$$\begin{aligned}\text{Var}(R_{p,t}) &= \boldsymbol{w}_p^T \left(\sigma_m^2 \boldsymbol{\beta}\boldsymbol{\beta}^T + \boldsymbol{\Sigma}_\epsilon\right) \boldsymbol{w}_p \\ &= \sigma_m^2 (\boldsymbol{w}_p^T \boldsymbol{\beta})(\boldsymbol{\beta}^T \boldsymbol{w}_p) + \boldsymbol{w}_p^T \boldsymbol{\Sigma}_\epsilon \boldsymbol{w}_p \\ &= \sigma^2 \beta_p^2 + \boldsymbol{w}_p^T \boldsymbol{\Sigma}_\epsilon \boldsymbol{w}_p.\end{aligned}$$

The goal in active portfolio management is to choose $\boldsymbol{w}_p$ so that the resulting portfolio outperforms the market portfolio.

The approach used here to construct such a portfolio is based on the following idea. We may view the market portfolio as a single asset in which we may place an investment; hence, we have, effectively, $N+1$ assets from which to form a portfolio. That is, we consider the market portfolio to be a tradeable asset. This is, in fact, essentially the case as there are a number of mutual funds constructed to track various market indices. We then choose the optimal weights for these $N+1$ assets. The result is a combination of the market portfolio and the N assets under consideration. This is known as the *Treynor–Black method.*

Adding a Single Asset to the Market Portfolio

We begin by considering the simplest case, in which we attempt to improve on the market portfolio by combining it with a single asset. Suppose we have an asset with return $R_{i,t}$ at time t and let $R_{m,t}$ denote the return on the market portfolio at time t. Under the assumption that the single-index model holds with respect to $R_{m,t}$ for asset i, we can write

$$R_{i,t} - R_{f,t} = \alpha_i + \beta_i(R_{m,t} - R_{f,t}) + \epsilon_{i,t}, \quad t = 1, 2, \ldots, T.$$

Here $R_{f,t}$ denotes the return on the risk-free asset at time t and $\epsilon_{i,1}, \epsilon_{i,2}, \ldots, \epsilon_{i,T}$ is a sequence of uncorrelated, mean-zero random variables that are uncorrelated with the $R_{m,t}$. Note that because we are only considering one asset, in this case, the single-index model is equivalent to the market model.

Consider the tangency portfolio formed from asset i and the market portfolio. Recall that, when forming the tangency portfolio from two assets, the weight given to asset 1 is given by

$$\frac{(\mu_1 - \mu_f)\sigma_2^2 - (\mu_2 - \mu_f)\rho_{12}\sigma_1\sigma_2}{(\mu_2 - \mu_f)\sigma_1^2 + (\mu_1 - \mu_f)\sigma_2^2 - [(\mu_1 - \mu_f) + (\mu_2 - \mu_f)]\rho_{12}\sigma_1\sigma_2}, \tag{9.18}$$

where μ_1, μ_2 are the mean returns on the two assets, σ_1, σ_2 are the return standard deviations, and ρ_{12} is the correlation of the returns.

Taking the asset with return $R_{i,t}$ to be asset 1 and the market portfolio to be asset 2, here

$$\mu_1 - \mu_f = \text{E}(R_{i,t}) - \mu_f = \alpha_i + \beta_i(\mu_m - \mu_f),$$

where $\mu_m = \mathrm{E}(R_{m,t})$,

$$\sigma_1^2 = \mathrm{Var}(R_{i,t}) = \beta_i^2\sigma_m^2 + \sigma_{\epsilon,i}^2$$

where $\sigma_m^2 = \mathrm{Var}(R_{m,t})$ and $\sigma_{\epsilon,i}^2 = \mathrm{Var}(\epsilon_{i,t})$,

$$\rho_{12} = \mathrm{Cov}(R_{i,t}, R_{m,t})\mathrm{Cov}(\beta_i R_{m,t}, R_{m,t}) = \beta_i\sigma_m^2,$$

$\mu_2 = \mu_m$ and $\sigma_2 = \sigma_m$.

Using these values in (9.18), and simplifying, leads to the expression

$$w_i^* = \frac{\alpha_i/\sigma_{\epsilon,i}^2}{(\mu_m - \mu_f)/\sigma_m^2 + (1-\beta_i)\alpha_i/\sigma_{\epsilon,i}^2} \tag{9.19}$$

for the weight given to asset 1, which in this case is asset i. The weight given to the market portfolio is therefore

$$1 - w_i^* = \frac{(\mu_m - \mu_f)/\sigma_m^2 - \beta_i\alpha_i/\sigma_{\epsilon,i}^2}{(\mu_m - \mu_f)/\sigma_m^2 + (1-\beta_i)\alpha_i/\sigma_{\epsilon,i}^2}.$$

We will refer to the portfolio placing weight w_i^* on asset i and weight $1 - w_i^*$ on the market portfolio as the *Treynor–Black portfolio* of asset i and the market portfolio.

If asset i is priced correctly relative to the market portfolio, in the sense that $\alpha_i = 0$, then the optimal weight to give asset i is 0; that is, we cannot improve the market portfolio by including more or less of asset i. However, if $\alpha_i \neq 0$, then the market portfolio may be improved by combining it with asset i.

The Treynor–Black portfolio of asset i and the market portfolio has mean excess return

$$\begin{aligned}
w_i^*(\mu_i - \mu_f) + (1 - w_i^*)(\mu_m - \mu_f) &= w_i^*\,(\alpha_i + \beta_i(\mu_m - \mu_f)) + (1 - w_i^*)(\mu_m - \mu_f)\\
&= w_i^*\alpha_i + (w_i^*\beta_i + 1 - w_i^*)(\mu_m - \mu_f)\\
&= \frac{1}{c}\left(\alpha_i^2/\sigma_{\epsilon,i}^2 + (\mu_m - \mu_f)^2/\sigma_m^2\right)
\end{aligned}$$

where

$$c = (\mu_m - \mu_f)/\sigma_m^2 + (1-\beta_i)\alpha_i/\sigma_{\epsilon,i}^2$$

is the denominator in the expressions for w_i^* and $1 - w_i^*$.

The variance of the return on this portfolio is given by

$$\begin{aligned}
&(w_i^*)^2\sigma_i^2 + (1 - w_i^*)^2\sigma_m^2 + 2w_i^*(1 - w_i^*)\mathrm{Cov}(R_i, R_m)\\
&\quad = (w_i^*)^2(\beta_i^2\sigma_m^2 + \sigma_{\epsilon,i}^2) + (1 - w_i^*)^2\sigma_m^2 + 2w_i(1 - w_i)\beta_i\sigma_m^2\\
&\quad = (w_i^*)^2\sigma_{\epsilon,i}^2 + (w_i\beta_i + 1 - w_i))\,\sigma_m^2\\
&\quad = \frac{1}{c^2}\left(\alpha_i^2/\sigma_{\epsilon,i}^2 + \left(\frac{\mu_m - \mu_f}{\sigma_m^2}\right)^2\sigma_m^2\right)\\
&\quad = \frac{1}{c^2}\left(\alpha_i^2/\sigma_{\epsilon,i}^2 + \frac{(\mu_m - \mu_f)^2}{\sigma_m^2}\right).
\end{aligned}$$

It follows that the squared Sharpe ratio of the Treynor–Black portfolio is given by

$$\frac{\left(\alpha_i^2/\sigma_{\epsilon,i}^2 + (\mu_m - \mu_f)^2/\sigma_m^2\right)^2}{\alpha_i^2/\sigma_{\epsilon,i}^2 + (\mu_m - \mu_f)^2/\sigma_m^2} = \alpha_i^2/\sigma_{\epsilon,i}^2 + (\mu_m - \mu_f)^2/\sigma_m^2. \tag{9.20}$$

Note that $(\mu_m - \mu_f)^2/\sigma_m^2$ is the squared Sharpe ratio of the market portfolio.

Therefore, if $\alpha_i = 0$, the Sharpe ratio of the Treynor–Black portfolio is equal to the Sharpe ratio of the market portfolio. However, if $\alpha_i \neq 0$, the Sharpe ratio of the Treynor–Black portfolio is greater than that of the market portfolio. The quantity $\alpha_i/\sigma_{\epsilon,i}$ is known as the *appraisal ratio* of the asset; recall that we considered the appraisal ratio as a measure of portfolio performance in Section 8.9. In the present context, the magnitude of the appraisal ratio indicates the possible improvement in the Sharpe ratio of the market portfolio that may be achieved by including more or less of the asset.

Example 9.7 Suppose that R_m, the return on the market portfolio, has mean 0.025 and standard deviation 0.04 and that the risk-free rate is $\mu_f = 0.005$. Consider an asset with return R_i with mean 0.02, standard deviation 0.08, and suppose that the correlation of R_i and R_m is $\rho_i = 0.30$ so that $\beta_i = 0.60$ and

$$\sigma_{\epsilon,i}^2 = (0.08)^2 - (0.60)^2(0.04)^2 = 0.00582.$$

In Example 7.6, it was shown that $\alpha_i = 0.008$ so that the price of asset i is mispriced, with a price that is too low.

Therefore, we can improve on the market portfolio by constructing a portfolio based on the market portfolio and asset i. The weight given to asset i in such a portfolio is

$$\begin{aligned}&\frac{\alpha_i/\sigma_{\epsilon,i}^2}{(\mu_m - \mu_f)/\sigma_m^2 + (1-\beta_i)\alpha_i/\sigma_{\epsilon,i}^2}\\ &\quad = \frac{(0.008)/(0.00582)}{(0.025-0.005)/(0.04)^2 + (1-0.60)(0.008)/(0.00582)}\\ &\quad = 0.105;\end{aligned}$$

the market portfolio receives weight $1 - 0.105 = 0.895$.

The squared Sharpe ratio of this portfolio is

$$(0.008)^2/(0.00582) + (0.025 - 0.005)^2/(0.04)^2 = 0.261$$

corresponding to a Sharpe ratio of $\sqrt{0.261} = 0.511$. This can be compared to the Sharpe ratio of the market portfolio, $(0.025 - 0.005)/(0.04) = 0.5$. □

Of course, in practice, these quantities must be estimated based on observed return data.

Example 9.8 Consider the stock in Apple Inc. (symbol AAPL); here we analyze five years of monthly returns for the period ending December 31, 2014. The variables aapl.alpha and aapl.s contain the estimates of α_i and $\sigma_{\epsilon,i}$, respectively, for this asset. Then the estimated appraisal ratio for Apple is

```
> aapl.alpha/aapl.s
(Intercept)
      0.233
```

The estimated Sharpe ratio of the S&P 500 index is 0.290. Thus, according to (9.20), the estimated Sharpe ratio of the Treynor–Black portfolio based on Apple stock together with the S&P 500 index is

$$\left((0.233)^2 + (0.290)^2\right)^{\frac{1}{2}} = 0.372.$$

The weight placed on Apple in this portfolio is estimated to be

```
> cc<-mean(sp500)/var(sp500) + (1-aapl.beta)*aapl.alpha/aapl.s^2
> (aapl.alpha/aapl.s^2)/cc
0.442
```

Here, aapl.beta is the estimate of beta for Apple stock. □

Of course, these results are estimates based on observed data; hence, it is important to keep in mind the sampling properties of the results. We have discussed two approaches to assessing such properties—the Monte Carlo simulation method used in Section 6.7 to study the sampling distribution of an estimator and the bootstrap method used in Section 8.10 to calculate standard errors and to estimate bias. Although either approach could be used here, for simplicity, we consider only the bootstrap method, as in the following example.

Example 9.9 Consider estimation of the appraisal ratio of Apple stock, as discussed in Example 9.8. Recall that the estimate obtained using five years of monthly data for the period ending December 31, 2014, is 0.233; the return data for Apple stock are stored in the variable aapl.

To obtain the standard error of this estimate, we use the procedure described in Section 8.10; specifically, we follow the method described in Example 8.22 for calculating the standard error of an estimated Treynor ratio.

Define a function appraisal that may be used to compute the appraisal ratio of a stock:

```
> appraisal<-function(rmat, ind){
+   ret<-rmat[ind, 1]
+   mkt<-rmat[ind, 2]
+   mm<-lm(ret~mkt)
+   alpha<-mm$coefficients[1]
+   sighat<-summary(mm)$sigma
+   alpha/sighat
+ }
```

Then the standard error of the estimated appraisal ratio for Apple may be calculated using

```
> library(boot)
> boot(cbind(aapl, sp500), appraisal, 10000)
ORDINAR{Y} NONPARAMETRIC BOOTSTRAP


Bootstrap Statistics :
    original  bias    std. error
t1*    0.233 0.00571       0.141
```

Thus, an approximate 95% confidence interval for the true appraisal ratio of Apple stock is

$$0.233 \pm 1.96(0.141) = (-0.043, 0.509).$$

Hence, although there is some evidence to suggest that increasing the investment in Apple stock leads to a portfolio with a larger Sharpe ratio, a formal test of the hypothesis that the true appraisal ratio is zero does not reject the null hypothesis at the 5% level.

A similar approach may be used to calculate a standard error for the estimated weight given to Apple stock in the Treynor–Black portfolio. The standard error based on a bootstrap sample size of 10,000 was calculated to be 6.84. An extremely large value such as this should be interpreted as an indication that there is large variability in bootstrap replications of the Treynor–Black weight.

In order to investigate this variability, the vector of bootstrap estimates of the Treynor–Black weight may be saved to a variable using

```
> aapl.boot<-boot(cbind(aapl, sp500), tb.wgt, 10000)
```

Here `tb.wgt` is a user-defined function, similar to `appraisal` defined earlier, that calculates the Treynor–Black weight. The bootstrap replications of the statistic specified in the `boot` function are stored in the component `$t` of the result, in this case `aapl.boot$t`.

The sample quantiles of the 10,000 values in `aapl.boot$t` may be calculated using

```
> quantile(aapl.boot$t, prob=c(0.01, 0.05, 0.10, 0.50, 0.90,
+ 0.95, 0.99))
      1%        5%       10%       50%       90%       95%       99%
-0.54860   0.00923   0.09006   0.42768   1.23355   1.77756   5.23613
```

These results show that there is considerable variability in these values, suggesting that the Treynor–Black weight is not accurately estimated, at least with the amount of data considered here.

It is worth noting that this same high degree of variability is observed if other assets are analyzed in place of Apple stock. Also, recall that in

Example 6.20, which considered a Monte Carlo study of the properties of estimated weights for the tangency portfolio, this same large variability was observed. Thus, given that the Treynor–Black portfolio is a type of tangency portfolio, the results obtained here are not surprising. □

The Treynor–Black Portfolio of N Assets Together with the Market Portfolio

We now consider the case in which there are N assets available for investment, together with the market portfolio, which plays the role of the $(N+1)$st asset.

Assume that the single-index model holds for all $N+1$ assets. Because the $(N+1)$st asset is the market index, $\alpha_{N+1} = 0$, $\beta_{N+1} = 1$, and $\sigma^2_{\epsilon,N+1} = 0$. We then construct a tangency portfolio from these $N+1$ assets. Therefore, instead of deriving the tangency portfolio weights for the N assets, as in Proposition 9.2, we start with the market index as one asset and derive an expression for how the market index must be modified when forming the tangency portfolio.

Note that the set of N assets need not contain all the assets in the market. In fact, if the set of N assets includes all assets in the market, then the single-index model cannot hold. To see this, note that in this case, the return on the market portfolio at time t would be of the form $\sum_{j=1}^{N} w_{m,j} R_{j,t}$ for some market portfolio weights $w_{m,1}, w_{m,2}, \ldots, w_{m,N}$; hence, the error term in the market model for the market portfolio is of the form $\sum_{j=1}^{N} w_{m,j}\epsilon_{j,t}$. Because the error term in the market model for the market portfolio must be zero, the variance of $\sum_{j=1}^{N} w_{m,j}\epsilon_{j,t}$ must be zero, contradicting the single-index-model assumption of uncorrelated error terms. Hence, we assume that the set of N assets is a subset of the assets in the market, for which the single-index model holds.

Let $\boldsymbol{\Sigma}_+$ denote the covariance matrix of the returns on the $N+1$ assets. Then $\boldsymbol{\Sigma}_+$ may be written as a partitioned matrix,

$$\boldsymbol{\Sigma}_+ = \begin{pmatrix} \sigma_m^2 \boldsymbol{\beta}\boldsymbol{\beta}^T + \boldsymbol{\Sigma}_\epsilon & \sigma_m^2 \boldsymbol{\beta} \\ \sigma_m^2 \boldsymbol{\beta}^T & \sigma_m^2 \end{pmatrix}. \tag{9.21}$$

Similarly, we may write the mean vector of the $N+1$ assets as

$$\boldsymbol{\mu}_+ = \begin{pmatrix} \boldsymbol{\mu} \\ \mu_m \end{pmatrix},$$

where

$$\boldsymbol{\mu} = \begin{pmatrix} \mu_1 \\ \mu_2 \\ \vdots \\ \mu_N \end{pmatrix}$$

denotes the mean-vector of the original N assets, and μ_m is the expected return on the market index.

According to Proposition 5.7, the weight vector of the tangency portfolio of the $N+1$ assets is proportional to

$$\boldsymbol{\Sigma}_+^{-1}(\boldsymbol{\mu}_+ - \mu_f \mathbf{1}_{N+1}). \tag{9.22}$$

The following lemma gives a simple expression for $\boldsymbol{\Sigma}_+^{-1}$; it is based on the well-known formula for the inverse of a partitioned matrix.

Lemma 9.3. *Define* $\boldsymbol{\Sigma}_+$ *by* (9.21). *Then*

$$\boldsymbol{\Sigma}_+^{-1} = \begin{pmatrix} \boldsymbol{\Sigma}_\epsilon^{-1} & -\boldsymbol{\Sigma}_\epsilon^{-1}\boldsymbol{\beta} \\ -\boldsymbol{\beta}^T\boldsymbol{\Sigma}_\epsilon^{-1} & 1/\sigma_m^2 + \boldsymbol{\beta}^T\boldsymbol{\Sigma}_\epsilon^{-1}\boldsymbol{\beta} \end{pmatrix}.$$

Proof. The result may be established by showing that

$$\boldsymbol{\Sigma}_+^{-1}\boldsymbol{\Sigma}_+ = \boldsymbol{\Sigma}_+\boldsymbol{\Sigma}_+^{-1} = \boldsymbol{I}_{N+1}.$$

Here, we consider $\boldsymbol{\Sigma}_+^{-1}\boldsymbol{\Sigma}_+$; the analysis of $\boldsymbol{\Sigma}_+\boldsymbol{\Sigma}_+^{-1}$ follows along similar lines.

Write

$$\boldsymbol{\Sigma}_+^{-1}\boldsymbol{\Sigma}_+ = \begin{pmatrix} \boldsymbol{A} & \boldsymbol{B} \\ \boldsymbol{B}^T & C \end{pmatrix}$$

for an $N \times N$ matrix $\boldsymbol{A}$, an $N \times 1$ matrix $\boldsymbol{B}$, and a scalar C. Then $\boldsymbol{\Sigma}_+^{-1}\boldsymbol{\Sigma}_+ = \boldsymbol{I}_{N+1}$ provided that $\boldsymbol{A} = \boldsymbol{I}_N$, $\boldsymbol{B} = \mathbf{0}$, and $C = 1$.

Using the expression for $\boldsymbol{\Sigma}_+^{-1}$ given in the statement of the lemma and the expression for $\boldsymbol{\Sigma}_+$ given in (9.21), it follows that

$$\boldsymbol{A} = \boldsymbol{\Sigma}_\epsilon^{-1}(\boldsymbol{\Sigma}_\epsilon + \sigma_m^2\boldsymbol{\beta}\boldsymbol{\beta}^T) - \sigma_m^2\boldsymbol{\Sigma}_\epsilon^{-1}\boldsymbol{\beta}\boldsymbol{\beta}^T = \boldsymbol{I}_N,$$

$$\boldsymbol{B} = \sigma_m^2\boldsymbol{\Sigma}_\epsilon^{-1}\boldsymbol{\beta} - \sigma_m^2\boldsymbol{\Sigma}_\epsilon^{-1}\boldsymbol{\beta} = \mathbf{0},$$

and

$$C = -\sigma_m^2\boldsymbol{\beta}^T\boldsymbol{\Sigma}_\epsilon^{-1}\boldsymbol{\beta} + (1 - \sigma_m^2\boldsymbol{\beta}^T(\boldsymbol{\Sigma}_\epsilon + \sigma_m^2\boldsymbol{\beta}\boldsymbol{\beta}^T)^{-1}\boldsymbol{\beta})^{-1} = 1,$$

verifying that $\boldsymbol{\Sigma}_+^{-1}\boldsymbol{\Sigma}_+ = \boldsymbol{I}_{N+1}$. □

We can now use the result in Lemma 9.3, along with the expression for the weights of the tangency portfolio given in (9.22), to derive the optimal modification to the market portfolio.

Proposition 9.3. *Consider a set of N assets with return vector $\boldsymbol{R}_t$ at time t and suppose that $\boldsymbol{R}_t$ follows the single-index model.*

Let $(\boldsymbol{w}_0, w_m)^T$ denote the weight vector of the tangency portfolio of these N assets, together with the portfolio corresponding to the market index; here $\boldsymbol{w}_0 = (w_{0,1}, \ldots, w_{0,N})^T$ is the $N \times 1$ weight vector for the N assets and w_m is the weight for the market index.

Then

$$w_{0,j} = \frac{1}{c}\frac{\alpha_j}{\sigma_{\epsilon,j}^2}, \quad j = 1, 2, \ldots, N$$

and

$$w_m = \frac{1}{c}\left(\frac{\mu_m - \mu_f}{\sigma_m^2} - \sum_{j=1}^{N} \alpha_j \beta_j / \sigma_{\epsilon,j}^2\right)$$

where c is chosen so that

$$\sum_{j=1}^{N} w_{0,j} + w_m = 1.$$

Proof. The expression for the weights of the tangency portfolio of the $N+1$ assets is given in (9.22). Write

$$\Sigma_+^{-1}(\boldsymbol{\mu}_+ - \mu_f \mathbf{1}_{N+1}) = c\begin{pmatrix} \boldsymbol{w}_0 \\ w_m \end{pmatrix}$$

where c is a constant, chosen so that the weights sum to 1.

Using the result in Lemma 9.3, along with the fact that

$$\boldsymbol{\mu}_+ - \mu_f \mathbf{1}_{N+1} = \begin{pmatrix} \boldsymbol{\mu} - \mu_f \mathbf{1} \\ \mu_m - \mu_f \end{pmatrix},$$

it follows that

$$\begin{aligned} c\boldsymbol{w}_0 &= \boldsymbol{\Sigma}_\epsilon^{-1}(\boldsymbol{\mu} - \mu_f \mathbf{1}) - \boldsymbol{\Sigma}_\epsilon^{-1}\boldsymbol{\beta}(\mu_m - \mu_f) \\ &= \boldsymbol{\Sigma}_\epsilon^{-1}(\boldsymbol{\mu} - \mu_f \mathbf{1} - \boldsymbol{\beta}(\mu_m - \mu_f)). \end{aligned}$$

Note that

$$\boldsymbol{\alpha} = \begin{pmatrix} \alpha_1 \\ \alpha_2 \\ \vdots \\ \alpha_N \end{pmatrix} = \boldsymbol{\mu} - \mu_f \mathbf{1} - \boldsymbol{\beta}(\mu_m - \mu_f)$$

so that

$$\boldsymbol{w}_0 = \frac{1}{c}\boldsymbol{\Sigma}_\epsilon^{-1}\boldsymbol{\alpha} = \frac{1}{c}\begin{pmatrix} \alpha_1/\sigma_{\epsilon,1}^2 \\ \alpha_2/\sigma_{\epsilon,2}^2 \\ \vdots \\ \alpha_n/\sigma_{\epsilon,N}^2 \end{pmatrix}.$$

The weight for the market index is given by

$$\begin{aligned} cw_m &= -\boldsymbol{\beta}^T\boldsymbol{\Sigma}_\epsilon^{-1}(\boldsymbol{\mu} - \mu_f \mathbf{1}) + \frac{1 + \sigma_m^2 \boldsymbol{\beta}^T \boldsymbol{\Sigma}_\epsilon^{-1}\boldsymbol{\beta}}{\sigma_m^2}(\mu_m - \mu_f) \\ &= \frac{\mu_m - \mu_f}{\sigma_m^2} - \boldsymbol{\beta}^T\boldsymbol{\Sigma}_\epsilon^{-1}(\boldsymbol{\mu} - \mu_f \mathbf{1} - \boldsymbol{\beta}(\mu_m - \mu_f)) \\ &= \frac{\mu_m - \mu_f}{\sigma_m^2} - \boldsymbol{\beta}^T\boldsymbol{\Sigma}_\epsilon^{-1}\boldsymbol{\alpha} = \frac{\mu_m - \mu_f}{\sigma_m^2} - \sum_{j=1}^{N} \alpha_j\beta_j/\sigma_{\epsilon,j}^2 \end{aligned}$$

as stated in the proposition. □

As in the case of a single asset, we will refer to the portfolio defined in Proposition 9.3 as the *Treynor–Black portfolio.*

Note that if $\alpha_1 = \alpha_2 = \cdots = \alpha_N = 0$, in agreement with the CAPM, then $\boldsymbol{w}_0 = \boldsymbol{0}$; hence, the Treynor–Black portfolio reduces to the market portfolio. If not all α_j are zero, then we may improve on the market portfolio by including more or less of some assets.

We may describe the Treynor–Black portfolio in terms of a portfolio constructed from the N assets combined with the market portfolio. Define a weight vector $\bar{\boldsymbol{w}}_0 = (\bar{w}_{0,1}, \bar{w}_{0,2}, \ldots, \bar{w}_{0,N})^T$ by

$$\bar{w}_{0,j} = \frac{\alpha_j/\sigma_{\epsilon,j}^2}{\sum_{j=1}^N \alpha_j/\sigma_{\epsilon,j}^2}, \quad j = 1, 2, \ldots, N,$$

assuming that $\sum_{j=1}^N \alpha_j/\sigma_{\epsilon,j}^2 \neq 0$. Note that $\sum_{j=1}^N \bar{w}_{0,j} = 1$.

To form the Treynor–Black portfolio, the market index is given weight

$$w_m^* = \frac{(\mu_m - \mu_f)/\sigma_m^2 - \sum_{j=1}^N \alpha_j\beta_j/\sigma_{\epsilon,j}^2}{\sum_{j=1}^N \alpha_j/\sigma_{\epsilon,j}^2 + (\mu_m - \mu_f)/\sigma_m^2 - \sum_{j=1}^N \alpha_j\beta_j/\sigma_{\epsilon,j}^2} \tag{9.23}$$

and the portfolio with weight vector $\bar{\boldsymbol{w}}_0$ is given weight $1 - w_m^*$.

Example 9.10 Consider the stocks of the five companies listed in Example 9.2 and analyzed in several examples in this chapter. Recall that the estimates of α_j for these stocks are stored in the variable `stks.alpha` and estimates of $\sigma_{\epsilon,j}$ are stored in the variable `stks.s`. The weights $\bar{w}_{0,j}$, $j = 1, 2, 3, 4, 5$ are calculated as follows.

```
> wbar0<-(stks.alpha/stks.s^2)/sum(stks.alpha/stks.s^2)
> wbar0
    CVC      EIX     EXPE      HUM      WMT
-0.0012  0.3587  0.0796  0.2752  0.2877
```

These results suggest that the best combination of the five stocks to combine with the market index consists primarily of Expedia, Humana, and Wal-Mart stocks, in roughly equal weights. The other stocks have weights that are relatively small in magnitude.

The weight given to the market index using this approach is given by

```
> c1<-mean(sp500)/var(sp500) - sum(stks.alpha*stks.beta/stks.s^2)
> wm-c1/(c1 + sum(stks.alpha/stks.s2))
> wm
      V1
V1 0.0781
```

with the remainder, $1 - 0.0781 = 0.922$, invested in the portfolio of the five stocks, with the weights given in the variable `wbar0` calculated earlier.

Alternatively, the Treynor–Black portfolio may be described in terms of the weight vector for the six assets (the five stocks plus the market portfolio); such a vector is given by

```
> c((1-wm)*wbar0, wm)
[1] -0.0011  0.3307  0.0734  0.2537  0.2652  0.0781
```

□

Properties of the Treynor–Black Portfolio

Proposition 9.4. *Consider the framework of Proposition 9.3. Let μ_{TB} and σ_{TB} denote the mean and standard deviation, respectively, of the return on the Treynor–Black portfolio, and let β_{TB} denote the value of beta in the market model for the Treynor–Black portfolio. Then*

$$\mu_{TB} - \mu_f = \frac{1}{c}\left(\frac{(\mu_m - \mu_f)^2}{\sigma_m^2} + \sum_{j=1}^{N} \frac{\alpha_j^2}{\sigma_{\epsilon,j}^2}\right),$$

$$\sigma_{TB}^2 = \frac{1}{c^2}\left(\frac{(\mu_m - \mu_f)^2}{\sigma_m^2} + \sum_{j=1}^{N} \frac{\alpha_j^2}{\sigma_{\epsilon,j}^2}\right),$$

and

$$\beta_{TB} = \frac{1}{c}\frac{\mu_m - \mu_f}{\sigma_m^2}.$$

Here c is as defined in the statement of Proposition 9.3:

$$c = \sum_{j=1}^{N} \alpha_j/\sigma_{\epsilon,j}^2 + \frac{\mu_m - \mu_f}{\sigma_m^2} - \sum_{j=1}^{N} \alpha_j \beta_j/\sigma_{\epsilon,j}^2.$$

Proof. For $j = 1, 2, \ldots, N$, let $\mu_j = \mathrm{E}(R_{j,t})$. Then, using the expression for the portfolio weights given in Proposition 9.3,

$$\begin{aligned}
\mu_{\mathrm{TB}} - \mu_f &= \frac{1}{c}\sum_{j=1}^{N} \frac{\alpha_j}{\sigma_{\epsilon,j}^2}(\mu_j - \mu_f) + \frac{1}{c}\frac{\mu_m - \mu_f}{\sigma_m^2}(\mu_m - \mu_f) + \frac{1}{c}\sum_{j=1}^{N} \frac{\alpha_j\beta_j}{\sigma_{\epsilon,j}^2}(\mu_m - \mu_f) \\
&= \frac{1}{c}\frac{(\mu_m - \mu_f)^2}{\sigma_m^2} + \frac{1}{c}\sum_{j=1}^{N} \frac{\alpha_j}{\sigma_{\epsilon,j}^2}(\mu_j - \mu_f - \beta_j(\mu_m - \mu_f)) \\
&= \frac{1}{c}\left(\frac{(\mu_m - \mu_f)^2}{\sigma_m^2} + \sum_{j=1}^{N} \frac{\alpha_j^2}{\sigma_{\epsilon,j}^2}\right),
\end{aligned}$$

as given in the statement of the proposition.

Note that because the market index has a beta of 1,

$$\beta_{\text{TB}} = \frac{1}{c}\left(\frac{\mu_m - \mu_f}{\sigma_m^2} - \sum_{j=1}^{N} \frac{\alpha_j \beta_j}{\sigma_{\epsilon,j}^2}\right) + \frac{1}{c}\sum_{j=1}^{N} \frac{\alpha_j}{\sigma_{\epsilon,j}^2}\beta_j = \frac{1}{c}\frac{\mu_m - \mu_f}{\sigma_m^2}.$$

Now consider σ_{TB}^2. It is convenient to consider separately the market and nonmarket components of the return variance. The market index has zero nonmarket variance; therefore, under the single-index model, the nonmarket component of σ_{TB}^2 is given by

$$\frac{1}{c^2}\sum_{j=1}^{N}\left(\frac{\alpha_j}{\sigma_{\epsilon,j}^2}\right)^2 \sigma_{\epsilon,j}^2 = \frac{1}{c^2}\left(\sum_{j=1}^{N}\frac{\alpha_j^2}{\sigma_{\epsilon,j}^2}\right) \tag{9.24}$$

and the market component of σ_{TB}^2 is $\beta_{\text{TB}}^2\sigma_m^2$, where β_{TB} is the value of beta for the Treynor–Black portfolio.

Using the expression for β_{TB} derived previously, the market component of σ_{TB}^2 is given by

$$\frac{1}{c^2}\frac{(\mu_m - \mu_f)^2}{\sigma_m^2}. \tag{9.25}$$

Adding (9.24) and (9.25) shows that

$$\sigma_{\text{TB}}^2 = \frac{1}{c^2}\left(\frac{(\mu_m - \mu_f)^2}{\sigma_m^2} + \sum_{j=1}^{N}\frac{\alpha_j^2}{\sigma_{\epsilon,j}^2}\right). \qquad \square$$

Let $\text{SR}_{\text{TB}} = (\mu_{\text{TB}} - \mu_f)/\sigma_{\text{TB}}$ denote the Sharpe ratio of the Treynor–Black portfolio. By construction, it is at least as large as the Sharpe ratio of the market index; an expression for the difference in the squared Sharpe ratios is given in the following corollary to Proposition 9.4.

Corollary 9.2.

$$(\text{SR}_{\text{TB}})^2 - \frac{(\mu_m - \mu_f)^2}{\sigma_m^2} = \sum_{j=1}^{N}\frac{\alpha_j^2}{\sigma_{\epsilon,j}^2}. \tag{9.26}$$

Example 9.11 The Sharpe ratio for the market index corresponding to the S&P 500 index is given by

```
> mean(sp500)/sd(sp500)
[1] 0.290
```

For the stocks represented in the data matrix `stks`, the estimated difference between the squared Sharpe ratio of the Treynor–Black portfolio described in Example 9.10 and the squared Sharpe ratio of the market portfolio is given by

```
> sum((stks.alpha^2)/stks.s^2)
[1] 0.125
```

Thus, the estimated Sharpe ratio of the Treynor–Black portfolio is

$$\left((0.290)^2 + 0.125\right)^{\frac{1}{2}} = 0.457. \qquad \square$$

Bias in the Estimator of $\sum_{j=1}^{N} \alpha_j^2/\sigma_{\epsilon,j}^2$

The quantity $\sum_{j=1}^{N} \alpha_j^2/\sigma_{\epsilon,j}^2$ measures the difference between the squared Sharpe ratio of the Treynor–Black portfolio and the squared Sharpe ratio of the market portfolio; hence, it gives a measure of the possible improvement in the market portfolio by combining it with a portfolio of the assets under consideration. Of course, in practice, $\sum_{j=1}^{N} \alpha_j^2/\sigma_{\epsilon,j}^2$ must be estimated using parameter estimators from the market models for the assets under consideration.

However, such an estimator tends to overestimate the true value of $\sum_{j=1}^{N} \alpha_j^2/\sigma_{\epsilon,j}^2$, giving an overly optimistic assessment of the benefit from modifying the market portfolio. The reason for this is that the estimator $\sum_{j=1}^{N} \hat{\alpha}_j^2/\hat{\sigma}_{\epsilon,j}^2$ is a sum of squared random variables $\hat{\alpha}_j/\hat{\sigma}_{\epsilon,j}$, $j = 1, 2, \ldots, N$. Recall that, for a random variable X, $\mathrm{E}(X^2) = \mathrm{E}(X)^2 + \mathrm{Var}(X)$. Therefore, even if each $\hat{\alpha}_j/\hat{\sigma}_{\epsilon,j}$ is an unbiased estimator of $\alpha_j/\sigma_{\epsilon,j}$, so that

$$\mathrm{E}\left(\hat{\alpha}_j/\hat{\sigma}_{\epsilon,j}\right) = \alpha_j/\sigma_{\epsilon,j},$$

it follows that

$$\begin{aligned}
\mathrm{E}\left(\sum_{j=1}^{N} \hat{\alpha}_j^2/\hat{\sigma}_{\epsilon,j}^2\right) &= \sum_{j=1}^{N} \mathrm{E}\left(\hat{\alpha}_j^2/\hat{\sigma}_{\epsilon,j}^2\right) \\
&= \sum_{j=1}^{N} \left(\mathrm{E}\left(\hat{\alpha}_j/\hat{\sigma}_{\epsilon,j}\right)^2 + \mathrm{Var}\left(\hat{\alpha}_j/\hat{\sigma}_{\epsilon,j}\right)\right) \\
&= \sum_{j=1}^{N} \alpha_j^2/\sigma_{\epsilon,j}^2 + \sum_{j=1}^{N} \mathrm{Var}\left(\hat{\alpha}_j/\hat{\sigma}_{\epsilon,j}\right) \\
&> \sum_{j=1}^{N} \alpha_j^2/\sigma_{\epsilon,j}^2.
\end{aligned}$$

One way to correct for this bias is to use the bootstrap method, as we did in Section 8.9 when estimating measures of portfolio performance. This is illustrated in the following example.

Example 9.12 Consider estimating the bias in $\sum_{j=1}^{N} \hat{\alpha}_j^2/\hat{\sigma}_{\epsilon,j}^2$ as an estimator of $\sum_{j=1}^{N} \alpha_j^2/\sigma_{\epsilon,j}^2$ using the function `boot` in the package `boot`. Recall that a function to be used in `boot` must take two arguments: the data, in the form of a vector or matrix, and the indices of the data values to be used in the computation. Define the function shrp.sq.diff

```
> shrp.sq.diff<-function(x, ind){
+ stks<-x[ind, -1*ncol(x)]
+ mkt<-x[ind, ncol(x)]
+ mm<-lm(stks~mkt)
+ shrp.m<-mean(mkt)/sd(mkt)
+ alpha<-mm$coefficients[1,]
+ beta<-mm$coefficients[2, ]
+ s.hat<-(apply(mm$residuals^2, 2, sum)/mm$df.residual)^.5
+ sum((alpha^2)/(s.hat^2))
+ }
```

This function assumes that the argument x is a matrix of returns, with the last column corresponding to the market portfolio and the remaining columns corresponding to the assets to be used in forming the Treynor–Black portfolio. Calculation of s.hat, the vector of estimates of $\sigma_{\epsilon,j}$, is carried out using the component $residuals of the output from the function lm; this component consists of a matrix of residuals corresponding to the different response variables used in the lm function, in this case, the returns on the different assets. Adding the squares of these residuals for each asset and dividing by the degrees of freedom, yields estimates of $\sigma^2_{\epsilon,j}$.

Therefore, shrp.sq.diff takes the values in x corresponding to the indices in the vector ind and uses those values to compute the difference in the estimated Sharpe ratios of the Treynor–Black and market portfolios. For example, taking x = cbind(stks, sp500) and ind = 1:60 returns the difference of the estimated squared Sharpe ratios calculated in Example 9.11

```
> shrp.sq.diff(cbind(stks, sp500), 1:60)
[1] 0.125
```

that agrees with the value obtained earlier.

The estimated bias of the estimator of $\sum_{j=1}^{N} \alpha_j^2/\sigma^2_{\epsilon,j}$ can now be obtained using the function boot.

```
> boot(cbind(stks, sp500), shrp.sq.diff, 10000)
ORDINAR{Y} NONPARAMETRIC BOOTSTRAP

Bootstrap Statistics :
     original  bias    std. error
t1*     0.125   0.111        0.129
```

Therefore, the estimated bias is 0.111 and the bias-corrected estimate is only $0.125 - 0.111 = 0.004$, considerably smaller than the original estimate.

The output from the boot also gives the standard error of the estimate. Although this value is useful for getting a rough idea of the variability in the estimates, in this case, it is not useful for constructing an approximate confidence interval for the true difference value of $\sum_{j=1}^{N} \alpha_j^2/\sigma_{\epsilon,j}^2$.

Note that the parameter $\sum_{j=1}^{N} \alpha_j^2/\sigma_{\epsilon,j}^2$ is nonnegative and the corresponding estimator is a nonnegative random variable. Hence, if the true value of $\sum_{j=1}^{N} \alpha_j^2/\sigma_{\epsilon,j}^2$ is close to zero, then the distribution of the estimator is not well approximated by a normal distribution; therefore, an approximate confidence interval of the form of the estimate plus or minus 1.96 times the standard error will not have the usual coverage property. One way to see that in the present example is to note that such an interval will include negative values, which are impossible. However, given that the standard error is 0.129 and the bias-corrected estimate is 0.004, it is clear that there is little evidence that the Treynor–Black portfolio has a larger Sharpe ratio than does the portfolio based on the market index. □

Numerical Computation of the Treynor–Black Portfolio Weights

In Sections 5.7 and 5.8, it was shown that the weights of the tangency portfolio may be found using quadratic programming. Given that the Treynor–Black portfolio is a type of tangency portfolio, it is clear that the same approach can be used to calculate the weights of the Treynor–Black portfolio. One advantage of calculating the weights in this way is that it is a simple matter to impose certain types of constraints on the weights; see Section 5.8 for a discussion of the type of constraints that may be handled in this manner. The following example illustrates the calculation when the weights are constrained to be nonnegative.

Example 9.13 Consider the stocks of the five companies listed in Example 9.2; the portfolio weights of the Treynor–Black portfolio were calculated in Example 9.10; here we repeat this calculation using the function solve.QP in the package quadprog.

Recall that the covariance matrix for these stocks based on the single-index model is stored in the variable stks.Sig; to calculate the portfolio weights, we need to extend this matrix to a covariance matrix for the returns on the five stocks, along with the returns on the market portfolio, in this case taken to be the returns on the S&P 500 index. Note that the covariance of the returns on a stock and the returns on the market portfolio is given by $\beta_i^2\sigma_m^2$, where β_i is the value of beta for the stock and σ_m^2 is the variance of the return on the market portfolio. Here, the values of beta for the five stocks are stored in the variable stks.beta. Hence, the extended covariance matrix is given by

```
> stks.Sig.plus<-rbind(cbind(stks.Sig, stks.beta*var(sp500)),
+ c( (stks.beta)*var(sp500), var(sp500)))
> stks.Sig.plus
```

```
           CVC     EIX     EXPE    HUM     WMT     SP500
CVC    0.00839 0.00072 0.00135 0.00099 0.00069 0.00151
EIX    0.00072 0.00241 0.00060 0.00044 0.00031 0.00067
EXPE   0.00135 0.00060 0.01226 0.00082 0.00057 0.00126
HUM    0.00099 0.00044 0.00082 0.00543 0.00042 0.00092
WMT    0.00069 0.00031 0.00057 0.00042 0.00198 0.00064
SP500  0.00151 0.00067 0.00126 0.00092 0.00064 0.00141
```

Recall that to compute the tangency portfolio weights numerically, we minimize the variance of the portfolio, subject to the constraint that the portfolio mean is one; the resulting weights are then normalized to sum to 1. Hence, let `A1` denote the 6×1 matrix of asset means, which may be calculated by

```
A1<-as.matrix(c(apply(stks, 2, mean), mean(sp500)))
```

Then the weights of the Treynor–Black portfolio may be calculated using

```
> w_tb1<-solve.QP(Dmat=stks.Sig.plus, dvec=rep(0, 6), Amat=A1,
+  bvec=c(1), meq=1)$solution
> w_tb<-w_tb1/sum(w_tb1)
> w_tb
[1] -0.0011  0.3307  0.0734  0.2537  0.2653  0.0781
```

Note that these weights match those calculated in Example 9.10.

To include the condition that all weights are nonnegative, we modify the arguments `Amat` and `bvec`.

```
> w_tb2<-solve.QP(Dmat=stks.Sig.plus, dvec=rep(0, 6),
+ Amat=cbind(A1,diag(6)),
+ bvec=c(1, rep(0, 6)), meq=1)$solution
> w_tb_nonneg<-w_tb2/sum(w_tb2)
> w_tb_nonneg
[1] 0.0000 0.3307 0.0734 0.2537 0.2653 0.0769
```

As expected, the constrained weights are very similar to the unconstrained weights, with the weight given to CVC set to zero and slight modifications to the other weights. □

9.7 Suggestions for Further Reading

The single-index model is one of the most commonly used statistical models in finance; it is also the starting point for many of the more sophisticated models used in financial modeling. Good detailed discussions of the single-index model are available from Francis and Kim (2013, Chapter 8) and Elton et al. (2007, Chapter 7); see also Sharpe (1963). Partial correlation is discussed by Agresti and Finlay (2009, Section 11.7). A number of useful results on matrix inverses, including the one used in Lemma 9.1, are available from Henderson and Searle (1981); the inverses of partitioned matrices are discussed by Lu and Shiou (2002).

Active portfolio management uses a variety of methods in an attempt to outperform the benchmark portfolio; see Grinold and Kahn (2000) and Chincarini and Kim (2006) for book-length treatments of this area. The Treynor–Black method is attributed to Treynor and Black (1973); see also Francis and Kim (2013, Section 17.2) and Kane et al. (2012). Optimal active portfolios based on the properties of their residual returns are considered by Grinold and Kahn (2000, Chapter 5); see Qian et al. (2007, Section 2.2.4) for an alternative approach.

Given that active portfolio management relies on the inefficiency of the market portfolio, it is not surprising that some analysts are skeptical of its benefits; see, for example, Samuelson (1974) and Sharpe (1991).

9.8 Exercises

1. Let $\boldsymbol{\beta}$ denote a vector in $\Re^N$, let $\boldsymbol{\Sigma}_\epsilon$ denote an $N \times N$ matrix, and suppose $\sigma_m^2 > 0$. Show that

$$\sigma_m^2 \boldsymbol{\beta}\boldsymbol{\beta}^T + \boldsymbol{\Sigma}_\epsilon \tag{9.27}$$

is a positive-definite matrix provided that $\boldsymbol{\Sigma}_\epsilon$ is positive definite.

Suppose that $\boldsymbol{\Sigma}_\epsilon$ is nonnegative definite but not positive definite. Is it possible that (9.27) is positive definite? Why or why not?

2. Consider a set of four assets with betas given by 1.1, 0.9, 0.4, 1.5, respectively, and error standard deviations $\sigma_{\epsilon,i}$ given by 0.2, 0.4, 0.5, 0.8 respectively; suppose that $\sigma_m = 0.1$. Find the covariance matrix of the return vector for these assets under the assumption that the single-index model holds and give the corresponding correlation matrix.

3. Calculate the monthly excess returns for the period ending December 31, 2015, for three stocks, Papa John's International, Inc. (symbol PZZA), Bed Bath & Beyond, Inc. (BBBY), and Time Warner, Inc. (TWX); for the risk-free rate, use the return on the three-month Treasury Bill, available on the Federal Reserve website.

For each pair of two stocks, calculate the partial correlation coefficient of the stock excess returns given the excess return on the S&P 500 index. Based on these results, does it appear that the single-index model holds for the returns on these three stocks? Why or why not?

4. Consider five mutual funds, representing five different industries: Fidelity Select Semiconductors Portfolio (symbol FSELX),

Fidelity Select Energy Portfolio (FSENX), Fidelity Select Health Care Services Portfolio (FSHCX), Fidelity Real Estate Investment Portfolio (FRESX), and Fidelity Select Transportation Portfolio (FSRFX). Mutual funds like these, which focus on a single industry or *sector* of the market, are sometimes called *sector funds*.

For each fund, calculate the monthly excess returns for the period ending December 31, 2015; for the risk-free rate, use the return on the three-month Treasury Bill, available on the Federal Reserve website.

For each pair of two funds, calculate the partial correlation coefficient of stock excess returns given the excess return on the S&P 500 index. Based on these results, does it appear that the single-index model holds for the returns on these assets? Why or why not?

5. Consider the three stocks analyzed in Exercise 3. Assume that the single-index model holds for these stocks and estimate the vector of betas, $\boldsymbol{\beta}$, and the error covariance matrix, $\boldsymbol{\Sigma}_\epsilon$. Take the market index to be the S&P 500 index.

   Based on these estimates, together with an estimate of σ_m^2, estimate the stock return correlation matrix under the single-index model assumption and compare it to the sample correlation matrix.

6. Consider the five mutual funds analyzed in Exercise 4. Assume that the single-index model holds for these funds and estimate the vector of betas, $\boldsymbol{\beta}$, and the error covariance matrix, $\boldsymbol{\Sigma}_\epsilon$. Take the market index to be the S&P 500 index.

   Based on these estimates, together with an estimate of σ_m^2, estimate the stock return correlation matrix under the single-index model assumption and compare it to the sample correlation matrix.

7. Consider the three stocks analyzed in Exercise 3. Using the function `solve.QP` in the `quadprog` package, calculate the weight vector of the risk-averse portfolio with risk-aversion parameter λ=5, first using the sample covariance matrix and then using the estimate of the covariance matrix based on the single-index model. Compare the results.

   Repeat the calculations using the restriction that all asset weights must be nonnegative. Compare the results.

8. For the five mutual funds analyzed in Exercise 4, estimate the weights of the tangency portfolio using the estimate of the return covariance matrix based on the single-index model. Compare the results to the estimate based on the sample covariance matrix.

9. Estimate the appraisal ratios for the five mutual funds analyzed in Exercise 4. Based on these results, if you were to modify your investment in the market portfolio, as represented by the S&P 500 index,

by increasing the weight given to one of these funds and decreasing the weight given to the market portfolio, which one would you choose? In this new portfolio, what weight would you give to the market portfolio?

10. For the five mutual funds analyzed in Exercise 4, estimate the weights of the Treynor–Black portfolio. Estimate the weight for each of the five sector funds along with the weight given to the portfolio corresponding to the S&P 500 index.

11. Using the result in Corollary 9.2, estimate the Sharpe ratio of the Treynor–Black portfolio constructed from the five sector mutual funds in Exercise 10. Compare the result to the estimated Sharpe ratio of the S&P 500 index.

12. Consider the Treynor–Black portfolio for a set of N assets together with the market portfolio. Let α_{TB} denote the value of alpha in the market model for this portfolio. Show that α_{TB}^2 may be written in terms of $\alpha_1^2, \alpha_2^2, \ldots, \alpha_N^2$, the squared values of alpha for the N assets; $\sigma_{\epsilon,1}^2, \sigma_{\epsilon,2}^2, \ldots, \sigma_{\epsilon,N}^2$, the error variances in the market models for the N assets; and $\sigma_{\epsilon,\text{TB}}^2$, the error variance in the market model for the Treynor–Black portfolio.

13. For the five mutual funds analyzed in Exercise 4, estimate the weights of the Treynor–Black portfolio subject to the constraint that the weights are nonnegative. Estimate the Sharpe ratio of the resulting portfolio.

10

Factor Models

10.1 Introduction

Although there are many assets available to an investor, the returns on these assets are often correlated—in some cases, highly correlated. One reason for this correlation is that the returns on a set of assets may all be affected by certain changes in the economy; alternatively, the assets may correspond to firms with similar properties. A *factor model* describes the returns on a set of assets in terms of a few underlying "factors" potentially affecting all of the assets.

We have already seen one example of a factor model, the single-index model discussed in Chapter 9, which describes the returns on a set of assets in terms of the returns on a market index. An important implication of this model is that the covariance between the returns on two assets arises from the fact that both assets' returns are related to the return on the market index. Although it may be reasonable to assume that the behavior of the market as a whole is the most important factor affecting asset returns, empirical research has shown that there are other factors, in addition to the return on the market index, that have important effects on asset returns. These factors are useful for describing the correlation structure of a set of asset returns as well as for describing the behavior of the mean returns of the assets, extending the type of relationship described by the single-index model in Chapter 9.

The goal of this chapter is to present the statistical methodology underlying these factor models along with the implications of these models for understanding the behavior of asset returns and for constructing and analyzing portfolios.

10.2 Limitations of the Single-Index Model

One role of the single-index model is to model the covariance between the returns on two assets in terms of the relationship between each asset's returns on the returns on a market index. However, as noted in the introduction, in some cases, there may be important economic variables, in addition to the return on the market, that have important effects on the relationship between the returns on the assets. This is illustrated in the following example.

Example 10.1 Consider the returns on two stocks, JetBlue Airways Corp. (symbol JBLU) and EV Energy Partners, L.P. (EVEP), an oil and natural gas company. The variables `jblu` and `evep` contain 5 years of monthly excess returns on JBLU and EVEP stock, respectively, for the period ending December 31, 2014, and suppose that `sp500` contains the corresponding excess returns on the Standard & Poors (S&P) 500 index. Then the estimated correlation of the returns on these stocks is given by

```
> cor(jblu, evep)
[1] -0.150
```

The estimated correlations of each return with the return on the S&P 500 index are given by

```
> cor(jblu, sp500)
[1] 0.311
> cor(evep, sp500)
[1] 0.268
```

Therefore, each stock's returns are *positively* correlated with the return on the market index, but the returns are *negatively* correlated with each other. Note that relationships of this type are not possible under the single-index model. The estimates of beta for the two stocks are given by

```
> lm(jblu~sp500)$coef
(Intercept)       sp500
     0.0137      0.8770
> lm(evep~sp500)$coef
(Intercept)       sp500
   -0.00437     0.74013
```

The estimated return variance for the S&P 500 index is

```
> var(sp500)
[1] 0.00141
```

According to the single-index model, the estimated covariance of the returns on JBLU and EVEP stock is

$$(0.877)(0.740)(0.00141) = 0.000915,$$

corresponding to an estimated correlation of

```
> 0.000915/(sd(jblu)*sd(evep))
[1] 0.0832
```

Hence, although the sample correlation of returns on JBLU and EVEP stock is negative (-0.150), the estimated correlation based on the single-index model is positive (0.0832).

One reason for this behavior may be the presence of other economic variables that are affecting the returns on JBLU and EVEP stock. For example, JBLU, as an airline stock, is likely to be negatively affected by increasing oil prices; EVEP, on the other hand, as a gas and oil stock, is likely to be positively affected by increasing oil prices. Hence, oil prices might have an important effect on the relationship between the returns on these two stocks.

One commonly used benchmark for crude oil prices is the price of West Texas Intermediate (WTI) oil, which is generally refined in the United States. Historical prices for WTI oil are available on the Federal Reserve Economic Data (FRED) website at https://research.stlouisfed.org/fred2/series/DCOILWTICO/downloaddata; like stock prices, these data are available for different sampling frequencies, such as daily or monthly prices. Let the variable `oil` denote the proportional change in monthly prices of WTI oil for the 5-year period ending December 31, 2014; thus, `oil` is calculated the same way that asset returns are calculated, except that oil prices, rather than stock prices, are used.

Note that, as expected, the returns on JBLU stock are negatively correlated with the change in oil prices, while the returns on EVEP stock are positively correlated with the change in oil prices:

```
> cor(jblu, oil)
[1] -0.265
> cor(evep, oil)
[1] 0.528
```

Therefore, in modeling the relationship between JBLU and EVEP stock, it may be important to take into account changes in oil prices. This is likely to be true when analyzing the returns on other stocks thought to be related to oil prices. □

A second use of the single-index model is in understanding the role of an asset's relationship with the market index in the expected return on the asset. According to the single-index model or, equivalently, the market model, the expected excess return on asset i, $\mu_i - \mu_f$, is related to the expected excess return on the market index, $\mu_m - \mu_f$, by

$$\mu_i - \mu_f = \alpha_i + \beta_i(\mu_m - \mu_f),$$

where α_i and β_i are the parameters in the market model for asset i.

If the asset is priced correctly, in the sense described in Section 8.4, then $\alpha_i = 0$ and the expected excess return on an asset is proportional to its value of beta; see Section 7.7 for further details. This fact suggests that assets with greater values of beta will tend to have higher expected excess returns. The following example shows that this interpretation of beta is not always useful in practice.

Example 10.2 Consider stocks for firms represented in the S&P 500 index. Five years of monthly returns for the period ending December 31, 2014, were analyzed; 474 of the stocks had returns for that entire period.

For each stock, the parameters of the market model were estimated, along with the sample mean excess return. These results suggest that all 474 stocks are priced correctly; for instance, the minimum p-value for testing $\alpha_i = 0$ is 0.00236 so that, using the Bonferroni method, we fail to reject the hypothesis that all α_i are equal to zero at any level.

The estimates of beta for the 474 stocks are stored in the variable `sp474.mmbeta`. Note that there is considerable variation in the estimates of beta:

```
> summary(sp474.mmbeta)
   Min. 1st Qu.  Median    Mean 3rd Qu.    Max.
  0.004   0.732   1.060   1.080   1.390   2.870
```

Hence, according to the CAPM, we expect that stocks with large estimates of beta will tend to have higher sample mean excess returns.

Figure 10.1 contains a plot of the sample mean excess returns versus the estimated value of beta for the 474 stocks. Note that there is, at most, a very weak relationship between a stock's sample mean excess return and its estimate of beta. Furthermore, this plot does not support the idea that stocks with larger values of beta tend to have large mean excess returns. For instance, the sample correlation of the estimates of beta and the sample mean excess returns based on these data is only 0.0359. The standard error of this estimate when the true correlation is zero is $1/\sqrt{N}$, where N is the number of observations; here $N = 474$. Hence, the standard error of the estimate is 0.0459 and the correlation is not significantly different than zero. The squared sample

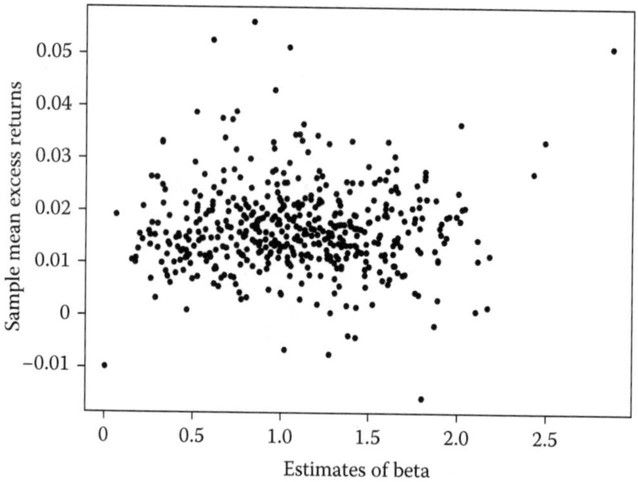

FIGURE 10.1
Plot of sample mean excess returns versus estimates of beta for stocks in the S&P 500 index.

correlation is about 0.0013, so only about 0.13% of the variation in the sample mean excess returns on the stocks may be explained by their estimates of beta. Thus, the theoretical relationship expressed in Figure 7.1 does not hold for these data. □

The results in the previous example suggest that, at least in some cases, the relationship between the returns on an asset and the returns on a market index is not sufficient to effectively describe the mean excess returns on an asset. That is, these results are consistent with the idea that it may be important to include factors other than the return on a market index in a model for asset returns.

10.3 The Model and Its Estimation

The single-index model relates the returns on a given asset, $R_{i,t}$, $t = 1, 2, \ldots, T$, to the returns on a market index, $R_{m,t}$, $t = 1, 2, \ldots, T$, through the model

$$R_{i,t} - R_{f,t} = \alpha_i + \beta_i(R_{m,t} - R_{f,t}) + \epsilon_{i,t}, \quad t = 1, 2, \ldots, T.$$

Here $R_{f,t}$ is the return on the risk-free asset at time t. The idea behind this model is that all assets are related to "the market" and volatility in the market induces volatility in the returns of individual assets. Furthermore, under this model, the correlation between the returns of two assets is a result of the fact that both assets are related to the market.

A general factor model extends the single-index model by including other risk factors, in addition to the market return, in the model. These factors may represent economic conditions that, like the return on a market index, affect all assets. Or the factors might reflect properties of the assets under consideration, such as the size of the company, in the case of a stock. This flexibility in the factors, which may be chosen to represent the analyst's beliefs and goals, is one of the strengths of factor models. In this section, we consider the form and properties of a factor model, along with parameter estimation; selection of the factors is considered in the following section.

Let $F_{1,t}, F_{2,t}, \ldots, F_{K,t}$ denote the values of K factors at time t, $t = 1, 2, \ldots, T$. For $i = 1, 2, \ldots, N$, let $R_{i,t}$ denote the return on asset i at time t. Then a factor model that describes $R_{i,t}$ in terms of $F_{1,t}, F_{2,t}, \ldots, F_{K,t}$ has the form

$$R_{i,t} - R_{f,t} = \alpha_i + \beta_{i,1}F_{1,t} + \beta_{i,2}F_{2,t} + \cdots + \beta_{i,K}F_{K,t} + \epsilon_{i,t}, \quad t = 1, 2, \ldots, T,$$

where $\epsilon_{i,1}, \epsilon_{i,2}, \ldots, \epsilon_{i,T}$ are unobserved mean-zero random variables that are uncorrelated with the factors. These terms represent the component of the asset's excess return not explained by the factors.

Note that the values of the factors $F_{1,t}, F_{2,t}, \ldots, F_{K,t}$ are the same for each asset and, hence, do not depend on i; in this sense, they may be viewed as *common factors.* The parameters $\beta_{i,1}, \beta_{i,2}, \ldots, \beta_{i,K}$, known as the *factor sensitivities* for asset i, measure how the factors affect a particular asset's returns. Hence, these parameters depend on i; however, they are assumed to be constant over the observation period, so that they do not depend on t. In the factor model, the factor sensitivities, like β in the single-index model, are unknown parameters that must be estimated.

It is assumed that the same factor model applies to all assets under consideration so that, for $t = 1, 2, \ldots, T$,

$$
\begin{aligned}
R_{1,t} - R_{f,t} &= \alpha_1 + \beta_{1,1}F_{1,t} + \beta_{1,2}F_{2,t} + \cdots + \beta_{1,K}F_{K,t} + \epsilon_{1,t} \\
R_{2,t} - R_{f,t} &= \alpha_2 + \beta_{2,1}F_{1,t} + \beta_{2,2}F_{2,t} + \cdots + \beta_{2,K}F_{K,t} + \epsilon_{2,t} \\
&\vdots \\
R_{N,t} - R_{f,t} &= \alpha_N + \beta_{N,1}F_{1,t} + \beta_{N,2}F_{2,t} + \cdots + \beta_{N,K}F_{K,t} + \epsilon_{N,t}
\end{aligned}
$$

or, in matrix notation,

$$\boldsymbol{R}_t - R_{f,t}\mathbf{1} = \boldsymbol{\alpha} + \boldsymbol{\beta}\boldsymbol{F}_t + \boldsymbol{\epsilon}_t, \tag{10.1}$$

where $\boldsymbol{\alpha} = (\alpha_1, \ldots, \alpha_N)^T$, $\boldsymbol{\beta}$ is the $N \times K$ matrix of factor sensitivities,

$$\boldsymbol{\beta} = \begin{pmatrix} \beta_{1,1} & \beta_{1,2} & \cdots & \beta_{1,K} \\ \beta_{2,1} & \beta_{2,2} & \cdots & \beta_{2,K} \\ \vdots & \vdots & \cdots & \vdots \\ \beta_{N,1} & \beta_{N,2} & \cdots & \beta_{N,K} \end{pmatrix},$$

$\boldsymbol{F}_t = (F_{1,t}, F_{2,t}, \ldots, F_{K,t})^T$ is the $K \times 1$ vector of factor values at time t, and $\boldsymbol{\epsilon}_t = (\epsilon_{1,t}, \epsilon_{2,t}, \ldots, \epsilon_{N,t})^T$ is an $N \times 1$ random vector of unobserved model errors at time t.

We assume that the stochastic process

$$\{((\boldsymbol{R}_t - R_{f,t}\mathbf{1})^T, \boldsymbol{F}_t^T)^T : t = 1, 2, \ldots\}$$

is a weakly stationary process, so any linear function of $\boldsymbol{R}_t - R_{f,t}\mathbf{1}, \boldsymbol{F}_t$ is a weakly stationary process; in particular, $\boldsymbol{\epsilon}_1, \boldsymbol{\epsilon}_2, \ldots$ is a weakly stationary process. Furthermore, we assume that

$$\text{Cov}(\boldsymbol{\epsilon}_t, \boldsymbol{F}_t) = \mathbf{0}, \quad t = 1, 2, \ldots, T,$$

so that $\boldsymbol{\Sigma}$, the covariance matrix of $\boldsymbol{R}_t$, may be written as

$$\boldsymbol{\Sigma} = \boldsymbol{\beta}\boldsymbol{\Sigma}_F\boldsymbol{\beta}^T + \boldsymbol{\Sigma}_\epsilon, \tag{10.2}$$

where $\mathbf{\Sigma}_F$ denotes the covariance matrix of $\boldsymbol{F}_t$ and $\mathbf{\Sigma}_\epsilon$ denotes the covariance matrix of $\boldsymbol{\epsilon}_t$; by the weak stationarity assumption, these parameters do not depend on t.

As with the single-index model, the errors for different assets are assumed to be uncorrelated so that $\mathbf{\Sigma}_\epsilon$ is a diagonal matrix of the form

$$\mathbf{\Sigma}_\epsilon = \begin{pmatrix} \sigma^2_{\epsilon,1} & 0 & \dots & 0 \\ 0 & \sigma^2_{\epsilon,2} & \ddots & \vdots \\ \vdots & \ddots & \ddots & 0 \\ 0 & \dots & 0 & \sigma^2_{\epsilon,N} \end{pmatrix}$$

where $\sigma^2_{\epsilon,i} = \text{Var}(\epsilon_{i,t})$, $t = 1, 2, \dots, N$. Under this assumption, any correlation among asset returns for different assets is attributable to the common factors that affect all assets.

Thus, the factor model may be viewed as an extension of the single-index model in which the excess return on the market index, $R_{m,t} - R_{f,t}$, is replaced by the factors $F_{1,t}, F_{2,t}, \dots, F_{K,t}$ and the vector $\boldsymbol{\beta} = (\beta_1, \beta_2, \dots, \beta_N)^T$ of the single-index model, which gives the value of beta for each asset, is replaced by the matrix $\boldsymbol{\beta}$. The ith row of $\boldsymbol{\beta}$, $(\beta_{i,1}, \beta_{i,2}, \dots, \beta_{i,K})$, gives the factor sensitivities for asset i; the jth column of $\boldsymbol{\beta}$, $(\beta_{1,j}, \beta_{2,j}, \dots, \beta_{N,j})^T$ gives the sensitivities to factor j for each of the N assets.

Portfolios

Consider a portfolio of the N assets under consideration based on a weight vector $\boldsymbol{w} = (w_1, w_2, \dots, w_N)^T$. Then the return on the portfolio at time t, $R_{p,t}$, may be written as

$$R_{p,t} = \boldsymbol{w}^T \boldsymbol{R}_t, \quad t = 1, 2, \dots, T.$$

It is straightforward to show that under the model (10.1)

$$R_{p,t} - R_{f,t} = \alpha_p + \boldsymbol{\beta}_p^T \boldsymbol{F}_t + \epsilon_{p,t}, \quad t = 1, 2, \dots, T,$$

where $\alpha_p = \boldsymbol{w}^T \boldsymbol{\alpha}$ and $\boldsymbol{\beta}_p$ denotes the $K \times 1$ vector factors sensitivities for the portfolio,

$$\boldsymbol{\beta}_p^T \equiv (\beta_{p,1}, \beta_{p,2}, \dots, \beta_{p,K}) = \boldsymbol{w}^T \boldsymbol{\beta},$$

and $\epsilon_{p,t} = \boldsymbol{w}^T \boldsymbol{\epsilon}_t$.

That is, the factor model applies to the portfolio as well, with the factor sensitivities for the portfolio given by weighted sums of the asset factor sensitivities:

$$\beta_{p,j} = \sum_{i=1}^{N} w_i \beta_{i,j}, \quad j = 1, 2, \dots, K.$$

The expected excess return on the portfolio at time t is given by

$$\begin{aligned} \mathrm{E}(R_{p,t}) - \mu_{f,t} &= \boldsymbol{w}^T \boldsymbol{\alpha} + \boldsymbol{w}^T \boldsymbol{\beta} \mathrm{E}(\boldsymbol{F}_t) \\ &= \alpha_p + \boldsymbol{\beta}_p^T \mathrm{E}(\boldsymbol{F}_t) \end{aligned}$$

and the variance of the return at time t is given by

$$\mathrm{Var}(R_{p,t}) = \boldsymbol{\beta}_p^T \boldsymbol{\Sigma}_F \boldsymbol{\beta}_p + \boldsymbol{w}^T \boldsymbol{\Sigma}_\epsilon \boldsymbol{w}.$$

The first term in this expression, $\boldsymbol{\beta}_p^T \boldsymbol{\Sigma}_F \boldsymbol{\beta}_p$, represents a measure of the systematic risk of the portfolio, that is, the risk explained by the factors, while the second term, $\boldsymbol{w}^T \boldsymbol{\Sigma}_\epsilon \boldsymbol{w}$, is a measure of the portfolio's specific risk. The systematic risk depends on the variation in the factors, as measured by $\boldsymbol{\Sigma}_F$, along with factor sensitivities of the portfolio, $\boldsymbol{\beta}_p$.

Estimation

Given a set of factors thought to be relevant to the asset returns under consideration, the factor sensitivities are estimated from the available data. Let $F_{1,t}, F_{2,t}, \ldots, F_{K,t}$ denote the values of K factors at time t, $t = 1, 2, \ldots, T$, and let $R_{i,t}$, $t = 1, 2, \ldots, T$ denote the returns on a given asset. Then, according to the factor model,

$$R_{i,t} - R_{f,t} = \alpha_i + \beta_{i,1} F_{1,t} + \beta_{i,2} F_{2,t} + \cdots + \beta_{i,K} F_{K,t} + \epsilon_{i,t}, \quad t = 1, 2, \ldots, T,$$

where $\epsilon_{i,t}$ is uncorrelated with $F_{1,t}, F_{2,t}, \ldots, F_{K,t}$; here, $R_{f,t}$ is the return on the risk-free asset at time t.

Hence, as in the case of the single-index model, the parameter estimates for asset i may be obtained using least-squares regression based on the returns on asset i; that is, all parameter estimates may be obtained from N regression analyses, one for each asset. Specifically, the parameters $\alpha_i, \beta_{i,1}, \beta_{i,2}, \ldots, \beta_{i,K}$ may be estimated using least-squares regression with response variable $R_{i,t} - R_{f,t}$ at time t and predictor variables $F_{1,t}, F_{2,t}, \ldots, F_{K,t}$ at time t. The error variance for asset i, $\sigma_{\epsilon,i}^2 = \mathrm{Var}(\epsilon_{it})$, may be estimated using the usual estimator from the regression analysis.

Example 10.3 Consider the returns on JetBlue and EV Energy Partners stock, as discussed in Example 10.1. According to that example, oil prices apparently have an important effect on the returns on these stocks. Consider a factor model with two factors, the return on the S&P 500 index and the change in oil prices; recall that these data are stored in the variables `sp500` and `oil`, respectively. Thus, using the general notation of this section, $K = 2$, with $F_{1,t}$ given by the excess returns on the S&P 500 index at time t and $F_{2,t}$ given by the change in West Texas Intermediate oil at time t.

The excess returns on JetBlue stock are stored in the variable `jblu`. To estimate the parameters of the factor model for JetBlue stock, we use the `lm` function to fit a regression model with predictor variables `sp500` and `oil`:

```
> summary(lm(jblu~sp500+oil))
Coefficients:
             Estimate Std. Error t value Pr(>|t|)
(Intercept)  0.00965    0.01282    0.75   0.4547
sp500         1.15544    0.34078    3.39   0.0013 **
oil          -0.60598    0.19626   -3.09   0.0031 **

Residual standard error: 0.0948 on 57 degrees of freedom
Multiple R-squared:  0.226,     Adjusted R-squared:  0.199
F-statistic: 8.33 on 2 and 57 DF,  p-value: 0.000672
```

Therefore, the parameter estimates for the factor model for JetBlue stock, which we take to be asset 1, are $\hat{\alpha}_1 = 0.00965$, $\hat{\beta}_{1,1} = 1.155$, $\hat{\beta}_{1,2} = -0.606$, and $\hat{\sigma}_{\epsilon,1} = 0.0948$.

These results may be compared to the estimates from the market model for JetBlue stock:

```
> summary(lm(jblu~sp500))
Coefficients:
             Estimate Std. Error t value Pr(>|t|)
(Intercept)   0.0137     0.0137     1.00    0.321
sp500         0.8770     0.3520     2.49    0.016 *

Residual standard error: 0.102 on 58 degrees of freedom
Multiple R-squared:  0.0967,     Adjusted R-squared:  0.0811
F-statistic: 6.21 on 1 and 58 DF,  p-value: 0.0156
```

Note that the estimated coefficient of the market index is different in the market model and the factor model. This is not unexpected; note that the coefficients have different interpretations. In the market model, the coefficient represents the change in the expected excess return on JetBlue stock corresponding to a change in the return on the market index, while in the factor model, it represents the change in the return on JetBlue stock corresponding to a change in the return on the market index *holding the change in oil prices constant.*

Similarly, the negative coefficient of `oil` in the factor model indicates that an increase in oil prices corresponds to a decrease in the expected excess return on JetBlue stock, holding the return on the S&P 500 constant.

Note that the value of R-squared in the factor model, 0.226, is larger than the value in the market model, 0.0967. Again, this is to be expected: the factors `sp500` and `oil` explain more of the variation in the returns on JetBlue stock than does `sp500` alone. In general, when adding predictor variables to a regression model, the R-squared value increases or stays the same.

Therefore, when comparing R-squared values from regression models with different numbers of predictors, it is generally preferable to use *adjusted R-squared.* As the name suggests, adjusted R-squared includes an adjustment for the number of predictors in the model. Adding a predictor to a model can lead to a decrease in adjusted R-squared. In the present case, the adjusted R-squared value for the market model is 0.0811, while for the factor model it is 0.199, so the basic conclusion does not change: Including the change in oil prices in the model explains much more of the variation in the returns on JetBlue stock.

The estimates of the factor model for EV Energy Partners stock are given by

```
> summary(lm(evep~sp500+oil))
Coefficients:
            Estimate Std. Error t value Pr(>|t|)
(Intercept) 0.000828   0.011979    0.07     0.95
sp500       0.380479   0.318424    1.19     0.24
oil         0.782726   0.183384    4.27  7.5e-05 ***

Residual standard error: 0.0886 on 57 degrees of freedom
Multiple R-squared:  0.297,     Adjusted R-squared:  0.272
F-statistic:    12 on 2 and 57 DF,  p-value: 4.43e-05
```

Here, the coefficient of `oil` in the estimated regression model is negative, so that an increase in the change in oil prices corresponds to an increase in the expected excess return on EVEP stock, holding the return on the S&P 500 constant.

An estimate of the covariance matrix of the returns on JetBlue and EV Energy Partners stock based on the factor model may be obtained using the relationship given in (10.2). Note that $\mathbf{\Sigma}_F$, the covariance matrix of the factors, may be estimated by the sample covariance matrix of observed factor values; the other parameters in (10.2) may be obtained from the factor model regressions. The relevant R commands are

```
> jblu.fm<-lm(jblu~sp500+oil)
> evep.fm<-lm(evep~sp500+oil)
> betamat<-rbind(jblu.fm$coef[2:3], evep.fm$coef[2:3])
> Sig.eps<-diag(c(summary(jblu.fm)$sigma,
+ summary(evep.fm)$sigma)^2)
> cov.fm<-betamat%*%cov(cbind(sp500, oil))%*%t(betamat) +
+ Sig.eps
```

Here `jblu.fm` and `evep.fm` contain the results from the factor model regressions. The matrix `betamat` is the matrix of coefficient estimates; note that

`$coef` extracts the coefficient estimates from the result of `lm`. Thus, here `betamat` contains the estimated factor sensitivities,

```
> betamat
     sp500    oil
[1,]  1.16 -0.606
[2,]  0.38  0.783
```

The matrix `Sig.eps` is the estimate of $\mathbf{\Sigma}_\epsilon$ from factor model regressions

```
> Sig.eps
        [,1]    [,2]
[1,] 0.00899 0.00000
[2,] 0.00000 0.00785
```

The matrix `cov.fm` is the estimate of $\mathbf{\Sigma}$ based on the factor model

```
> cov.fm
         [,1]     [,2]
[1,]  0.01153 -0.00096
[2,] -0.00096  0.01104
```

and the corresponding correlation matrix is given by

```
> cov2cor(cov.fm)
        [,1]    [,2]
[1,]  1.0000 -0.0851
[2,] -0.0851  1.0000
```

Thus, the factor model with two factors, the return on the S&P 500 index and the change in the price of WTI oil, captures the negative correlation between the returns on JetBlue and EV Energy Partners stock. □

10.4 Factors

As noted in the previous section, there is considerable freedom in the selection of the factors to include in a factor model. Factors are often divided into two categories. *Economic factors* are variables measuring the general state of the market or of the economy; for instance, the return on a market index and the unemployment rate are two examples of economic factors. *Fundamental factors* are based on the characteristics of a particular firm, such as the size of the firm, as measured by its market capitalization; however, because the factors in our model must apply to all assets, such characteristics must first be converted to a common factor.

An economic factor may be based on any macroeconomic variable thought to have an important influence on the asset returns under consideration.

As noted earlier, the return on a market index is one example of an economic factor; another, used in the previous section, is the change in oil prices. Other commonly used economic factors include the unemployment rate, measures of industrial production, inflation rate, and measures of consumer sentiment or confidence. The data needed to construct such factors are available from a variety of sources, such as the FRED website and the Bureau of Labor Statistics website.

Example 10.4 Consider the four stocks of four companies, Caterpillar, Inc. (symbol CAT), maker of construction equipment; Cintas Corp. (CTAS), which provides uniforms and related services to businesses; Exxon Mobile Corp. (XOM), and Reliance Steel and Aluminum Co. (RS). Five years of monthly data for the period ending December 31, 2014, were analyzed using a factor model. The data matrix of excess return data for these stocks is stored in the matrix `stks4`.

Three factors were used, the excess returns on the S&P 500 index, the proportional change in the Industrial Production Index, and the unemployment rate, expressed as a proportion, rather than as a percentage. The Industrial Production Index is a measure of the industrial output of U.S. facilities; it is available on the FRED website at https://research.stlouisfed.org/fred2/series/INDPRO. The series of proportional changes in the index are stored in the variable `indpro`. The unemployment rate is also taken from the FRED website at https://research.stlouisfed.org/fred2/series/UNRATE. These data are stored in the variable `unemp`.

The parameter estimates for the factor model may be calculated by using the `lm` function with the data matrix as the response.

```
> summary(lm(stks4~sp500+indpro+unemp))
Response CAT :
Coefficients:
            Estimate Std. Error t value Pr(>|t|)
(Intercept)  -0.1081     0.0471   -2.29    0.026 *
sp500         1.7498     0.1927    9.08  1.3e-12 ***
indpro        0.0290     0.0169    1.71    0.092 .
unemp         1.1782     0.5821    2.02    0.048 *

Residual standard error: 0.0549 on 56 degrees of freedom
Multiple R-squared:  0.605,     Adjusted R-squared:  0.583
F-statistic: 28.5 on 3 and 56 DF,  p-value: 2.49e-11

Response CTAS :
Coefficients:
            Estimate Std. Error t value Pr(>|t|)
(Intercept)   0.0517     0.0290    1.78    0.080 .
sp500         0.8775     0.1187    7.39  7.8e-10 ***
indpro        0.0203     0.0104    1.95    0.056 .
unemp        -0.5635     0.3587   -1.57    0.122
```

```
Residual standard error: 0.0339 on 56 degrees of freedom
Multiple R-squared:  0.512,     Adjusted R-squared:  0.485
F-statistic: 19.6 on 3 and 56 DF,  p-value: 8.46e-09

Response RS :
Coefficients:
            Estimate Std. Error t value Pr(>|t|)
(Intercept)  -0.0782     0.0416   -1.88    0.065 .
sp500         1.7668     0.1700   10.39  1.1e-14 ***
indpro        0.0304     0.0149    2.03    0.047 *
unemp         0.7603     0.5137    1.48    0.144

Residual standard error: 0.0485 on 56 degrees of freedom
Multiple R-squared:  0.662,     Adjusted R-squared:  0.644
F-statistic: 36.5 on 3 and 56 DF,  p-value: 3.27e-13

Response XOM :
Coefficients:
            Estimate Std. Error t value Pr(>|t|)
(Intercept) -0.03568    0.02590   -1.38   0.1738
sp500        0.84442    0.10591    7.97  8.6e-11 ***
indpro      -0.02685    0.00931   -2.88   0.0056 **
unemp        0.52112    0.31996    1.63   0.1090

Residual standard error: 0.0302 on 56 degrees of freedom
Multiple R-squared:  0.59,      Adjusted R-squared:  0.568
F-statistic: 26.8 on 3 and 56 DF,  p-value: 6.97e-11
```

Note that the statistical significance of the different factors varies considerably by stock. This is not surprising; although, in general, the factors are related to the stocks' returns, the nature of the relationship depends on the particular features of the stock under consideration. It is interesting to note that, for each stock considered, one of `indpro` and `unemp` is statistically significant, or close to significant, at the 0.05 level.

The estimate of the matrix of factor coefficients, $\boldsymbol{\beta}$, may be extracted from the result of the `lm` function and is stored in the R matrix `betamat`.

```
> betamat<- t(lm(stks4~sp500+indpro+unem)$coef[-1, ])
> betamat
     sp500  indpro     unem
CAT  1.750  0.0290  0.01178
CTAS 0.878  0.0203 -0.00564
RS   1.767  0.0304  0.00760
XOM  0.844 -0.0268  0.00521
```

The index [-1,] is used to drop the estimate of the intercept when constructing betamat.

The estimates of $\mathbf{\Sigma}_F$, the covariance matrix of the factors, and $\mathbf{\Sigma}_\epsilon$, the error covariance matrix, are stored in the matrices Sig_F and Sig_eps, respectively.

```
> Sig.F<-cov(cbind(sp500, indpro, unem))
> Sig.F
          sp500   indpro     unem
sp500   0.00141 -0.00235 -0.00283
indpro -0.00235  0.18510  0.06878
unem   -0.00283  0.06878  1.53847
> f.sig<-function(y){summary(lm(y~sp500+indpro+unem))$sigma}
> Sig.eps<-diag(apply(stks4, 2, f.sig)^2)
> Sig.eps
         CAT    CTAS      RS      XOM
CAT  0.00302 0.00000 0.00000 0.000000
CTAS 0.00000 0.00115 0.00000 0.000000
RS   0.00000 0.00000 0.00235 0.000000
XOM  0.00000 0.00000 0.00000 0.000912
```

The estimate of the covariance matrix of the assets based on the factor model, that is, using (10.2), is given by

```
> Sig<-betamat%*%Sig.F%*%t(betamat) + Sig.eps
```

and the corresponding correlation matrix is given by

```
> cov2cor(Sig)
       CAT  CTAS    RS   XOM
CAT  1.000 0.494 0.618 0.506
CTAS 0.494 1.000 0.535 0.420
RS   0.618 0.535 1.000 0.530
XOM  0.506 0.420 0.530 1.000
```

This can be compared to the estimate based on the single-index model

```
       CAT  CTAS    RS   XOM
CAT  1.000 0.499 0.578 0.528
CTAS 0.499 1.000 0.531 0.485
RS   0.578 0.531 1.000 0.561
XOM  0.528 0.485 0.561 1.000
```

and to the sample correlation matrix

```
> cor(stks4)
       CAT  CTAS    RS   XOM
CAT  1.000 0.432 0.729 0.484
CTAS 0.432 1.000 0.569 0.451
RS   0.729 0.569 1.000 0.579
XOM  0.484 0.451 0.579 1.000
```

The estimates based on the factor model and the single-index model are generally similar, but there are some differences. For instance, the correlation of the returns on XOM and the returns on the other stocks is smaller for the factor model than for the single-index model. This is likely because of the fact that the estimate of the coefficient of `indpro` for XOM is negative, while the estimates of this coefficient for the other stocks are positive, reducing the factor model estimate of correlation as compared to the estimate based on the single-index model. □

Using Fundamental Factors in a Factor Model

Variables measuring certain characteristics of the assets may also be used to construct a factor. However, because factors must be common to all assets, and not specific to individual assets, such variables must be converted to common factors in order to use them in a factor model. The method used for this is based on the pioneering work of Fama and French (1993).

For instance, Fama and French (1993) note that the size of a firm is related to its profitability. Therefore, it may be useful to include a "size" factor in a factor model, such as one based on a firm's market capitalization, the price per share of the firm's stock times the number of shares outstanding.

To construct a common factor capturing a size effect, the following procedure is used. A large set of stocks, all those traded on the New York Stock Exchange in the case of Fama and French (1993), is divided into three groups, "small" firms, "medium-sized" firms, and "big" firms. Stocks in the "small" group correspond to firms whose size is in the bottom one-third of the market capitalizations of the stocks under consideration; the "big" group corresponds to firms whose size is in the top one-third. Two portfolios are formed, one using stocks from the "small" group and one using stocks from the "big" group. For instance, we might consider equally-weighted portfolios of small and big stocks, respectively, although the method used by Fama and French (1993) to construct their portfolios is more sophisticated. The Fama–French "small minus big" (SMB) is the return on the "small" portfolio minus the return on the "big" portfolio. Therefore, SMB is the return on a zero-investment portfolio designed to capture risk factors related to the size of the firm.

The same basic approach may be used to convert any variable measuring some characteristic of a firm into a common factor. Let X denote a characteristic of a firm that can be measured for all assets under consideration. Rank all assets by their value of X and form "low-X" and "high-X" groups. For instance, when defining the factor SMB, the low-X group is taken to be the one-third with the smallest X-values and the high-X group is taken to be the one-third with the largest X-values; however, other criteria for choosing the groups can be used. Then, form two portfolios, one from the low-X group and one from the high-X group; equally weighted portfolios are often used but, as in Fama and French (1993), other approaches might be considered. The factor corresponding to X is then the difference in the returns of the low-X portfolio and those of the high-X portfolio. This procedure yields a common factor that applies to all assets; it may also be used in a factor model for a portfolio, for which it may not be possible to measure X, such as in the firm size example.

Example 10.5 Consider the stocks of the five companies, Cablevision Systems Corp. (symbol CVC), Edison International (EIX), Expedia, Inc. (EXPE), Humana, Inc. (HUM), and Wal-Mart Stores, Inc. (WMT), that were analyzed in Chapter 9, in Examples 9.2 and 9.3. The data matrix of 5 years of excess monthly returns for the period ending December 31, 2014, is stored in the variable `stks`.

In Example 9.3, the parameters of the single-index model, using the S&P 500 index as the market index, were estimated. The estimates of α and β for these stocks are given by

```
> stks.mm$coef
                  CVC     EIX   EXPE    HUM    WMT
(Intercept) -9.75e-05 0.00914 0.0108 0.0162 0.0059
sp500        1.07e+00 0.47407 0.8902 0.6516 0.4568
```

In this example, the returns on these stocks are analyzed using a factor model based on three factors. The first factor is the return on the S&P 500 index and the second factor is the SMB factor described earlier.

For the third factor, we use a factor based on the book-to-market ratio of the stock. The "book value" of a stock is the value of the stock based on accounting information for the firm, while the "market value" of the stock is based on the price of the stock. Thus, a large book-to-market ratio suggests that the stock is undervalued by the market; such a stock is often referred to as a *value stock.* Fama and French (1993) construct a factor based on the book-to-market ratio of the stocks, using the same general procedure used to construct the factor SMB. The factor based on the book-to-market ratio is known as HML, for "high minus low"; thus, this factor is based on a zero-investment portfolio contrasting stocks with high book-to-market ratios with stocks with low book-to-market ratios. The model with factors SMB and HML, together with the return on a market index, is known as the *Fama–French three-factor model.*

Data on the factors SMB and HML are available from the Kenneth R. French Data Library, on the website http://mba.tuck.dartmouth.edu/pages/faculty/ken.french/data_library.html. This site contains extensive data useful in analyzing financial data. The values for the factors SMB and HML are given in the file "Fama/French 3 Factors" found in the section on U.S. Research Returns Data.

Five years of monthly data on SMB and HML for the period ending December 31, 2014, are stored in the variables `smb` and `hml`, respectively. Note that the data in the French Data Library are generally in the form of percentage returns, while here we use proportional returns; hence, the variables `smb` and `hml` contain the values given in the French Data Library, divided by 100.

To fit the factor model with these three factors to the stocks represented in the data matrix `stks`, we may use the function `lm`:

```
> stks.fact<-lm(stks~sp500+smb+hml)
```

The coefficient estimates for these factors may be extracted by

```
> stks.fact$coef
                CVC    EIX    EXPE    HUM     WMT
(Intercept) 0.0026 0.0094  0.0072  0.016  0.0044
sp500       0.8379 0.4461  1.1570  0.535  0.6333
smb         0.5268 0.0637 -0.4224  0.625 -0.5659
hml         0.9994 0.1158 -1.5607 -0.389 -0.3341
```

Wal-Mart (WMT) has the largest market capitalization of the five stocks considered; hence, it is not surprising that its sensitivity to SMB is negative. CVC has the smallest market capitalization and its sensitivity to SMB is fairly large and positive. However, the sensitivities are not a direct measure of the firm's size; they measure a property of the relationship between the stock's returns and those of the zero-investment portfolio on which SMB is based. Hence, it is possible for the stock returns of a large firm to have a positive sensitivity to SMB or the returns of a small company to have a negative sensitivity to SMB; that is, in fact, the case for EXPE, which has a fairly small market capitalization but a negative sensitivity to SMB.

The `summary` function may be used to obtain the standard errors of the estimates along with other useful information such as the R-squared and adjusted R-squared values for the regressions; see Example 10.4.

Estimates of the residual standard deviations may be obtained by defining a function `f.sighat` by

```
> f.sighat<-function(y){summary(lm(y~sp500+smb+hml))$sigma}
```

and then using the `apply` function

```
> apply(stks, 2, f.sighat)
   CVC    EIX   EXPE    HUM    WMT
0.0811 0.0465 0.1033 0.0692 0.0398
```

□

Note that the factors in a factor model are, in general, correlated. For instance, for the three factors analyzed in the previous example, the estimated correlation matrix of the factors is

```
> cor(cbind(sp500, smb, hml))
      sp500    smb    hml
sp500 1.000 0.4380 0.2122
smb   0.438 1.0000 0.0614
hml   0.212 0.0614 1.0000
```

An important consequence of this is that the estimated sensitivity of an asset's returns to the excess returns on the S&P 500 index is different than the asset's estimated value of beta in the market model. For example,

```
> lm(wmt~sp500)$coef
(Intercept)        sp500
     0.0059       0.4568
```

10.5 Arbitrage Pricing Theory

Like the single-index model, a factor model may be used to model the risk of an asset as well as the covariance structure of a set of asset returns. In addition, it gives information regarding the relationship between the expected returns on an asset and the factors used in the model.

In this section, we consider the implications of a factor model for modeling the expected return on an asset using an approach known as *arbitrage pricing theory (APT)*. Because the ideas in this section apply to any single time period, for notational convenience, the subscript t will be omitted from the random variables representing returns, factors, and so on.

First consider the market model for a given asset, asset i,

$$R_i - R_f = \alpha_i + \beta_i(R_m - R_f) + \epsilon_i,$$

where R_i, R_m, and R_f are the returns on asset i, the returns on a market index, and the returns on the risk-free asset, respectively. The random variable ϵ_i is assumed to be uncorrelated with R_m and to have mean zero so that the expected excess return on the asset may be written

$$\mathrm{E}(R_i - R_f) = \mathrm{E}(R_i) - \mu_f = \alpha_i + \beta_i \mathrm{E}(R_m - R_f).$$

However, without any further assumptions, this relationship is not very useful—in fact, it is always true by simply defining α_i to be

$$\alpha_i = \mathrm{E}(R_i - R_f) - \beta_i \mathrm{E}(R_m - R_f).$$

According to the CAPM, if R_m is the return on a market portfolio that is efficient in the sense that it has the maximum Sharpe ratio, then $\alpha_i = 0$. Therefore, if the CAPM holds with respect to the market index used in the market model, then

$$\mathrm{E}(R_i) - \mu_f = \beta_i \mathrm{E}(R_m - R_f).$$

That is, under the efficiency of the portfolio with return R_m, the expected excess return on an asset is proportional to its value of beta, β_i. Thus, the difference in the expected returns on two assets may be described in terms of the difference in their values of beta:

$$E(R_i) - E(R_j) = (\beta_i - \beta_j)E(R_m - R_f).$$

See Section 7.3 for further discussion of the implications of the CAPM.

Now consider a factor model of the form

$$R_i - R_f = \alpha_i + \beta_{i,1}F_1 + \beta_{i,2}F_2 + \cdots + \beta_{i,K}F_K + \epsilon_i,$$

where $F_1, F_2, \ldots, F_K$ are the values of the factors. Then, under the usual assumptions on ϵ_i, the expected excess return on asset i may be written as

$$E(R_i) - \mu_f = \alpha_i + \beta_{i,1}E(F_1) + \beta_{i,2}E(F_2) + \cdots + \beta_{i,K}E(F_K).$$

However, like the corresponding relationship based on the market model, without some conditions on α_i, this equation is not useful for describing the expected return on an asset.

Thus, the goal of APT is to derive a result similar to the CAPM for a general factor model. Note that such a result does not need to conclude that $\alpha_i = 0$ for all i in order to be useful; it is enough that the difference in the expected returns on two assets can be described in terms of the differences in their factor sensitivities, $\beta_{i,k} - \beta_{j,k}$, $k = 1, 2, \ldots, K$:

$$E(R_i) - E(R_j) = (\beta_{i,1} - \beta_{j,1})\Gamma_1 + (\beta_{i,2} - \beta_{j,2})\Gamma_2 + \cdots + (\beta_{i,K} - \beta_{j,K})\Gamma_K,$$

for some constants $\Gamma_1, \Gamma_2, \ldots, \Gamma_K$. Note that such a result holds under the condition that all α_i are equal,

$$\alpha_1 = \alpha_2 = \cdots = \alpha_N \equiv \alpha,$$

for some constant α but, as we will see, it also holds under weaker conditions on $\alpha_1, \alpha_2, \ldots, \alpha_N$.

The key assumption of the CAPM is the efficiency of the market portfolio. Given the generality of a factor model—there is considerable flexibility in the specific factors used in the model—a more general approach is needed for factor models. Hence, we rely on a concept more fundamental than efficiency, *arbitrage*.

Roughly speaking, an arbitrage opportunity is one in which an investor makes no net investment, has no chance of losing money, and has at least some chance of making money.

Let $\boldsymbol{R}$ denote an $N \times 1$ vector of asset returns. Consider a zero-investment portfolio, that is, one based on a weight vector $\boldsymbol{v}$ satisfying $\boldsymbol{v}^T\mathbf{1} = 0$, with return $R_p = \boldsymbol{v}^T\boldsymbol{R}$. The portfolio corresponding to $\boldsymbol{v}$ is said to be an *arbitrage portfolio* if there is zero probability of a negative return, $P(R_p < 0) = 0$, and there is positive probability of a positive return $P(R_p > 0) > 0$. Thus, an arbitrage portfolio requires no investment, has zero probability of a negative return, and positive probability of a positive return. According to economic theory, a market in equilibrium does not contain any arbitrage portfolios, a condition that we will refer to as the *no-arbitrage assumption.*

Consider a zero-investment portfolio with return R_p and suppose $\text{Var}(R_p) = 0$. Then, under the no-arbitrage assumption, we must have $\text{E}(R_p) \leq 0$, for if $\text{E}(R_p) > 0$, then $\text{P}(R_p > 0) = 1$; in that case, there is zero probability of a negative return and positive probability of a positive return, violating the no-arbitrage assumption. That is, a zero-investment portfolio with return R_p for which $\text{Var}(R_p) = 0$ and $\text{E}(R_p) > 0$ contradicts the no-arbitrage assumption.

However, a zero-investment portfolio with return R_p for which $\text{Var}(R_p) = 0$ and $\text{E}(R_p) < 0$ also violates the no-arbitrage assumption because, if $\boldsymbol{v}$ denotes the weight vector corresponding to R_p, then the portfolio with weight vector $-\boldsymbol{v}$ is a zero-investment portfolio with zero variance and positive expected return. Therefore, we must also have $\text{E}(R_p) \geq 0$. Because $\text{E}(R_p)$ must satisfy both $\text{E}(R_p) \leq 0$ and $\text{E}(R_p) \geq 0$, it must satisfy $\text{E}(R_p) = 0$. That is, under the no-arbitrage assumption, a zero-investment portfolio with return R_p such that $\text{Var}(R_p) = 0$ must also have $\text{E}(R_p) = 0$.

Consider the factor model (10.1) for a set of N assets; as noted previously, because we are focusing on a single time period, the subscript t is dropped. Then we may write the vector of asset returns $\boldsymbol{R}$ as

$$\boldsymbol{R} - R_f\mathbf{1} = \boldsymbol{\alpha} + \boldsymbol{\beta}\boldsymbol{F} + \boldsymbol{\epsilon}.$$

Taking expectations,

$$\text{E}(\boldsymbol{R}) - \mu_f\mathbf{1} = \boldsymbol{\alpha} + \boldsymbol{\beta}\text{E}(\boldsymbol{F}).$$

Our goal is to show that, under the no-arbitrage assumption, the vector of expected excess returns $\text{E}(\boldsymbol{R}) - \mu_f\mathbf{1}$ is of the form

$$\alpha\mathbf{1} + \boldsymbol{\beta}\boldsymbol{\Gamma}$$

for some scalar α and vector $\boldsymbol{\Gamma} = (\Gamma_1, \Gamma_2, \ldots, \Gamma_K)^T$.

We begin by considering the case in which the residual returns are all zero, $\boldsymbol{\epsilon} = \mathbf{0}$, so that the excess returns of the assets are completely described by the factors; thus,

$$\boldsymbol{R} - R_f\mathbf{1} = \boldsymbol{\alpha} + \boldsymbol{\beta}\boldsymbol{F}.$$

Consider the following simple result from linear algebra.

Lemma 10.1. *Let $\mathcal{M}$ denote a linear subspace of $\Re^N$ and let $\boldsymbol{u}, \boldsymbol{v}$ be elements of $\Re^N$. If $\boldsymbol{v}$ is orthogonal to $\mathcal{M}$ implies that $\boldsymbol{v}$ and $\boldsymbol{u}$ are orthogonal, then $\boldsymbol{u} \in \mathcal{M}$.*

Proof. Let $\mathcal{M}^\perp$ denote the linear subspace of $\Re^N$ consisting of vectors that are orthogonal to $\mathcal{M}$. Note that $\mathcal{M}$ and $\mathcal{M}^\perp$ are disjoint linear subspaces so that they have only the zero vector in common. Also, $(\mathcal{M}^\perp)^\perp = \mathcal{M}$; that is, the set of vectors that are orthogonal to $\mathcal{M}^\perp$ is simply $\mathcal{M}$. If $\boldsymbol{v} \in \mathcal{M}^\perp$ implies that $\boldsymbol{u}$ and $\boldsymbol{v}$ are orthogonal, then $\boldsymbol{u}$ must be orthogonal to all vectors that are orthogonal to $\mathcal{M}$; that is,

$$\boldsymbol{u} \in (\mathcal{M}^{\perp})^{\perp},$$

so that

$$\boldsymbol{u} \in \mathcal{M},$$

proving the result. □

Lemma 10.1 may now be used to prove the following simple form of APT.

Proposition 10.1. *Suppose that the following factor model holds:*

$$\boldsymbol{R} - R_f\mathbf{1} = \boldsymbol{\alpha} + \boldsymbol{\beta}\boldsymbol{F} \tag{10.3}$$

for some vector of random variables $\boldsymbol{F}$ *representing the factors.*

Then, if the no-arbitrage assumption holds,

$$E(\boldsymbol{R}) - \mu_f\mathbf{1} = \alpha\mathbf{1} + \boldsymbol{\beta}\Gamma \tag{10.4}$$

for some scalar α *and some vector* $\Gamma \in \Re^K$.

Proof. Let $\mathcal{M}$ denote the $(K+1)$-dimensional subspace of $\Re^N$ spanned by the columns of $\boldsymbol{\beta}$ together with $\mathbf{1}$. Let $\boldsymbol{v} \in \mathcal{M}^{\perp}$. Then $\boldsymbol{v}$ is orthogonal to $\mathbf{1}$ and to each column of $\boldsymbol{\beta}$ so that

$$\mathbf{1}^T\boldsymbol{v} = 0 \quad \text{and} \quad \boldsymbol{\beta}^T\boldsymbol{v} = \mathbf{0}_K. \tag{10.5}$$

Consider a portfolio with return $\boldsymbol{v}^T\boldsymbol{R}$; according to (10.5) and (10.3), this is a zero-investment portfolio such that

$$\boldsymbol{v}^T\boldsymbol{R} = \boldsymbol{v}^T\boldsymbol{\alpha} + \boldsymbol{v}^T\boldsymbol{\beta}\boldsymbol{F} = \boldsymbol{v}^T\boldsymbol{\alpha}.$$

It follows that the portfolio return has zero variance and an expected value $\boldsymbol{v}^T\boldsymbol{\alpha}$. Under the no-arbitrage assumption, the expected return must be 0 so that $\boldsymbol{v}^T\boldsymbol{\alpha} = 0$.

That is, $\boldsymbol{v}$ is orthogonal to $\mathcal{M}$ implies that $\boldsymbol{v}$ is orthogonal to $\boldsymbol{\alpha}$. It now follows from Lemma 10.1 that $\boldsymbol{\alpha}$ lies in $\mathcal{M}$, the space spanned by $\mathbf{1}$ and the columns of $\boldsymbol{\beta}$; hence, $\boldsymbol{\alpha}$ must be of the form

$$\boldsymbol{\alpha} = \alpha\mathbf{1} + \boldsymbol{\beta}\tilde{\boldsymbol{\Gamma}}$$

for some scalar α and vector $\tilde{\boldsymbol{\Gamma}} \in \Re^K$. Therefore,

$$\mathrm{E}(\boldsymbol{R}) - \mu_f\mathbf{1} = \alpha\mathbf{1} + \boldsymbol{\beta}\tilde{\boldsymbol{\Gamma}} + \boldsymbol{\beta}\mathrm{E}(\boldsymbol{F});$$

the result now follows by writing this equation in the form (10.4) by defining $\boldsymbol{\Gamma} = \tilde{\boldsymbol{\Gamma}} + \mathrm{E}(\boldsymbol{F})$. □

According to this result, the expected return on asset i may be written as

$$\mathrm{E}(R_i) - \mu_f = \alpha + \beta_{i,1}\Gamma_1 + \beta_{i,2}\Gamma_2 + \cdots + \beta_{i,K}\Gamma_K$$

for some constants $\alpha, \Gamma_1, \Gamma_2, \ldots, \Gamma_K$. Furthermore, the difference in the expected returns of two assets, say asset i and asset j, can be written

$$\mathrm{E}(R_i) - \mathrm{E}(R_j) = (\beta_{i,1} - \beta_{j,1})\Gamma_1 + (\beta_{i,2} - \beta_{j,2})\Gamma_2 + \cdots + (\beta_{i,K} - \beta_{j,K})\Gamma_K.$$

That is, the difference in the expected returns of two assets can be described in terms of the differences in the factor sensitivities for the two assets.

Asymptotic Arbitrage

Of course, the assumption that the asset returns are completely determined by the factors is an unrealistic one. More general versions of APT are based on the idea that if a zero-investment portfolio has a "small" return variance, then it must have a "small" expected return. Note that this is a type of continuity assumption. Recall that a function $f(\cdot)$ is continuous at a point x_0 if $|f(x) - f(x_0)|$ is "small" whenever $|x - x_0|$ is "small." Definitions of continuity are generally based on the concept of convergence; for instance, in the case of a function, $f(\cdot)$ is continuous at x_0 if, for a sequence $x_1, x_2, \ldots$, the condition that $x_n \to x_0$ implies that $f(x_n) \to f(x_0)$.

Therefore, a more general treatment of APT is based on assumptions regarding sequences of portfolios. For $N = 1, 2, \ldots$, consider a set of N assets and let $\boldsymbol{R}^{(N)}$ denote the corresponding $N \times 1$ vector of asset returns. Let $\boldsymbol{v}_N$, $N = 1, 2, \ldots$ denote a sequence of vectors such that $\boldsymbol{v}_N$ is $N \times 1$ and $\boldsymbol{v}_N^T \mathbf{1}_N = 0$, $N = 1, 2, \ldots$. Thus, each $\boldsymbol{v}_N$ defines a zero-investment portfolio based on the set of N assets.

Let $R_p^{(N)} = \boldsymbol{v}_N^T \boldsymbol{R}^{(N)}$, $N = 1, 2, \ldots$ so that, for each N, $R_p^{(N)}$ is the return on a zero-investment portfolio of N assets. We say that the *no-asymptotic-arbitrage assumption* holds if

$$\mathrm{Var}(R_p^{(N)}) \to 0 \text{ as } N \to \infty$$

implies that

$$\mathrm{E}(R_p^{(N)}) \to 0 \text{ as } N \to \infty.$$

Thus, under the no-asymptotic-arbitrage assumption, a zero-investment portfolio with a small return variance must also have a small mean return.

Suppose that for each $N = 1, 2, \ldots$ the factor model

$$\boldsymbol{R}^{(N)} - R_f \mathbf{1}_N = \boldsymbol{\alpha}^{(N)} + \boldsymbol{\beta}^{(N)} \boldsymbol{F} + \boldsymbol{\epsilon}^{(N)}$$

holds, where $\boldsymbol{\alpha}^{(N)}$ is an $N \times 1$ vector of constants, $\boldsymbol{\beta}^{(N)}$ is an $N \times K$ matrix of factor sensitivities, $\boldsymbol{F}$ is a $K \times 1$ vector of factors, and $\boldsymbol{\epsilon}^{(N)}$ is an $N \times 1$ random vector with zero mean vector.

Then, under the no-asymptotic-arbitrage assumption, it may be shown that

$$E(\boldsymbol{R}^{(N)}) - \mu_f \doteq \alpha\mathbf{1} + \boldsymbol{\beta}^{(N)}\boldsymbol{\Gamma} \tag{10.6}$$

where α is a scalar and $\boldsymbol{\Gamma}$ is a $K \times 1$ vector, both of which depend on N; that is, they depend on the assets under consideration. Here "$\doteq$" means that the difference between $E(\boldsymbol{R}^{(N)}) - \mu_f\mathbf{1}$ and an expression of the form

$$\alpha\mathbf{1}_N + \boldsymbol{\beta}^{(N)}\boldsymbol{\Gamma}$$

is small when N is large.

A formal proof of this result is quite technical and will not be presented here. The basic idea is that, by constructing a portfolio from a large number of assets, the residual risk of the portfolio may be made small, so that the situation is similar to the one in Proposition 10.1 in which there is no residual risk. Thus, an important condition is that we may form portfolios such that the residual returns on the portfolios have small standard deviation. If the residual returns for different assets are uncorrelated, it is straightforward to show that such a condition holds; however, it may also hold in cases in which the residual returns are correlated.

The interpretation of the result is the same as the interpretation of the result given in Proposition 10.1. The difference in the expected returns of two assets may be described in terms of the differences in their factor sensitivities. That is, the factor sensitivities are useful in describing the expected returns on the assets. In particular, the expected excess return on an asset may be written as a sum of a constant, not depending on the asset, plus a weighted sum of its factor sensitivities.

10.6 Factor Premiums

The APT discussed in the previous section shows that factor models can be used to describe the vector of asset mean returns. Consider a factor model for an asset return vector $\boldsymbol{R}_t$ of the form

$$\boldsymbol{R}_t - R_{f,t}\mathbf{1} = \boldsymbol{\alpha} + \boldsymbol{\beta}\boldsymbol{F}_t + \boldsymbol{\epsilon}_t, \quad t = 1, 2, \ldots, T, \tag{10.7}$$

where $\boldsymbol{F}_t$ is a vector of factor values and the residual return vector $\boldsymbol{\epsilon}_t$ satisfying $E(\boldsymbol{\epsilon}_t) = 0$, in addition to the other conditions described in Section 10.3.

According to APT, we may write

$$E(\boldsymbol{R}_t - R_{f,t}\mathbf{1}) = \alpha\mathbf{1} + \boldsymbol{\beta}\boldsymbol{\Gamma}, \quad t = 1, 2, \ldots, T \tag{10.8}$$

for a scalar α and a vector $\boldsymbol{\Gamma}$. Thus, estimates of α and $\boldsymbol{\Gamma}$ may be used to estimate the expected excess returns on the assets.

More importantly, this expression gives a relationship between the factor sensitivities of an asset and its expected excess return that is useful for understanding the roles played by various economic variables in describing the expected returns of the different assets. In particular, the factor premiums tell us how an asset's sensitivities to various factors are compensated by a higher expected return.

For $i = 1, 2, \ldots, N$, let

$$\bar{R}_i - \bar{R}_f = \frac{1}{T}\sum_{t=1}^{T}(R_{i,t} - R_{f,t})$$

denote the sample mean excess return on the asset. It follows from (10.8) that

$$E(\bar{R}_i - \bar{R}_f) = \alpha + \Gamma_1\beta_{i,1} + \Gamma_2\beta_{i,2} + \cdots + \Gamma_K\beta_{i,K} \tag{10.9}$$

where $\beta_{i,1}, \beta_{i,2}, \ldots, \beta_{i,K}$ are the factor sensitivities of asset i and $\alpha, \Gamma_1, \Gamma_2, \ldots, \Gamma_K$ are unknown parameters; Γ_k is known as the *risk premium* or, simply, the *premium* of factor k. The premium of a factor represents the reward, in terms of a greater expected return, for assuming the risk associated with a factor. Thus, our goal is to estimate the factor premiums $\Gamma_1, \Gamma_2, \ldots, \Gamma_K$ or, equivalently, the vector $\boldsymbol{\Gamma} = (\Gamma_1, \Gamma_2, \ldots, \Gamma_K)^T$ along with the scalar parameter α.

Role of Arbitrage Pricing Theory

Before considering the estimation of α and $\boldsymbol{\Gamma}$, it is useful to reiterate some of the important aspects of APT and how they relate to the estimation of factor premiums.

Taking expectations in the model (10.7), the expected excess returns on the assets may be written as

$$E(\boldsymbol{R}_t - R_{f,t}\mathbf{1}) = \boldsymbol{\alpha} + \boldsymbol{\beta}E(\boldsymbol{F}_t), \quad t = 1, 2, \ldots, T. \tag{10.10}$$

This equation gives an expression for the excess mean return vector $E(\boldsymbol{R}_t - R_f\mathbf{1})$ in terms of the factor sensitivities for the assets, as given by the matrix $\boldsymbol{\beta}$, the vector $\boldsymbol{\alpha}$, which is a property of the assets under consideration, and the vector of factor means $E(\boldsymbol{F}_t)$.

APT tells us that the vector $\boldsymbol{\alpha}$ is approximately of the form

$$\alpha\mathbf{1} + \boldsymbol{\beta}\tilde{\boldsymbol{\Gamma}} \tag{10.11}$$

for a scalar α and a vector $\tilde{\boldsymbol{\Gamma}}$. Using this fact in (10.10), and defining $\boldsymbol{\Gamma}$ to be $\tilde{\boldsymbol{\Gamma}} + E(\boldsymbol{F}_t)$, leads to (10.8). Thus, it is important to keep in mind that, in the expression (10.8), the parameter $\boldsymbol{\Gamma}$ is different than the vector of factor means, $E(\boldsymbol{F}_t)$. In particular, we cannot estimate $\boldsymbol{\Gamma}$ by the sample means of the factors.

A second important fact to keep in mind is that the result of APT is valid only if the factor model correctly describes the asset returns under consideration. There are at least two aspects to this. First, we assume that the error term $\boldsymbol{\epsilon}_t$ in (10.7) is uncorrelated with the factor vector $\boldsymbol{F}_t$. In particular, if we inadvertently omit an important factor from the factor model, and that factor is correlated with the factors in the model, then the least-squares estimators of the factor sensitivities will be biased, a case of *omitted-variable bias.*

A second aspect of model misspecification is its effect on the correlation structure of $\boldsymbol{\epsilon}_t$. A key part of the proof of (10.6) is that we can form a portfolio in which the variance of residual returns of the portfolio is negligible. If the covariance matrix of $\boldsymbol{\epsilon}_t$ is diagonal, then this is generally possible when N is large. However, if, after accounting for the factors, the asset returns are still correlated, it might not be possible. Thus, it is important that the factor model includes those factors that play an important role in describing the correlation between the returns of different assets. That is, there is a second potential problem related to an omitted factor, in addition to possible omitted variable bias—the omitted variable may lead to correlation in the residual returns for different assets, so that the expression in (10.11) is not an accurate approximation to $\boldsymbol{\alpha}$.

Two-Stage Least-Squares Estimation

To estimate the parameters $\alpha, \Gamma_1, \Gamma_2, \ldots, \Gamma_K$, we use a two-stage approach. In the first stage, we estimate the factor sensitivities $\beta_{i,1}, \beta_{i,2}, \ldots, \beta_{i,K}$ for each asset using the least-squares method described in Section 10.3. This procedure yields estimates $\hat{\beta}_{i,1}, \hat{\beta}_{i,2}, \ldots, \hat{\beta}_{i,K}$ for each $i = 1, 2, \ldots, N$.

In the second stage, we estimate $\alpha, \Gamma_1, \Gamma_2, \ldots, \Gamma_K$ using the relationships

$$\mathrm{E}(\bar{R}_i - \bar{R}_f) = \alpha + \beta_{i,1}\Gamma_1 + \beta_{i,2}\Gamma_2 + \cdots + \beta_{i,N}\Gamma_N$$

for $i = 1, 2, \ldots, N$ but with the estimates $\hat{\beta}_{i,k}$ replacing the parameters $\beta_{i,k}$. Specifically, we use least-squares regression with the sample mean excess returns, $\bar{R}_i - \bar{R}_f$, $i = 1, 2, \ldots, N$, as the observed response variable and the first-stage estimates of the factor sensitivities, $\hat{\beta}_{i,1}, \hat{\beta}_{i,2}, \ldots, \hat{\beta}_{i,K}$, as the predictor variables corresponding to $\bar{R}_i - \bar{R}_f$. The methodology is illustrated in the following example.

Example 10.6 Recall that in Example 10.2, stock return data for firms represented in the S&P 500 index were analyzed; 474 of the stocks had returns for the period under consideration. The data matrix for these 474 stocks is stored in the variable `sp474.data`.

The first step is to estimate the parameters of a factor model for each stock. Here, we use four factors: the return on the S&P 500 index (`sp500`), the factors SMB and HML described in Example 10.5 (variables `smb` and `hml`, respectively), and a "momentum" factor, denoted by MOM. Like SMB and HML, the momentum factor is based on constructing portfolios from two sets of stocks, those that have performed well in recent months and those that have performed poorly; the factor MOM is the return on a zero-investment portfolio based on the difference in the returns on these two portfolios. Thus, MOM is different than SMB and HML in the sense that it is based on previous returns of the assets rather than on properties of the firms issuing the stock. The momentum factor is originally attributed to Carhart (1997); values of MOM are available in the Kenneth R. French Data Library, in the section "Sorts involving Prior Returns." The variable `mom` contains the values of MOM from the French Data Library, divided by 100.

The factor model estimates for the 474 stocks may be calculated using

```
> sp474.fm<-lm(sp474.data~sp500+smb+hml+mom)
```

The estimates of β_1, the coefficient of the S&P 500 index; β_2, the coefficient of SMB; β_3, the coefficient of HML; and β_4, the coefficient of MOM, for each of the 474 stocks are stored in the matrix `sp474.beta`, which has 474 rows and 4 columns and that may be obtained by the command

```
> sp474.beta<-t(sp474.fm$coef)[, -1]
> head(sp474.beta)
     sp500     smb     hml     mom
MMM   1.01  0.3661 -0.0988 -0.0107
ABT   0.80 -0.6508 -0.0792  0.2731
ACN   1.27 -0.0683 -0.6131 -0.4408
ATVI  1.19 -0.4622 -0.3889  0.4143
AYI   1.28  1.6768 -0.5286  0.3750
ADBE  1.23  0.1768  0.0784  0.0318
```

Note that the index `[, -1]` is used in defining `sp474.beta` in order to drop the column of intercept estimates; the transpose function `t` is used so that `sp474.beta` has the same form as the parameter $\boldsymbol{\beta}$.

To estimate the factor premiums, we fit a regression model with the sample mean excess returns for each of the 474 stocks as the response variable and beta estimates as the predictor variables. Note that, in the function `lm`, we may specify the model in terms of a matrix of predictor variables; then the columns of the matrix are taken as the predictors.

```
> sp474.mean<-apply(sp474.data, 2, mean)
> summary(lm(sp474.mean~sp474.beta))
Coefficients:
                 Estimate Std. Error t value Pr(>|t|)
(Intercept)      0.014464   0.000901   16.06  < 2e-16 ***
sp474.betasp500  0.000372   0.000825    0.45     0.65
sp474.betasmb    0.006519   0.000666    9.79  < 2e-16 ***
sp474.betahml   -0.004380   0.000615   -7.12  4.1e-12 ***
sp474.betamom    0.006333   0.000886    7.15  3.3e-12 ***
```

```
Residual standard error: 0.00726 on 469 degrees of freedom
Multiple R-squared:  0.254,     Adjusted R-squared:  0.248
F-statistic: 39.9 on 4 and 469 DF,  p-value: <2e-16
```

Therefore, the estimated mean excess return on a stock with beta estimates given by $\hat{\beta}_1$, $\hat{\beta}_2$, $\hat{\beta}_3$, and $\hat{\beta}_4$ is

$$0.0145 + 0.000372\hat{\beta}_1 + 0.00652\hat{\beta}_2 - 0.00438\hat{\beta}_3 - 0.00633\hat{\beta}_4.$$

According to the R-squared value for the regression, about 25.4% of the variation in the sample mean excess returns on the 474 stocks can be explained by their estimates of the factor sensitivities; a better estimate of this quantity is the adjusted R-squared value, which, in this case, is essentially the same at 24.8%.

For instance, for the stock of the 3M Company (symbol MMM), $\hat{\beta}_1 = 1.008$, $\hat{\beta}_2 = 0.00366$, $\hat{\beta}_3 = -0.000988$, and $\hat{\beta}_4 = -0.000107$. Therefore, the estimate of the mean excess return on 3M stock is

$$\begin{aligned} 0.0145 + 0.000372(1.008) + 0.00652(0.366) - 0.00438(-0.0988) \\ - 0.00633(-0.0107) = 0.0176; \end{aligned}$$

for comparison, the observed sample mean excess returns for MMM is 0.0148.

For the 474 stocks, the estimated mean excess returns are given by fitted values from the regression

```
> sp474.fit<-lm(sp474.mean~sp474.beta)$fitted.values
```

and the average error in the fitted values as estimates of sample mean excess return is

```
> mean(abs(sp474.fit-sp474.mean))
[1] 0.00535
```

The correlation of the fitted values and the sample mean excess returns is

```
> cor(sp474.fit, sp474.mean)
[1] 0.504
```

which is simply the square root of the R-squared value from the regression. □

Obtaining Standard Errors of the Premium Estimates

In the example, the factor premiums were estimated using ordinary least-squares regression with $\bar{R}_i - \bar{R}_f$, $i = 1, 2, \ldots, N$, as the response variable and the estimated factor sensitivities $\hat{\beta}_{i,k}$, $k = 1, 2, \ldots, K$, $i = 1, 2, \ldots, N$ as the predictor variables.

The least-squares procedure used in the example gives valid estimates of the factor premiums. However, the standard errors given by `lm` are based on the assumption that observations on the response variable are uncorrelated. In practice, the returns on different assets in a given time period are correlated; hence, $\bar{R}_i$ and $\bar{R}_j$ are correlated. It follows that the methods used by `lm` tend to overstate the amount of information available to estimate the factor premiums, and the reported standard errors are generally too small.

To obtain valid standard errors of the estimated factor premiums, we may use an alternative approach in which the factor premiums are estimated for each time period, using

$$R_{1,t} - R_{f,t}, R_{2,t} - R_{f,t}, \ldots R_{N,t} - R_{f,t}$$

as the dependent variable in the analysis for period t. Thus, if our data consist of 5 years of monthly returns, we would obtain 60 estimates of each factor premium—one estimate for each time period. These estimates are obtained using least-squares regression with the estimated factor sensitivities as the predictor variables; note that the same predictor variables are used in each regression. The final estimates of the factor premiums are given by the sample means of these 60 estimates, and the standard errors of the premiums estimates are given by the usual expression for the standard error of a sample mean.

Example 10.7 Consider estimation of the factor premiums for the factor model analyzed in Examples 10.2 and 10.6. Recall that the data matrix for the 474 stocks under consideration is stored in the variable `sp474.data` and the corresponding estimates of the factor sensitivities for the assets are stored in the matrix `sp474.beta`.

To estimate the factor premiums for each time period, we may use the command

```
> spcoef<-lm(t(sp474.data)~sp474.beta)$coef
```

Note that the data matrix must be transposed in order to use it as the response variable in `lm`. The result of this command, `spcoef`, is a matrix with five rows, one for the intercept of the model and one for each factor coefficient in the model, and 60 columns, one for each time period. For instance, the first column of the matrix

```
> spcoef[,1]
(Intercept)       sp500         smb         hml         mom
     0.0334     -0.0681      0.0118      0.0117     -0.0593
```

gives the estimated factor premiums based on the data in period 1.

To obtain the overall estimates of the factor premiums, we average the estimates in each row of `spcoef`

```
> apply(spcoef, 1, mean)
(Intercept)       sp500         smb         hml         mom
   0.014464    0.000372    0.006519   -0.004380    0.006333
```

The standard errors of the estimates are given by the sample standard deviations of each row, divided by $\sqrt{60}$

```
> apply(spcoef, 1, sd)/(60^.5)
(Intercept)        sp500          smb          hml          mom
    0.00276      0.00553      0.00324      0.00272      0.00366
```

These results may be compared to those based on the analysis in Example 10.6:

```
> summary(lm(sp474.mean~sp474.beta))
Coefficients:
                 Estimate Std. Error t value Pr(>|t|)
(Intercept)      0.014464   0.000901   16.06  < 2e-16 ***
sp474.betasp500  0.000372   0.000825    0.45     0.65
sp474.betasmb    0.006519   0.000666    9.79  < 2e-16 ***
sp474.betahml   -0.004380   0.000615   -7.12  4.1e-12 ***
sp474.betamom    0.006333   0.000886    7.15  3.3e-12 ***
```

Note that the parameter estimates based on averaging the 60 estimates are identical to the ones obtained in Example 10.6. This is a general property of least-squares estimates; it follows from the facts that the least-squares estimates of regression coefficients are linear functions of the response data and the same predictor variables are used in each of the 60 regressions. The standard errors of these estimates, using the sample standard deviations of the 60 estimates, are considerably larger than those obtained in Example 10.6. For instance, for estimating the premium of HML, the least-squares standard error is 0.000615, while the two-stage standard error is 0.00272. □

When interpreting estimates of the factor premiums, it is useful to keep in mind that the estimated factor sensitivities tend to be correlated across assets. For instance, in the previous example, the estimated correlation matrix of the factor sensitivities for the 474 stocks analyzed is

```
> cor(sp474.beta)
       sp500    smb    hml    mom
sp500  1.000  0.166  0.030 -0.279
smb    0.166  1.000  0.109 -0.451
hml    0.030  0.109  1.000 -0.152
mom   -0.279 -0.451 -0.152  1.000
```

For example, stocks with a large sensitivity to the MOM factor tend to have smaller sensitivities to the other factors.

Rolling Regressions

In the models used in this section to estimate factor premiums, the excess returns on the stocks under consideration are related to those stocks' estimated factor sensitivities. An important aspect of these analyses is that those excess returns used are from the same data used to estimate the factor sensitivities. However, in practice, a common goal of a factor-model analysis is to use current data to study the properties of *future* returns. Thus, it may be more appropriate to estimate factor premiums using a response variable based on return data for periods following the ones used to estimate the factor sensitivities.

Example 10.8 Consider estimation of the factor premiums for the factor model analyzed in Examples 10.2 and 10.6. In Example 10.6, estimates of the factor premiums were obtained using least-squares regression with the response variable taken to be the sample mean excess returns for the 474 assets and the predictor variables given by the estimated factor sensitivities. The response and predictor variables are all based on data in the matrix `sp474.data`, which contains the monthly excess returns on the 474 stocks for the 5-year period ending December 31, 2014.

Now suppose we are interested in using the estimated factor sensitivities to describe future asset returns. The variable `sp474.data.115` contains the excess returns for the 474 stocks for January 2015. Therefore, we can estimate the factor premiums for the four factors in the model using least-squares regression with response variable `sp474.data.115` and the predictor variables given in the matrix `sp474.beta`.

```
> summary(lm(sp474.data.115~sp474.beta))
Coefficients:
                Estimate Std. Error t value Pr(>|t|)
(Intercept)      0.04205    0.00821    5.12 4.5e-07 ***
sp474.betasp500 -0.06412    0.00752   -8.52 < 2e-16 ***
sp474.betasmb    0.00330    0.00607    0.54     0.59
sp474.betahml   -0.02486    0.00561   -4.43 1.2e-05 ***
sp474.betamom    0.00248    0.00808    0.31     0.76

Residual standard error: 0.0662 on 469 degrees of freedom
Multiple R-squared:  0.178,     Adjusted R-squared:  0.171
F-statistic: 25.4 on 4 and 469 DF,  p-value: <2e-16
```
□

The estimates obtained in the previous example have the drawback that the response variable in the regression is based on the stock returns from only a single month. One way to include additional months of returns in the analysis is to use *rolling regressions*. For instance, suppose that we have m months of asset returns, along with the corresponding factor values, and suppose that our goal is to estimate the factor premiums using estimates of the factor sensitivities based on 60 months of data.

To do this, we first estimate the factor sensitivities using data from months 1 through 60 and then use those results as the predictor variables in a regression model, with the response variable taken to be the excess returns for all stocks from month 61. This procedure may then be repeated, estimating the factor sensitivities using data from months 2 through 61 and estimating the factor premiums using a regression model with response variable based on data from month 62. We may continue in this way until the final regression with the returns from month m as the response variable and the factor sensitivities estimated using data from months $m-60$ through $m-1$ as the predictors. The result is $m-60$ estimates of the factor premiums, which can be averaged using an approach similar to the one used in Example 10.7 to obtain the final estimates.

Example 10.9 Consider estimation of the factor premiums as in Example 10.8, using rolling regressions as discussed previously. The data matrix `sp474.data6` contains 6 years of monthly excess returns on the 474 stocks described in Example 10.2, and the matrix `factor6` contains 6 years of monthly factor values for the factors described in Example 10.8. Hence, we will calculate 12 sets of factor premium estimates, each obtained using the same basic procedure used in Example 10.8. The first set of estimates uses data from months 1 through 60 to estimate the factor sensitivities, which are then used as predictor variables in a regression model for the returns in month 61. The second set of estimates uses data from months 2 to 61 to estimate the factor sensitivities, which are then used as predictor variables in a regression model for the returns in month 62, and so on.

To perform these rolling regressions, we may use the following commands.

```
> pmat<-matrix(0, 5, 12)
> for (j in 1:12){
+     dat<-sp474.data6[j:(j+59), ]
+     x.fact<-factor6[j:(j+59), ]
+     bet<-t(lm(dat~x.fact)$coef)[, -1]
+     pmat[, j]<-lm(sp474.data6[(j+60), ]~bet)$coef
+ }
```

The matrix `pmat` stores the 12 sets of factor premiums, one in each column. For a given value of the index `j`, `dat` contains the return data and `x.fact` contains the factor data for months j to $j+59$ that will be used to obtain the estimates of the factor sensitivities; these estimates are calculated using the command `lm(dat~x.fact)$coef` and are stored in the variable `bet`. Note that the transpose function `t` is used to put the estimates in the correct format to use in a subsequent function, and the index `[, -1]` is used to drop the estimate of the intercept term from the results.

The command `lm(sp474.data6[(j+60), ]~bet)` runs the regression with the returns in month $j+60$ as the response variable and the estimated factor sensitivities in `bet` as the predictor variables. The estimated coefficients from this regression are stored in the jth column of the matrix `pmat`. After the loop is completed, `pmat` contains 12 estimates of each factor premium.

Using the approach from Example 10.7, the factor premium estimates may then be obtained by averaging the 12 columns of the matrix pmat for each column

```
> apply(pmat, 1, mean)
[1]  0.00762 -0.00851  0.00235 -0.00878  0.00895
```

and the standard errors may be estimated using

```
> apply(pmat, 1, sd)/(12^.5)
[1] 0.00846 0.00889 0.00290 0.00391 0.00875
```

Of course, the first value in these vectors refers to the intercept, which is not a factor.

These results may be compared to those obtained in Example 10.7, in which the estimated factor premiums are given by

```
> apply(spcoef, 1, mean)
(Intercept)       sp500         smb         hml         mom
   0.014464    0.000372    0.006519   -0.004380    0.006333
```

with standard errors given by

```
> apply(spcoef, 1, sd)/(60^.5)
(Intercept)       sp500         smb         hml         mom
    0.00276     0.00553     0.00324     0.00272     0.00366
```

The two sets of premium estimates have several differences that are likely to be important in practice. For instance, the estimate of the premium for the return on the S&P 500 index was found to be 0.00372 in Example 10.7; using future returns as the predictor variable, the estimate is −0.00851. However, the standard errors of both estimates are relatively large so that the difference is unlikely to be statistically significant.

The average R-squared value for the 12 regressions used to calculate the estimates in pmat is roughly 0.11, which is considerably less than the R-squared value in Example 10.6. This is not surprising; we expect that estimates of the factor sensitivities will have a weaker relationship with future returns than they have with returns for the time period used for their estimation. □

Note that the standard errors of the estimates based on the rolling regressions are generally larger than those of the estimates calculated in Example 10.7, which is not surprising since they are based on the averages of only 12 rolling estimates as compared to the averages of 60 estimates used in Example 10.7. Of course, by going back further in time, we could obtain more premium estimates, thus reducing the standard errors; as always, such an approach is only useful if the relationships among the variables do not change in important ways over the time period considered. Thus, one drawback of the rolling-regression method is that the estimates must be based on a relatively long series of data in order to achieve the standard errors of the same general magnitude as those obtained by the method used in Example 10.7.

10.7 Applications of Factor Models

The ultimate goal of analyzing return data using a factor model is to provide information that is useful in the investment process. There are several ways in which a factor model may be used to better understand the properties of assets and to guide the selection of portfolio weights. The most straightforward of these is to use the factor model to estimate parameters of the distribution of the return vector on the assets under consideration, such as its covariance matrix.

Example 10.10 Consider stocks for firms represented in the S&P 100 index; see Example 8.5. The data matrix `sp96.data` contains 5 years of monthly returns for the period ending December 31, 2014, for each stock; only 96 of the 100 stocks had 5 years of monthly returns available, so `sp96.data` has 60 rows and 96 columns.

Consider the Fama–French three-factor model used in Example 10.5; using this model, the covariance matrix of the returns may be estimated using the same approach used in Example 10.4.

```
> sp96.ff<-lm(sp96.data~sp500+smb+hml)
> sp96.ff.beta<-sp96.ff$coef[-1,]
> Sig.FF<-cov(cbind(sp500, smb, hml))
> f.sighat.ff<-function(y){summary(lm(y~sp500+smb+hml))$sigma}
> sp96.Sig.fact<-t(sp96.ff.beta)%*%Sig.FF%*%sp96.ff.beta +
+  diag(apply(sp96.data, 2, f.sighat.ff)^2)
```

Hence, the matrix `sp96.Sig.fact` contains an estimate of the return covariance matrix for the 96 stocks represented in `sp96.data`.

Consider the minimum-variance portfolio of those stocks, subject to the constraint that all portfolio weights are nonnegative. Estimates of the portfolio weights may be obtained using the function `solve.QP` in the `quadprog` package.

```
> library(quadprog)
> sp96.mv<-solve.QP(Dmat=2*sp96.Sig.fact, dvec=rep(0, 96),
+      Amat=cbind(rep(1,96), diag(96)), bvec=c(1, rep(0, 96)),
+      meq=1)$solution
```

Note that many of the estimated weights are close to zero, but not exactly zero

```
> head(sp96.mv)
[1] -4.2e-18 -2.5e-17 -5.3e-17 -3.0e-17 -2.8e-17  1.0e-02
```

The function `round` may be used to round values to a specified number of decimal places; hence, the command

```
> sp96.mv<-round(sp96.mv, 4)
```

rounds the elements of `sp96.mv` to four decimal places:

```
> head(sp96.mv)
[1] 0.000 0.000 0.000 0.000 0.000 0.010.
```

After rounding, only 20 stocks have nonzero weights in the minimum-variance portfolio using the constraint that all weights are nonnegative.

Note that the sample covariance matrix cannot be used in this context.

```
> solve.QP(Dmat=2*cov(sp96.data), dvec=rep(0, 96),
+   Amat=cbind(rep(1, 96), diag(96)), bvec=c(1, rep(0, 96)),
+    meq=1)$solution
Error in solve.QP(Dmat = 2 * cov(sp96.data), dvec = rep(0, 96),
  Amat = cbind(rep(1, : matrix D in quadratic function is not
  positive definite!
```
□

The mean excess return on an asset may be estimated using the same general approach. Such an estimate is based on estimates of the factor premiums, as discussed in Section 10.6, along with estimates of the factor sensitivities for a given asset. The methodology is illustrated in the following example.

Example 10.11 Consider the factor model with factors SMB, HML, MOM, along with the excess return on the S&P 500 index. Estimates of the factor premiums were obtained in Example 10.6; they are stored in the variable `fact.prem`

```
> fact.prem
    sp500        smb        hml        mom
 0.000372  0.006519 -0.004380  0.006333
```

The constant in the equation for an asset's excess mean return in terms of its factor sensitivities is 0.01446.

For example, for 3M Company stock (symbol MMM), the factor sensitivities are

```
> sp474.beta["MMM",]
  sp500     smb     hml     mom
 1.0080  0.3661 -0.0988 -0.0107
```

It follows that the estimated mean excess return for 3M stock is

```
> 0.01446 + sum(fact.prem*sp474.beta["MMM", ])
[1] 0.0176
```

This may be compared to the sample mean excess return of 0.0148. □

Using Factor Sensitivities to Describe a Portfolio

Although it may be more natural to describe a portfolio in terms of the weights given to the assets in the portfolio, such an approach does not take into account the fact that many assets are closely related, with returns that are, in some cases, highly correlated. Hence, even a portfolio constructed from a large number of assets may not be truly diversified. An alternative approach is to describe a portfolio in terms of its sensitivities to the different factors used in a factor model. The estimated sensitivities of a portfolio's returns to the various factors give a succinct description of its properties.

Example 10.12 Consider the returns on three mutual funds, TIAA-CREF Small-Cap Equity Fund (TISEX), T. Rowe Price Small-Cap Value Fund (PRSVX), and Vanguard Value Index Fund (VIVAX). Like mutual funds in general, these three funds have holdings in a large number of stocks, with around 300 stocks in each fund; hence, it is difficult to obtain an understanding of a fund by analyzing the weights given to each stock. One way to summarize this information is to look at the composition of the fund by market sector, such as energy, financial services, health care, and so on; although such an analysis is often useful, not all stocks in a given sector are affected the same way by different economic conditions and returns for stocks in different sectors may still be correlated, making assessment of the diversification of the portfolio difficult. Analyzing a portfolio's returns using a factor model summarizes the relationship between the returns and the various underlying factors that affect their volatility.

Suppose that the returns on these funds are analyzed using the factor model described in Example 10.6, which has factors SMB, HML, and MOM, along with the excess return on the S&P 500 index. The data matrix `funds3.data` contains monthly excess returns on the three funds for the 5-year period ending December 31, 2014. The estimated factor sensitivities for the three mutual funds are given by

```
> funds3.fact<-lm(funds3.data~sp500+smb+hml+mom)
> t(funds3.fact$coef)[,-1]
      sp500     smb     hml     mom
tisex 1.040  0.9498 -0.0427 -0.0278
prsvx 0.893  0.9412  0.2335 -0.0494
vivax 0.950 -0.0161  0.2578 -0.0483
```

The adjusted R-squared values for the four-factor model regressions are given by

```
> f.rsq<-function(y)
+ {summary(lm(y~sp500+smb+hml+mom))$adj.r.squared}
> apply(funds3.data, 2, f.rsq)
tisex prsvx vivax
0.985 0.974 0.983
```

indicating that the returns of each of the funds are closely related to the four factors.

The estimated coefficients of the return on the market index are similar for the three funds; however, their sensitivities to the other factors are often quite different. For instance, TISEX and PRSVX both have a relatively large sensitivity to SMB of about 0.95, indicating that the returns on both funds have a positive relationship to the returns on small cap stocks, and their sensitivities to MOM are similar; however, PRSVX has a relatively large positive sensitivity to HML while TISEX has a negative sensitivity to this factor. This suggests that PRSVX tends to invest in more value stocks than does TISEX. The funds PRSVX and VIVAX have similar coefficients of the market index, HML, and MOM; however, the coefficient of SMB for VIVAX is small and negative, while for PRSVX it is large and positive. This suggests that the returns on PRSVX are taking advantage of a "small-cap stock effect" while the returns on VIVAX are approximately unrelated to the size factor. □

Imposing Factor-Sensitivity Constraints

As discussed in Section 10.3, the factor sensitivities of a portfolio are weighted sums of the factor sensitivities of the individual assets,

$$\beta_{p,j} = \sum_{i=1}^{N} w_i \beta_{i,j}, \quad j = 1, 2, \ldots, K,$$

where $w_1, w_2, \ldots, w_N$ are the portfolio weights. Hence, it is often possible to choose the portfolio weights in order to achieve a desired factor sensitivity for the portfolio. This is illustrated in the following example.

Example 10.13 Consider the stocks represented in the data matrix `big8` and consider a factor model based on five economic factors: the return on the S&P 500 index, stored in the variable `sp500`; the unemployment rate, stored in the variable `unemp`; the change in the Industrial Production Index stored in the variable `indpro`; the change in the Consumer Sentiment Index, stored in the variable `consum`; and the change in the Consumer Price Index, stored in the variable `cpi`.

The Industrial Production Index and the unemployment rate are described in Example 10.4. The Consumer Sentiment Index is based on the University of Michigan's Surveys of Consumers; these data are available on the FRED website at https://research.stlouisfed.org/fred2/series/UMCSENT. The Consumer Price Index is the Consumer Price Index for All Urban Consumers prepared by the Bureau of Labor Statistics; it is also available on the FRED website at https://research.stlouisfed.org/fred2/series/CPIAUCSL.

Some summary statistics of the factors for the time period considered are given by

```
> apply(cbind(sp500, unemp, indpro,consum, cpi), 2, summary)
          sp500 unemp indpro  consum      cpi
Min.     -0.082 0.056 -0.702 -0.1290 -0.00336
1st Qu. -0.014 0.073  0.012 -0.0132  0.00021
Median    0.018 0.082  0.263  0.0119  0.00161
Mean      0.011 0.080  0.266  0.0057  0.00141
3rd Qu.   0.033 0.090  0.485  0.0386  0.00256
Max.      0.108 0.099  1.560  0.1060  0.00599
```

The estimated factor sensitivities for the eight stocks are stored in the variable `big8.fact.beta`:

```
>big8.fact<-lm(big8~sp500+unemp+indpro+consum+cpi)
>big8.fact.beta<-t(big8.fact$coefficients)[, -1]
> big8.fact.beta
     sp500  unemp indpro consum    cpi
AAPL 0.940  0.282  0.011  0.000  0.988
BAX  0.678 -0.304 -0.013  0.017  2.888
KO   0.507  0.092  0.013 -0.119 -0.275
CVS  1.096 -0.446  0.015  0.036 -1.989
XOM  0.846  0.496 -0.022  0.142  2.298
IBM  0.617  0.950 -0.008  0.030  2.030
JNJ  0.504 -0.138 -0.022 -0.023 -1.107
DIS  1.196  0.083 -0.014  0.079 -5.806
```

The estimate of the return covariance matrix based on the factor model is stored in the variable `big8.Sig.fact`

```
> f.sighat.fact<-function(y){summary(lm(y~sp500+unem+indpro+
+ consum+cpi))$sigma^2}
> Sig.FF1<-cov(cbind(sp500, unem, indpro, consum, cpi))
> big8.Sig.fact<-t(big8.fact.beta)%*%Sig.FF1%*%big8.fact.beta +
+        diag(apply(big8, 2, f.sighat.fact))
```

Using this estimate, along with the sample mean excess returns, the estimated weight vector of the risk-averse portfolio with the risk-aversion parameter taken to be $\lambda = 10$ is given by

```
> big8.mean<-apply(big8, 2, mean)
> solve.QP(Dmat=10*big8.Sig.fact, dvec=big8.mean,
+ A=cbind(rep(1,8)), bvec=c(1), meq=1)$solution
[1]  0.282 -0.078  0.180  0.337 -0.296 -0.105  0.359  0.320
```

If we enforce the restriction that all asset weights must be nonnegative, the estimated weights are given by

```
> big8.ra10<-solve.QP(Dmat=10*big8.Sig.fact, dvec=big8.mean,
+   A=cbind(rep(1,8),diag(8)), bvec=c(1, rep(0, 8)),
+   meq=1)$solution
> big8.ra10
[1] 0.244 0.000 0.084 0.270 0.000 0.000 0.195 0.207
```

The estimated factor sensitivities for the portfolio with weight vector `big8.ra10` are given by

```
> big8.fact.beta%*%big8.ra10
          [,1]
sp500   0.9144
unemp  -0.0538
indpro  0.0004
consum  0.0119
cpi    -1.7397
```

Suppose we wish to construct a portfolio that is insensitive to the factors `unemp` and `cpi`. This may be done using the function `solve.QP`, including the constraint that the portfolio sensitivities to those factors are both zero. Define a matrix `A` by

```
>  A<-cbind(rep(1, 8), t(big8.fact.beta[c(2, 5),]), diag(8))
```

Then the second and third rows of the transpose of `A` contain the factor sensitivities for `unemp` and `cpi`, respectively. Thus, the command

```
> big8.ra10.con<-solve.QP(Dmat=10*big8.Sig.fact, dvec=big8.mean,
+ Amat=A, bvec=c(1, 0, 0, rep(0, 8)), meq=3)$solution
```

computes the weights of the risk-averse portfolio based on the risk-aversion parameter $\lambda = 10$ subject to the constraints that all portfolio weights are nonnegative and that the factor sensitivities for `unemp` and `cpi` are zero. The weights are given by

```
> big8.ra10.con
[1] 0.317 0.080 0.139 0.226 0.028 0.035 0.175 0.000
```

and the estimated factor sensitivities are

```
> big8.fact.beta%*%big8.ra10.con
          [,1]
sp500   0.8039
unemp   0.0000
indpro  0.0026
consum -0.0059
cpi     0.0000
```

Thus, we expect that the returns on the portfolio with weight vector `big8.ra10.con` will be less sensitive to the economic conditions reflected in `unemp` and `cpi` than is the portfolio with weight vector `big8.ra10`.

The estimated mean excess returns and return standard deviations for the two portfolios are given by

```
> sum(big8.ra10*big8.mean)
[1] 0.0193
> sum(big8.ra10.con*big8.mean)
[1] 0.0173
> (big8.ra10%*%big8.Sig.fact%*%big8.ra10)^.5
       [,1]
[1,] 0.0411
> (big8.ra10.con%*%big8.Sig.fact%*%big8.ra10.con)^.5
       [,1]
[1,] 0.0392
```

Thus, the constrained portfolio has a smaller estimated mean excess return but, because its return does not depend on the factors `unemp` and `cpi`, its estimated return standard deviation is smaller as well. In terms of the risk-aversion criterion function, the value for the unconstrained portfolio is 0.0109, while the value for the constrained portfolio is slightly less, at 0.0096. □

10.8 Suggestions for Further Reading

Factor models are among the core methodologies for understanding the behavior of asset returns; hence, there are many books and research papers covering their properties. Chincarini and Kim (2006) present an in-depth discussion of the application of factor models to portfolio management; this book is particularly useful for its detailed treatment of the possible factors to include in such a model. Alexander (2008, Chapter II.1), Elton et al. (2007, Chapter 8), Fabozzi et al. (2006, Chapter 8), Francis and Kim (2013, Chapter 8), and Ruppert (2004, Section 7.8) provide introductions to a number of different aspects of factor models. A more technical treatment of these topics is available in Campbell et al. (1997, Chapter 6). The momentum factor used in Sections 10.6 and 10.7 is attributed to Carhart (1997).

The analysis in Example 10.2 of Section 10.2, showing that the market model beta is not closely related to the average returns on an asset, is a simplified version of the discussion in Fama and French (1993). The procedure, discussed in Section 10.4, for constructing a factor based on the characteristics of the various firms, is covered in detail by Chincarini and Kim (2006, Section 7.4); a more comprehensive treatment is presented by Bali et al. (2016, Chapter 5), where this methodology is called "portfolio analysis."

The proposition on APT, along with the corresponding discussion, given in Section 10.5 is based on the work of Francis and Kim (2013, Chapter 16), which provides an excellent introduction to this topic; see also Reilly and Brown (2009, Chapter 9). Shumway (2000) outlines further details on asymptotic arbitrage, presenting important results without using extensive mathematics; see also Huberman and Wang (2008).

The use of factor models to estimate expected returns and the estimation of factor premiums are important topics in financial statistics, and the discussion in Section 10.6 provides just an overview of this topic. The two-stage approach, in which the factor sensitivities are first estimated and these estimates are then the predictor variables in a model used to estimate the factor premiums, is generally known as "Fama–MacBeth regression" after Fama and MacBeth (1973). See Cochrane (2000, Chapter 12) for a useful introduction to this method; a more general treatment is presented by Bali et al. (2016, Chapter 6). In particular, Bali et al. (2016) discuss the use of "future" returns as the response variable in the model used to estimate factor premiums. Because different assets have different error variances, some analysts use weighted least-squares when estimating factor premiums; see Draper and Smith (1981, Section 2.11) for a general discussion of weighted least-squares estimation and Shanken and Zhou (2007) for a comparison of weighted least-squares with other methods of estimating factor premiums.

The application of factor models to the investment process is often described by the term *factor-based investing*; useful nontechnical introductions to the ideas of factor-based investing are given by Bender et al. (2013) and Pappas and Dickson (2015). The use of factor models in describing the properties of a portfolio is described by Francis and Kim (2013, Chapter 19); factor-sensitivity constraints are discussed by Chincarini and Kim (2006, Section 9.7).

The type of factor model discussed in this chapter is often described as an "economic factor model." Two other types of factor models that are often used are "fundamental factor models" and "statistical factor models." A fundamental factor model has the same basic form as the models considered in this chapter. However, in such models, the goal is to describe the returns on an asset in terms of asset-specific variables, such as those related to a firm's value or its profitability, rather than in terms of common factors. Because these variables depend on the properties of each asset, they correspond to the factor sensitivities in the factor model, and the factor values are treated as unknown parameters to be estimated. In a statistical factor model, the goal is to model the covariance matrix of the asset return vector directly by describing it in terms of the covariances of certain linear combinations of the asset returns, using a technique known as *principal components analysis*. See, for example, Chincarini and Kim (2006, Chapter 3) and Qian et al. (2007, Chapter 3) for general discussions of the different types of factor models.

10.9 Exercises

1. Let $\boldsymbol{\beta}$ denote a $N \times K$ matrix, let $\boldsymbol{\Sigma}_F$ denote a $K \times K$ symmetric matrix, and let $\boldsymbol{\Sigma}_\epsilon$ denote an $N \times N$ symmetric matrix. Show that

$$\boldsymbol{\beta}\boldsymbol{\Sigma}_F\boldsymbol{\beta}^T + \boldsymbol{\Sigma}_\epsilon \tag{10.12}$$

is a positive-definite matrix provided that $\boldsymbol{\Sigma}_F$ and $\boldsymbol{\Sigma}_\epsilon$ are positive definite.

Suppose that one of $\boldsymbol{\Sigma}_F$ and $\boldsymbol{\Sigma}_\epsilon$ is nonnegative definite but not positive definite. Does it follow that (10.12) is positive definite? Why or why not?

2. Consider a factor model with three factors. Suppose that, for a set of four assets, the matrix of factor sensitivities is given by

$$\boldsymbol{\beta} = \begin{pmatrix} 1.0 & 0.3 & -0.2 \\ 0.9 & 0.7 & 0.5 \\ 1.1 & -0.1 & 1.0 \\ 0.5 & 1.1 & 0 \end{pmatrix}$$

and that the residual standard deviations for these assets are given by $0.4, 0.5, 0.2, 0.6$, respectively. Let

$$\boldsymbol{\Sigma}_F = \begin{pmatrix} 0.15 & 0.035 & 0.015 \\ 0.035 & 0.050 & 0.002 \\ 0.015 & 0.002 & 0.030 \end{pmatrix}.$$

Find the covariance matrix of the return vector for these assets under the assumption that the factor model holds and give the corresponding correlation matrix.

3. Consider a factor model with three factors applied to a set of four assets. For the parameter values given in Exercise 2, find the factor sensitivities and residual standard deviation for the equally weighted portfolio of the four assets.

4. Calculate 5 years of monthly excess returns for the period ending December 31, 2015, for five stocks, Papa John's International, Inc. (symbol PZZA), Bed Bath & Beyond, Inc. (BBBY), Netflix, Inc. (NFLX), Time Warner, Inc. (TWX), and Verizon Communications, Inc. (VZ); for the risk-free rate, use the return on the 3-month Treasury Bill, available on the Federal Reserve website.

Consider a factor model with three factors, the excess return on the S&P 500 index (symbol ^GSPC) and the Fama–French factors SMB and HML. Data on the factors SMB and HML are available from the Kenneth R. French Data Library, available on the website http://mba.tuck.dartmouth.edu/pages/faculty/ken.french/data_library.html in the file "Fama/French 3 Factors" found in the section on U.S. Research Returns Data. Note that the data in the French Data Library are generally in the form of percentage returns; hence, you should divide the values by 100.

Estimate the factor sensitivities for these five stocks, along with the residual standard deviations.

5. Is a portfolio consisting entirely of an investment in the risk-free asset, which has a positive return with probability one, considered an arbitrage portfolio? Why or why not?

6. Show that, under the no-arbitrage assumption, all risk-free assets must have the same return.

7. Consider a factor model with three factors. Suppose that the factor premiums for the factors are 0.002, 0.001, and 0.0025, respectively, and that the value of α in the equation for expected return on an asset is 0.005.

   Suppose that a given asset has factor sensitivities $\beta_1 = 1.0$, $\beta_2 = 0.75$, and $\beta_3 = -0.10$. Find the expected return on the asset.

8. Consider the returns on the five stocks and the factor model described in Exercise 4. Using these data, estimate the factor premiums for the three factors using the procedure described in Example 10.7. Provide standard errors of the premium estimates.

9. Consider the returns on the five stocks and the factor model described in Exercise 4.

   a. Using the estimated factor sensitivities for that model, estimate the covariance function of the assets' returns. Calculate the corresponding correlation matrix.
   b. Using the estimated covariance function based on the factor model and the sample mean vector for the assets, estimate the weights of the risk-averse portfolio with risk-aversion parameter $\lambda = 10$.

10. Consider the returns on the five stocks and the factor model described in Exercise 4 and analyzed in Exercise 9. Find the risk-averse portfolio with risk-aversion parameter $\lambda = 10$ subject to the following constraints: all portfolio weights are nonnegative; the sensitivity of the portfolio to the factor based on the S&P 500 index is one; the sensitivity of the portfolio to the factor SMB is at least 0.25.

Use the estimated covariance matrix of the assets' returns based on the factor model together with the sample mean vector of the assets.

11. Consider the following five large mutual funds: American Funds Income Fund of America Class A (symbol AMECX), Dodge & Cox Stock Fund (DODGX), Fidelity Contrafund (FCNTX), Franklin Income Fund Class A (FKINX), and Vanguard Wellington Fund (VWENX). For each fund, calculate 5 years of monthly excess returns for the period ending December 31, 2015; for the risk-free rate, use the return on the 3-month Treasury Bill, available on the Federal Reserve website.

    Consider a factor model with three factors, the excess return on the S&P 500 index (symbol ^GSPC) and the Fama–French factors SMB and HML. Data on the factors SMB and HML are available from the Kenneth R. French Data Library, available on the website http://mba.tuck.dartmouth.edu/pages/faculty/ken.french/data_library.html in the file "Fama/French 3 Factors" found in the section on U.S. Research Returns Data. Note that the data in the French Data Library are generally in the form of percentage returns; hence, you should divide the values by 100.

    a. Estimate the factor sensitivities for these five funds. Based on these results, identify the two funds that appear to be the most similar and identify the two funds that appear to be the most different.
    b. Estimate the correlation matrix of the funds' returns. Are the returns of the "similar" funds identified in Part (a) highly correlated compared to other pairs of funds? Do the "different" funds identified in Part (a) have a low return correlation compared to the other funds?
    c. Compute the sample mean excess returns and sample return standard deviations for the five funds. Are the sample mean excess returns and the sample return standard deviations of the "similar" funds identified in Part (a) generally similar, as compared to the results for the other funds? Are the sample mean excess returns and the sample return standard deviations of the "different" funds identified in Part (a) generally different, as compared to the results for the other funds?
    d. Based on your results, would you conclude that the factor sensitivities are useful for identifying portfolios with similar properties? Why or why not?

References

Agresti, A. and Finlay, B. (2009). *Statistical Methods for the Social Sciences.* Pearson, Upper Saddle River, NJ, fourth edition.

Alexander, C. (2008). *Practical Financial Econometrics.* Wiley, Chichester.

Bali, T. G., Engle, R. F., and Murray, S. (2016). *Empirical Asset Pricing: The Cross Section of Stock Returns.* Wiley, Hoboken, NJ.

Bender, J., Briand, R., Melas, D., and Subramanian, R. A. (2013). Foundations of Factor Investing. Technical report. MSCI Index Research, New York, NY.

Benninga, S. (2008). *Financial Modeling.* The MIT Press, Cambridge, MA, third edition.

Bernstein, W. J. (2001). *The Intelligent Asset Allocator: How to Build Your Portfolio to Maximize Returns and Minimize Risk.* McGraw-Hill, New York, NY.

Best, M. J. and Grauer, R. R. (1991). On the Sensitivity of Mean-Variance-Efficient Portfolios to Changes in Asset Means: Some Analytical and Computational Results. *The Review of Financial Studies*, 4:315–342.

Black, F. (1972). Capital Market Equilibrium with Restricted Borrowing. *Journal of Business*, 45:444–455.

Blattberg, R. C. and Gonedes, N. J. (1974). A Comparison of the Stable and Student Distributions as Statistical Models for Stock Prices. *Journal of Business*, 47:244–280.

Blitzstein, J. K. and Hwang, J. (2015). *Introduction to Probability.* CRC Press, Boca Raton, FL.

Bouchaud, J.-P. and Potters, M. (2011). Financial Applications of Random Matrix Theory: A Short Review. In Akemann, G., Baik, J., and Francesco, P. D., editors, *Oxford Handbook on Random Matrix Theory*, chapter 40, pages 824–850. Oxford University Press, Oxford, UK.

Bradfield, D. (2003). Investment Basics XLVI. On Estimating the Beta Coefficient. *Investment Analysts Journal*, 57:47–53.

Caeiro, F. and Mateus, A. (2014). *randtests: Testing randomness in R.* R package version 1.0.

Campbell, J. Y., Lo, A. W., and MacKinlay, A. C. (1997). *The Econometrics of Financial Markets.* Princeton University Press, Princeton, NJ.

Canty, A. and Ripley, B. (2015). *boot: Bootstrap R (S-plus) Functions.* R package version 1.3-17.

Carhart, M. M. (1997). On Persistence in Mutual Fund Performance. *Journal of Finance*, 52:57–82.

Carlin, B. P. and Louis, T. A. (2000). *Bayes and Empirical Bayes Methods for Data Analysis.* CRC Press, Boca Raton, FL, second edition.

Chambers, J. M., Cleveland, W. S., Keiner, B., and Tukey, P. A. (1983). *Graphical Methods for Data Analysis.* Wadsworth, Belmont, CA.

Chincarini, L. B. and Kim, D. (2006). *Quantitative Equity Portfolio Management.* McGraw-Hill, New York, NY.

Cochrane, J. H. (2000). *Asset Pricing.* Princeton University Press, Princeton, NJ.

Cowpertwait, P. S. P. and Metcalfe, A. V. (2009). *Introductory Time Series with R.* Springer, New York, NY.

Dalgaard, P. (2008). *Introductory Statistics with R.* Springer, New York, NY, second edition.

Damodaran, A. (1999). Estimating Risk Parameters. Report No. S-CDM-99-02. Faculty Digital Archive, New York University. http://hdl.handle.net/2451/26789.

Davison, A. C. and Hinkley, D. V. (1997). *Bootstrap Methods and Their Application.* Cambridge University Press, Cambridge, UK.

DeMiguel, V., Garlappi, L., and Uppal, R. (2009). Optimal versus Naive Diversification: How Inefficient Is the 1/N Portfolio Strategy? *The Review of Financial Studies*, 22:1915–1953.

DeMiguel, V., Martin-Utrerab, A., and Nogalesb, F. J. (2013). Size Matters: Optimal Calibration of Shrinkage Estimators for Portfolio Selection. *Journal of Banking and Finance*, 37:3018–3034.

Draper, N. R. and Smith, H. (1981). *Applied Regression Analysis.* Wiley, New York, NY, second edition.

Efron, B. and Morris, C. (1970). Stein's Paradox in Statistics. *Scientific American*, 236:119–127.

Efron, B. and Tibshirani, R. J. (1993). *An Introduction to the Bootstrap.* Chapman & Hall, New York, NY.

Elton, E. J., Gruber, M. J., Brown, S. J., and Goetzmann, W. N. (2007). *Modern Portfolio Theory and Investment Analysis.* Wiley, Hoboken, NJ, seventh edition.

Evans, M. J. and Rosenthal, J. S. (2004). *Probability and Statistics: The Science of Uncertainty.* Freeman, New York, NY.

Fabozzi, F. J., Focardi, S. M., and Kolm, P. N. (2006). *Financial Modeling of the Equity Market: From CAPM to Cointegration.* Wiley, Hoboken, NJ.

Fama, E. F. (1965). Random Walks in Stock Market Prices. *Financial Analysts Journal*, 51:75–80.

Fama, E. F. (1970). Efficient Capital Markets: A Review of Theory and Empirical Work. *Journal of Finance*, 25:383–417.

Fama, E. F. (1976). *Foundations of Finance: Portfolio Decisions and Securities Prices.* Basic Books, New York, NY.

Fama, E. F. and French, K. R. (1993). Common Risk Factors in the Returns on Stocks and Bonds. *Journal of Financial Economics*, 33:3–56.

Fama, E. F. and MacBeth, J. D. (1973). Risk, Return, and Equilibrium: Empirical Tests. *Journal of Political Economy*, 81:607–636.

Foulkes, A. S. (2009). *Applied Statistical Genetics with R.* Springer, New York, NY.

Francis, J. C. and Kim, D. (2013). *Modern Portfolio Theory: Foundation, Analysis, and New Developments.* Wiley, Hoboken, NJ.

Genz, A., Bretz, F., Miwa, T., Mi, X., Leisch, F., Scheipl, F., and Hothorn, T. (2016). *mvtnorm: Multivariate Normal and t Distributions.* R package version 1.0-5.

Granger, C. W. J. (1963). A Quick Test for Serial Correlation Suitable for Use with Non-Stationary Time Series. *Journal of the American Statistical Association*, 58:728–736.

Gray, J. B. and French, D. W. (1990). Empirical Comparisons of Distributional Models for Stock Index Returns. *Journal of Business Finance & Accounting*, 17:451–459.

Grinold, R. C. and Kahn, R. N. (2000). *Active Portfolio Management.* McGraw-Hill, New York, NY, second edition.

Hammersley, J. M. and Handscomb, D. C. (1964). *Monte Carlo Methods.* Metheun, London.

Henderson, H. V. and Searle, S. R. (1981). On Deriving the Inverse of a Sum of Matrices. *SIAM Review*, 23:53–60.

Hogg, R. V. and Tanis, E. A. (2006). *Probability and Statistical Inference.* Pearson, Upper Saddle River, NJ, seventh edition.

Huberman, G. and Wang, Z. (2008). Arbitrage Pricing Theory. In Durlauf, S. N. and Blume, L. E., editors, *The New Palgrave Dictionary of Economics*, pages 197–205. Palgrave Macmillan, Basingstoke, UK.

Hult, H., Lindskog, F., Hammarlid, O., and Rehn, C. J. (2012). *Risk and Portfolio Analysis: Principles and Methods.* Springer, New York, NY.

Jobson, J. D. and Korkie, B. (1980). Estimation for Markowitz Efficient Portfolios. *Journal of the American Statistical Association*, 75:544–554.

Jobson, J. D. and Korkie, B. (1981). Putting Markowitz Theory to Work. *Journal of Portfolio Management*, 7:70–74.

Johnson, R. A. and Wichern, D. W. (2007). *Applied Multivariate Statistical Analysis.* Pearson, Upper Saddle River, NJ, sixth edition.

Jorion, P. (1986). Bayes-Stein Estimation for Portfolio Analysis. *Journal of Financial and Quantitative Analysis*, 21:279–292.

Kane, A., Kim, T., and White, H. (2012). Active Portfolio Management : The Power of the Treynor-Black Model. In Kyrtsou, C. and Vorlow, C. editors, *Progress in Financial Markets Research*, pages 311–332. Nova Publishers, New York, NY.

Larson, R. and Edwards, B. H. (2014). *Calculus.* Brooks Cole, Boston, MA, tenth edition.

Ledoit, O. and Wolf, M. (2004). Honey, I Shrunk the Sample Covariance Matrix. *Journal of Portfolio Management*, 30:110–119.

Lintner, J. (1952). The Valuation of Risk Assets and the Selection of Risky Investments in Stock Portfolios and Capital Budget. *Review of Economics & Statistics*, 47:13–37.

Lo, A. W. (1991). Long-Term Memory in Stock Market Prices. *Econometrica*, 59:1279–1313.

Lo, A. W. (1997). A Nonrandom Walk Down Wall Street: Recent Advances in Financial Technology. *TIAA-CREF Research Dialogues*, 52:1–7.

Lo, A. W. and MacKinlay, A. C. (2002). *A Non-Random Walk Down Wall Street.* Princeton University Press, Princeton, NJ.

Lu, T.-T. and Shiou, S.-H. (2002). Inverses of 2×2 Block Matrices. *Computers & Mathematics with Applications*, 43:119–129.

Malkiel, B. G. (1973). *A Random Walk Down Wall Street.* W. W. Norton & Company, New York, NY.

Malkiel, B. G. (2003). *A Random Walk Down Wall Street: The Time-Tested Strategy for Successful Investing.* W. W. Norton, New York, NY.

Markowitz, H. (1952). Portfolio Selection. *Journal of Finance*, 7:77–91.

Markowitz, H. M. (1987). *Mean-Variance Analysis in Portfolio Choice and Capital Markets.* Wiley, New York, NY.

Martin, R. D. and Simin, T. T. (2003). Outlier-Resistant Estimates of Beta. *Financial Analysts Journal*, 59:56–69.

Merton, R. C. (1972). An Analytic Derivation of the Efficient Portfolio Frontier. *Journal of Financial and Quantitative Analysis*, 7:1851–1872.

Michaud, R. O. (1989). The Markowitz Optimization Enigma: Is 'Optimized' Optimal? *Financial Analysts Journal*, 45:31–42.

Miller, M. B. (2012). *Mathematics and Statistics for Financial Risk Management.* Wiley, Hoboken, NJ.

Modigliani, F. and Pogue, G. A. (1974). An Introduction to Risk and Return: Concepts and Evidence. *Financial Analysts Journal*, 30:68–80.

Montgomery, D. C., Jennings, C. L., and Kulahci, M. (2008). *Introduction to Time Series Analysis and Forecasting.* Wiley, Hoboken, NJ.

Mossin, J. (1966). Equlibrium in a Capital Asset Market. *Econometrica*, 34:768–783.

Newbold, P., Carlson, W. L., and Thorne, B. M. (2013). *Statistics for Business and Economics.* Pearson, Upper Saddle River, NJ, eighth edition.

Pappas, S. N. and Dickson, J. M. (2015). Factor-Based Investing. Technical report. Vanguard Research, Springfield, VA.

Praetz, P. D. (1972). The Distribution of Share Price Changes. *Journal of Business*, 45:49–55.

Qian, E. E., Hua, R. H., and Sorensen, E. H. (2007). *Quantitative Equity Portfolio Management: Modern Techniques and Applications.* CRC Press, Boca Raton, FL.

Reilly, F. K. and Brown, K. C. (2009). *Investment Analysis and Portfolio Management.* South-Western, Mason, OH, ninth edition.

Rice, J. A. (2007). *Mathematical Statistics and Data Analysis.* Brooks/Cole, Belmont, CA, third edition.

Roll, R. (1977). A Critique of the Asset Pricing Theory's Tests: Part I: On Past and Potential Testability of the Theory. *Journal of Financial Economics*, 4:129–176.

Ross, S. (2006). *A First Course in Probability*. Pearson, Upper Saddle River, NJ, seventh edition.

Ross, S. A. (1977). The Capital Asset Pricing Model (CAPM), Short-Sale Restrictions and Related Issues. *Journal of Finance*, 32:177–183.

Ross, S. M. (2013). *Simulation*. Academic Press, San Diego, CA, fifth edition.

Ruppert, D. (2004). *Statistics and Finance: An Introduction*. Springer-Verlag, New York, NY.

Samuelson, P. (1965). Proof that Properly Anticipated Prices Fluctuate Randomly. *Industrial Management Review*, 6:41–49.

Samuelson, P. (1974). Challenge to Judgement. *Journal of Portfolio Management*, 1:17–19.

Sclove, S. L. (2013). *A Course on Statistics for Finance*. CRC Press, Boca Raton, FL.

Severini, T. A. (2005). *Elements of Distribution Theory*. Cambridge University Press, Cambridge, UK.

Severini, T. A. (2016). A Nonparametric Approach to Measuring the Sensitivity of an Asset's Return to the Market. *Annals of Finance*, 12:179–199.

Shanken, J. and Zhou, G. (2007). Estimation and Testing Beta Pricing Models: Alternative Methods and their Performance in Simulations. *Journal of Financial Economics*, 84:40–86.

Sharpe, W. F. (1963). A Simplified Model for Portfolio Analysis. *Management Science*, 9:277–293.

Sharpe, W. F. (1964). Capital Asset Prices: A Theory of Market Equilibrium under Conditions of Risk. *Journal of Finance*, 19:425–442.

Sharpe, W. F. (1991). The Arithmetic of Active Management. *Financial Analysts Journal*, 47:7–9.

Shumway, T. (2000). Course Notes for Bus Admin 855. http://www-personal.umich.edu/~shumway/courses.dir/ba855.dir/Notes1.pdf.

Stewart, J. (2015). *Calculus*. Brooks Cole, Boston, MA, eighth edition.

Tamhane, A. C. and Dunlop, D. D. (2000). *Statistics and Data Analysis: From Elementary to Intermediate*. Prentice-Hall, Upper Saddle River, NJ.

Touloumis, A. (2015). Nonparametric Stein-Type Shrinkage Covariance Matrix Estimators in High-Dimensional Settings. *Computational Statistics and Data Analysis*, 83:251–261.

Trapletti, A. and Hornik, K. (2016). *tseries: Time Series Analysis and Computational Finance.* R package version 0.10-35.

Treynor, J. L. and Black, F. (1973). How to Use Security Analysis to Improve Portfolio Selection. *Journal of Business*, 46:66–86.

Turlach, B. A. and Weingessel, A. (2013). *quadprog: Functions to Solve Quadratic Programming Problems.* R package version 1.5-5.

Vasicek, O. A. (1973). A Note on Using Cross-Sectional Information in Bayesian Estimation of Betas. *Journal of Finance*, 28:1233–1239.

Venables, W. N. and Ripley, B. D. (2002). *Modern Applied Statistics with S.* Springer, New York, fourth edition.

Warnes, G. R., Bolker, B., and Lumley, T. (2015). *gtools: Various R Programming Tools.* R package version 3.5.0.

Wei, W. W. S. (2006). *Time Series Analysis: Univariate and Multivariate Methods.* Pearson, Boston, MA, second edition.

Woolridge, J. M. (2013). *Introductory Econometrics: A Modern Approach.* South-Western, Mason, OH, fifth edition.

Index

M

N

S

W

Z

9781138198371